TITANIUM FOR CONSUMER APPLICATIONS

TITANIUM FOR CONSUMER APPLICATIONS

Real World Use of Titanium

FRANCIS FROES

MA QIAN

MITSUO NIINOMI

ELSEVIER

TITANIUM
titanium.org

Elsevier
Radarweg 29, PO Box 211, 1000 AE Amsterdam, Netherlands
The Boulevard, Langford Lane, Kidlington, Oxford OX5 1GB, United Kingdom
50 Hampshire Street, 5th Floor, Cambridge, MA 02139, United States

Notices
Knowledge and best practice in this field are constantly changing. As new research and experience broaden our understanding, changes in research methods, professional practices, or medical treatment may become necessary.

Practitioners and researchers must always rely on their own experience and knowledge in evaluating and using any information, methods, compounds, or experiments described herein. In using such information or methods they should be mindful of their own safety and the safety of others, including parties for whom they have a professional responsibility.

To the fullest extent of the law, neither the Publisher nor the authors, contributors, or editors, assume any liability for any injury and/or damage to persons or property as a matter of products liability, negligence or otherwise, or from any use or operation of any methods, products, instructions, or ideas contained in the material herein.

Library of Congress Cataloging-in-Publication Data
A catalog record for this book is available from the Library of Congress

British Library Cataloguing-in-Publication Data
A catalogue record for this book is available from the British Library

ISBN 978-0-12-815820-3

For information on all Elsevier publications
visit our website at https://www.elsevier.com/books-and-journals

Publisher: Matthew Deans
Acquisition Editor: Christina Gifford
Editorial Project Manager: Andrae Akeh
Production Project Manager: Kamesh Ramajogi
Cover Designer: Christian J. Bilbow

Typeset by SPi Global, India

Contents

Contributors

Milan Brandt
Centre for Additive Manufacturing, School of Engineering, RMIT University, Melbourne, VIC, Australia

A. du Plessis
Physics Department, University of Stellenbosch, Stellenbosch, South Africa

T. Ebel
Helmholtz-Zentrum Geesthacht, Institute of Materials Research, Geesthacht, Germany

Yong Feng
Western Superconducting Technologies Co., Ltd. (WST); National Engineering Laboratory for Superconducting Materials Preparation, Xi'an, China

F.H. (Sam) Froes
Titanium Industry, Tacoma, WA, United States

Tadahiko Furuta
Applied Powder Metallurgy and Additive Manufacturing Laboratory, Totoya Central R&D Labs., Inc., Nagakute, Japan

N. Gui
Centre for Additive Manufacturing, School of Engineering, RMIT University, Melbourne, VIC, Australia

Qiang Guo
Western Superconducting Technologies Co., Ltd. (WST); National Engineering Laboratory for Superconducting Materials Preparation, Xi'an, China

Takuji Horie
HORIE Corporation, Niigata, Japan

M. Ikeda
Faculty of Chemistry, Materials, and Bioengineering, Kansai University, Osaka, Japan

M. Krzywicka
Department of Fundamentals of Technology, University of Life Sciences in Lublin, Lublin, Poland

Jianfeng Li
Western Superconducting Technologies Co., Ltd. (WST); National Engineering Laboratory for Superconducting Materials Preparation, Xi'an, China

Xianghong Liu
Western Superconducting Technologies Co., Ltd. (WST); National Engineering Laboratory for Superconducting Materials Preparation, Xi'an, China

Bing Liu
Research Institute for New Materials Technology, Chongqing Powder Metallurgy Institute, Chongqing University of Arts and Sciences, Chongqing, China

Shudong Luo
RMIT University, School of Engineering, Centre for Additive Manufacturing; Centre for Additive Manufacturing, School of Engineering, RMIT University, Melbourne, VIC, Australia

Scott Mayson
School of Design, RMIT University, Melbourne, VIC, Australia

Mitsuo Niinomi
Institute for Materials Research, Tohoku University, Sendai; Department of Materials and Manufacturing Science, Graduate School of Engineering, Osaka University, Osaka; Department of Materials Science and Engineering, Graduate School of Science and Technology, Meijo University; Institute of Materials and Systems for Sustainability, Nagoya University, Nagoya; Faculty of Chemistry, Materials, and Bioengineering, Kansai University, Osaka, Japan

K. Pałka
Department of Materials Engineering, Lublin University of Technology, Lublin, Poland

R. Pokrowiecki
Department of Head and Neck Surgery-Maxillofacial Surgery, Otolaryngology and Ophtalmology, Olsztyn, Poland

Ma Qian
RMIT University, School of Engineering, Centre for Additive Manufacturing; Centre for Additive Manufacturing, School of Engineering, RMIT University, Melbourne, VIC, Australia

Ellen A. Semeniuta
Grandis Titanium, Rancho Santa Margarita, CA, United States

Tingting Song
Centre for Additive Manufacturing, School of Engineering, RMIT University, Melbourne, VIC, Australia

Hiroyuki Tada
Charmant Co. Ltd, Technology Development Section, Sabae, Japan

Kazuhiro Takahashi
Nippon Steel Corporation, Chiba, Japan

M. Ueda
Faculty of Chemistry, Materials, and Bioengineering, Kansai University, Osaka, Japan

Ruilong Wang
Western Superconducting Technologies Co., Ltd. (WST); National Engineering Laboratory for Superconducting Materials Preparation, Xi'an, China

I. Yadroitsava
Department of Mechanical and Mechatronics Engineering, Central University of Technology, Bloemfontein, South Africa

I. Yadroitsev
Department of Mechanical and Mechatronics Engineering, Central University of Technology, Bloemfontein, South Africa

Kaijuan Yan
Western Superconducting Technologies Co., Ltd. (WST); National Engineering Laboratory for Superconducting Materials Preparation, Xi'an, China

M. Yan
Department of Materials Science and Engineering, and Shenzhen Key Laboratory for Additive Manufacturing of High-performance Materials, Southern University of Science and Technology, Shenzhen, China

Huan Yang
Panzhihua Sunshine Poetry Culture Propagation Co. Ltd., Panzhihua, China

Shulong Ye
Department of Materials Science and Engineering, Southern University of Science and Technology, Shenzhen; Institute of Applied Physics and Materials Engineering, University of Macau, Macau, China

Peng Yu
Department of Materials Science and Engineering, Southern University of Science and Technology, Shenzhen, China

Pingxiang Zhang
Western Superconducting Technologies Co., Ltd. (WST); National Engineering Laboratory for Superconducting Materials Preparation; Northwest Institute for Non-ferrous Metal Research (NIN), Xi'an, China

Kailin Zhang
Western Superconducting Technologies Co., Ltd. (WST); National Engineering Laboratory for Superconducting Materials Preparation; Northwest Institute for Non-ferrous Metal Research (NIN), Xi'an, China

Yongyun Zhang
Department of Materials Science and Engineering, Southern University of Science and Technology, Shenzhen, China

Yanmin Zhu
Western Superconducting Technologies Co., Ltd. (WST); National Engineering Laboratory for Superconducting Materials Preparation, Xi'an, China

Editors Biography

Dr. Ma Qian is a Distinguished Professor of the School of Engineering at Royal Melbourne Institute of Technology (RMIT University), Melbourne, Australia. His current research interests include metal additive manufacturing, powder metallurgy of light metals and alloys, solidification processing, metallic biomaterials, high entropy and medium entropy alloys, biomimetic design, and lattice materials. He has published 215 peer-reviewed journal papers (nearly 100 papers on titanium), which have attracted more than 7200 Scopus citations as of March 2019. He initiated the biennial international conference on *Titanium Powder Metallurgy* in 2011 (cosponsored by Materials Australia, TiDA, TMS, JSPM, and CSM). He serves as an editorial member for a number of journals, including being appointed as an associate editor for both *Acta Materialia* and *Scripta Materialia* in April 2017.

Francis H. Froes, Ph.D. has been involved in the Titanium field with an emphasis on Powder Metallurgy (P/M) for more than 40 years. He was employed by a primary Titanium producer-Crucible Steel Company-where he was leader of the Titanium group. He was the program manager on a multi-million dollar US Air Force (USAF) contract on Titanium P/M. He then spent time at the USAF Materials Lab where he was supervisor of the Light Metals group (which included Titanium). This was followed by 17 years at the University of Idaho where he was a Director and Department Head of the Materials Science and Engineering Department. He has over 800 publications, in excess of 60 patents, and has edited almost 30 books-the majority on various aspects of Titanium again with an emphasis on P/M. He gave the key-note presentation at the first TDA (ITA) Conference. In recent years he has co-sponsored four TMS Symposia on Cost Effective Titanium featuring numerous papers on P/M. He is a Fellow of ASM, is a member of the Russian Academy of Science, and was awarded the Service to Powder Metallurgy by the Metal Powder Association. Recently he has been a co-author of three comprehensive papers on the Additive Manufacturing of Titanium.

Dr. Mitsuo Niinomi received his PhD from Faculty of Engineering, Nagoya University in 1980. He received his Doctor of Science from School of Dentistry, Aichi-Gakuin University in 2003. From 1980 to 2005, he was a research associate, Associate Professor, and Professor at Toyohashi University of Technology. In 2005, he moved to Institute for Materials Research, Tohoku University as a Professor. He was a Director at Institute for Materials Research, Tohoku University from 2009 to 2014. He was retired from Institute of Materials Research, Tohoku University. He has been an Honorary Professor at Tianjin University (China) from 2011. He has been a Professor Emeritus at Tohoku University, a Vising Professor at Institute for Materials Research, Tohoku University, a Specially Appointed Professor at Osaka University and Meijo University, and a Visiting Professor at Nagoya University from 2016. He has been a visiting professor at Kansai University from 2017. He has been an Honorary Professor at North Easter University (China) from 2018. He was a Visiting Associate Professor at Carnegie-Mellon University (USA) from 1988 to 1989, and a Visiting Professor at The University of Dayton (USA) from 1997 to 1997. His research is focused on designing, development, and evaluation of biological and mechanical biocompatibility of titanium alloys for medical and dental applications.

Some other rainbows

James Brent Ward
Fife, Scotland, UK

The pleasure of the how, when, and where my adventures with refractory metals began are interwoven with my questing attitude to life. I was blessed with a good British education, which encouraged a sense of wonder in the world around me. This was certainly compounded by finding myself in the London Arts school system of the late 1960s. Like many of the "fortunate few," we were looking out into a rapidly changing world and collectively anxious to do something "different." I came from a traditional gold and silversmithing background where craftsmanship was often measured against the achievements of the 19th and 18th centuries. The trade, as such, was conservative, and our main ideology as young aspirant artists was both to provoke and shake the establishment.

Remember, the late 1960s was a time of the moon landing, drugs, innovative music, and artistic experimentation, so eloquently described by everyone from Kurt Vonnegut to Anthony Burgess.

As a bunch of students, we were collectively bursting with energy, ideas, and attitude. Albert, one of our eccentric and venerable technicians, asked me, "Have you seen this metal? Apparently if you pass electricity through it, it changes color!" I was absolutely mesmerized. It stimulated me to be both creative and take risks at the same time.

The gentleman concerned had some of this new metal from his covert British government work with the "Blue Streak Missile" project, and he handed it to me for examination. Having only a cursory knowledge of physics and chemistry, my reaction was to attach the metal to two bare wires from the mains supply and flick the switch, resulting in a huge BANG, which fused all the lights in the jewelry department, with thankfully no major injuries. A spectacular blaze of colors appeared as by magic across the previously dull grey metal.

Once I recovered from a major reprimand from the head of the department, I was entranced and in awe in equal measure. My natural hedonism contained a modicum of respect for human life, so artistic passion was tempered by a desire to both know more and to control this phenomena. This was indeed my artistic Damascene moment; I was converted and enthralled by both the magic of electricity, and the fact that I knew virtually nothing about it. This was undoubtedly the beginning of a major and everlasting romance circa 1969.

I became a refractory evangelist, a zealot with a passion for new knowledge. Essentially, I had a quest to learn and know more about the extraordinary possibilities of this lightweight dull grey metal, and the journey of being a pioneer. There was little to no information about titanium (primarily used only in warfare) and its application to the arts. Availability of the raw material was a complex engagement with bureaucracy because, when I tried to buy titanium, all the suppliers were military institutions who decidedly wanted to know just what I intended to do with the metal. Once in London I actually had to sign a document saying that I was only using the material for creative purposes and not for any military application.

Even at that time, the word "titanium" was virtually unknown, and most of my fellow students were at best quizzical. Frankly, there was nothing technically available about how to saw drill, bend, hammer, or polish it, and certainly, after trying several times, no chance of soldering it!

All of these factors obviously changed my attitude toward design. I constantly thought about rivets and the press fit accuracy of joints, all design work leading to the visually liberating moment of coloration, when the metal revealed its true magnificence. From 1969–1972, I experimented with various forms of controlled mains electricity. I was introduced to the "Variac" rectifier, which controlled A/C mains voltage but was often unstable and prone to fusing. Thanks to the unselfish attitudes of Terry Green and Brian Hanson at IMI Birmingham, both knowledge and availability of raw materials became easier.

The very concept of connecting the metal ferule of an artist's paintbrush to a Variac is very much in the tradition British Boffins, with often explosive results. Some of my tutors at the time would nervously approach the workbench with cautious optimism and sometimes even rubber gloves. However the results opened a veritable Pandora's box of color potential and ultimately lead to a greater desire to understand the metallurgy of the metals. When heated or colored straight from the workbench, the colors were pastel and often dull, but when etched with a diabolical brew of hydrofluoric and nitric acid, the colors shimmered with radiance and an almost iridescent glow. Inevitably, this lead to further extensive research into the actual chemistry and metallurgy of the metals.

The color or apparent color was not pigmented but an optical phenomena known as "interference" similar to oil on water, a peacock's feather, or a butterfly's wing. From a designer's perspective, this was absolutely captivating, and the genie really was "let out of the bottle." My thirst for knowledge was unleashed, and days were spent texturing, analyzing, and documenting the subtleties of tonality, hue, and reflectivity. The same voltage could create multitudinous differences in visual impact; the work of Manet and Monet took on a whole new meaning. The Holy Grail of Refractory Metals became the "Why no red?" question. How come, if the colors are similar to a pyramidal sequence, do anodized colors not follow the ROYGBIV phenomena? Not a rainbow, but similar! No doubt this lead to the Royal College of Art and the tremendous contributions made by The Goldsmiths Company in London, particularly Chris Walton, Peter Gainsbury, and my subsequent report on "Working with Refractory Metals."

Perhaps the most significant change in my determination to spread the titanium word was the understanding and cocreation of a safe power source for D/C anodizing whilst the variable voltage Variac system is OK. It is not until a power source of less than 1% ripple is utilized that the subtleties of coloration can truly be revealed, especially when using tantalum and niobium at higher voltages, and I still heartily recommend the use of rubber gloves.

My commitment to creativity lead to the Royal College to India where the vivacity of the country renewed my enthusiasm for experimentation and the search for creating harmonious mandalas. In 1978, I joined "Dust Jewelry "in Scotland to create a range of commercially available titanium jewelry. A particular innovation was the photoetching range, which broadened the visual availability of the metal and certainly made a significant contribution to the modern jewelry movement. From Scotland, I moved to the U.S. and joined Edward Rosenberg as a founder of Spectore Corporation, which continues to the present day. In 1986, I joined Comyns of London as production director and created a range of cups and drinking vessels for Mitsukoshi and a variety of retailers around the world. In 1993, I joined Royal Selangor Malaysia and set up their gold and silver manufacturing facility in Kuala Lumpur, returning in 1998. I have spent the last 20 years working as a craftsman creating works in gold, silver and the refractory metals of titanium, niobium, tantalum.

These wondrous elements, so diverse and often metallurgically baffling, are renowned for their inert properties; they are considered hypoallergenic by both the British and American Health authorities, as they are never rejected by the human body and are now considered biocompatible. I am indeed proud to say that, for more than 30 years, I have been actively involved with a number of neurosurgeons in maxillofacial and cranial reconstruction, particularly utilizing the strength/weight ratio of commercially pure titanium (Figs. 1–5).

Over the last 49 years, I have come to love their individual "coat of many colors," but recently I have been experimenting with the reflectivity of the nascent grey state, and the subtleties of tones and shades that can be achieved by diligent engraving. This new development has proven fantastically, creatively stimulating and continues to provide inspiration for many

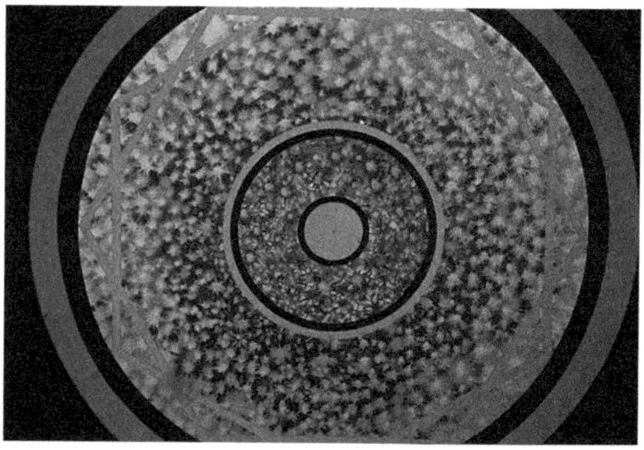

Fig. 1 Painting with electricity.

Fig. 2 Butterfly photo etching.

Fig. 3 Dragonfly photo etching.

Fig. 4 Collection photo etching silver and titanium.

Fig. 5 Sun face mandala carving of large titanium circle.

new works. I can even after all this time state honestly that I am just beginning to fully understand the artistic potential of these unique materials, who knows what extraordinary technology the future will provide, "sputtering" vapor deposition and casting or even 3D printing on a jewelry scale are all being developed but all come with individual technical problems.

Recently there have been scientific breakthroughs in to what color and light actually are and how we and those around us perceive it. The chromatophores and dichromic perception of the cephalopods is enough wonder to fill a hundred pages, but for now we should just relax, admire, enjoy and share the universal sense of wonder provided by our glorious prismatic planet Earth.

CHAPTER 1

An introduction to titanium in consumer applications

F.H. (Sam) Froes[a], Ma Qian[b], Mitsuo Niinomi[c,d,e,f,g]

[a]Titanium Industry, Tacoma, WA, United States
[b]RMIT University, School of Engineering, Centre for Additive Manufacturing, Melbourne, VIC, Australia
[c]Institute for Materials Research, Tohoku University, Sendai, Japan
[d]Department of Materials and Manufacturing Science, Graduate School of Engineering, Osaka University, Osaka, Japan
[e]Department of Materials Science and Engineering, Graduate School of Science and Technology, Meijo University, Nagoya, Japan
[f]Institute of Materials and Systems for Sustainability, Nagoya University, Nagoya, Japan
[g]Faculty of Chemistry, Materials, and Bioengineering, Kansai University, Osaka, Japan

Contents

The discovery of titanium in 1791 in Cornwall, England, is attributed to William Gregor, a clergyman. A few years later, Martin Heinrich Klaproth named it for the Titans of Greek mythology. Titanium is found as a number of mineral deposits, predominantly as rutile (TiO_2) and ilmenite (which contains iron as an impurity), which are found in the Earth's crust, with rutile occurring as a beach sand, for example, in southeast Australia. Metal extraction is accomplished by the Kroll (magnesium reduction) and Hunter (sodium reduction) processes [1–32].

Titanium is alloyed with elements such as iron, aluminium, vanadium, and molybdenum, resulting in strong, low-density alloys used in aerospace (jet engines, missiles, and spacecraft), military, industrial applications

Titanium for Consumer Applications
https://doi.org/10.1016/B978-0-12-815820-3.00001-0

(chemicals and petrochemicals, desalination plants, pulp, and paper), automotive, agri-food, medical and dental uses, sporting goods, jewelry, cellular phones, and a variety of other applications (see later in this chapter) [1–32].

Titanium exhibits a number of attractive features including corrosion resistance (where unalloyed titanium is generally used), high-strength-to-density ratio, and good fracture resistance (where the titanium alloys are generally used). In its alloyed condition, the metal is as strong as many forms of steel but exhibits lower density. Titanium has two allotropic forms: the high-temperature beta phase (having a body-centered cubic structure) and the low-temperature alpha phase (which has a closely packed hexagonal structure), below 1625°F (885°C) in commercially pure titanium. Titanium can be strengthened by precipitating the alpha phase in the beta matrix [1–32].

In this chapter, we define consumer applications as automotive use, sporting applications, backpacking, architecture, firearms and laptops, tools and jewelry, as well as watches and drones.

1 Consumer applications

1.1 Automotive use

Automotive and motorcycle use of titanium is concentrated in racing applications where reduced weight and high strength along with rigidity can provide the racer's edge. However even in the automotive family, decreased automobile weight can noticeably improve fuel economy (Fig. 1). The metal is generally too expensive for extensive use in automotive applications, although some niche use has occurred (Table 1). A number of late model Corvettes feature titanium exhausts, and a Corvette Z06's LT4 supercharged engine has low-density solid titanium intake valves to achieve

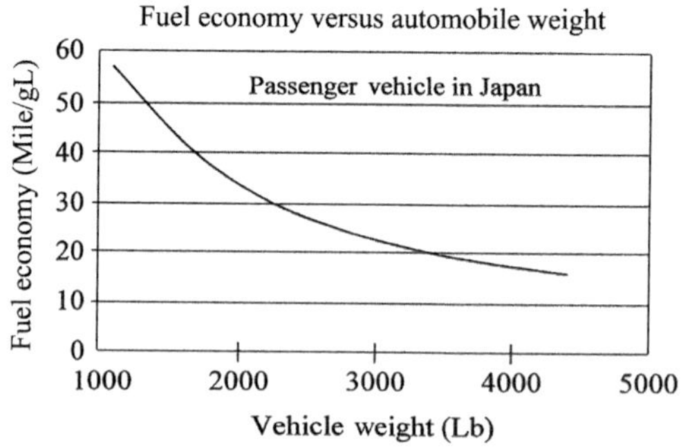

Fig. 1 Fuel economy versus automobile weight for a family automobile in Japan.

Table 1 Some examples of niche titanium applications in the family automobile

Component	Material	Manufacturer	Model
Connecting rods	Ti-3Al-2V-rare earth	Honda	Acura NSX
Connecting rods	Ti-6Al-4V	Ferrari	All 12-cyl.
Wheel rim screws	Ti-6Al-4V	Porsche	Sport wheel option
Brake pad guide pins	Ti grade 2	Daimler	S-Class
Brake sealing washers	Ti grade 1s	Volkswagen	All
Gearshift knob	Ti grade 1	Honda	S2000 Roadster
Connecting rods	Ti-6Al-4V	Porsche	GT3
Valves	Ti-6Al-4V & PM-Ti	Toyota	Altezza 6-cyl.
Turbo charger wheel	Ti-6Al-4V	Daimler	Truck diesel
Suspension springs	TIMETAL LCB	Volkswagen	Lupo FSI
Wheel rim screws	Ti-6Al-4V	BMW	M-Techn. option
Valve spring retainers	β-Titanium alloys	Mitsubishi	All 1.8 1—4-cyl.
Turbo charger wheel	γ-TiAl	Mitsubishi	Lancer
Exhaust system	Ti grade 2	General Motors	Corvette Z06
Wheel rim screws	Ti-6Al-4V	Volkswagen	Sport package GTI
Valves	Ti-6Al-4V & PM-Ti	Nissan	Infiniti Q45
Suspension springs	TIMETAL LCB	Ferrari	360 Stradale

greater strength and heat resistance. A notable use is in the Thrust SSC, or Thrust supersonic car, a British jet-propelled titanium car developed by Richard Noble, Glynne Bowsher, Ron Ayers, and Jeremy Bliss. In October 1997, the Thrust SSC set a World Land Speed Record when it achieved a speed of 1228 km/h (763 mph), driven by Andy Green, becoming the first car to officially break the sound barrier at close to sea level [1–32].

1.2 Sporting goods

Many sporting goods make use of titanium, including tennis rackets; golf clubs (Fig. 2); lacrosse stick shafts; cricket, hockey, lacrosse, and football helmet grills; and bicycle frames and components. Although it is not a material in wide-spread use for bicycles, titanium bikes have seen use in racing events.

Fig. 2 A titanium golf club head, which allows greater distance and accuracy.

1.3 Backpacking

Titanium equipment, including cookware, eating utensils, lanterns, and tent stakes (Fig. 3), are used extensively by backpackers. Although somewhat more expensive than competing materials such as steel or aluminum, titanium components are lighter without reducing strength. Titanium horseshoes (Fig. 4) are chosen over steel by farriers due to lightness and durability.

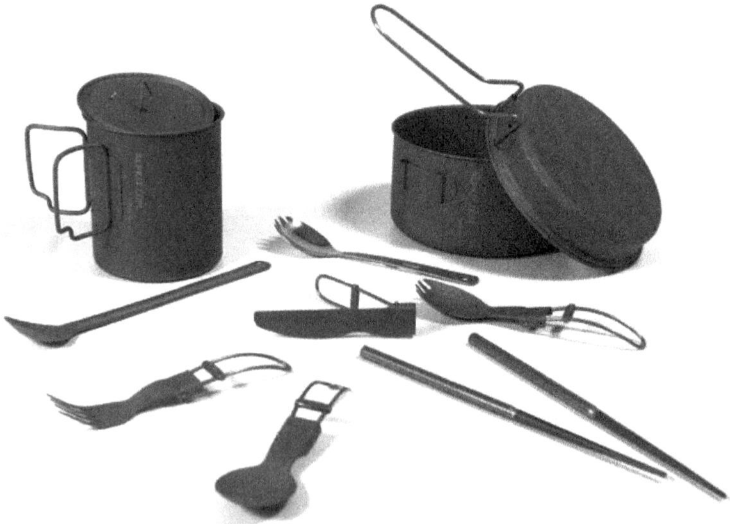

Fig. 3 Titanium backpacking equipment.

Fig. 4 Titanium horseshoes.

1.4 Architecture

Architectural use of titanium is quite extensive. For example, the 42.5-m (139-foot) monument to Yuri Gagarin, the first space traveller (Fig. 5), located in Moscow, Russia, is fabricated from titanium. Additionally, the 110-m (360.9-foot) monument to the Conquerors of Space, topping the Cosmonaut Museum also in Moscow is also made of titanium. The choice here is based on titanium's attractive color and close relationship with rockery. The first constructions in Europe and North America, respectively, to be sheathed in titanium were the Guggenheim Museum in Bilbao (Fig. 6) and the Cerritos Millennium Library. Sheathing with titanium was also used on the Frederic C. Hamilton Building in Denver, Colorado. Temples in Japan have titanium roof tiles.

1.5 Firearms and laptop computers

The high strength and low density of titanium compared to competing metals (steel, stainless steel, and aluminium), combined with enhanced fabrication methods, has led to increased use in firearm production. Uses include pistols and rifles. For the same reasons, titanium has also shown increased use in the body of laptop computers (for example, in Apple's PowerBook line; Fig. 7).

Fig. 5 A titanium statue of Yuri Gurganin, the first person in space, located in Moscow, Russia.

Fig. 6 The Guggenheim Museum in Bilbao, Northern Spain, is sheathed in titanium sheet. *From https://en.wikipedia.org/wiki/Guggenheim_Museum_Bilbao#/media/File:Bilbao_-_Guggenheim_aurore.jpg.*

Fig. 7 An Apple computer case fabricated from titanium.

1.6 Tools and kitchenware

Titanium has shown increased use in high-end, low-density, and corrosion-resistant tools, such as shovels, flashlights, knife handles, and hammers (Fig. 8). Bolts are made of titanium or titanium alloys (Fig. 9). Also high-end kitchenware is fabricated from titanium (Fig. 10).

1.7 Jewelry including watches and eyeglass frames

The durability of titanium has led to increased use for designer jewelry (particularly titanium rings; Fig. 11). The inertness of the metal makes it a good choice for those with allergies or for use of jewelry in environments such as swimming pools. Titanium can be alloyed with gold to give an alloy that can be marketed as 24-carat gold because the 1% of alloying is below the additive amount requiring a change in designation. The Ti-Au alloy has approximately the hardness of 14-carat gold with durability greater than pure 24-carat gold.

The durability, low density, and dent and corrosion resistance of titanium makes it a good choice for watch cases. Titanium has also been used to produce pictures, sculptures, decorative objects, and furniture. Titanium can be anodized to change the thickness of the surface oxide layer, giving a

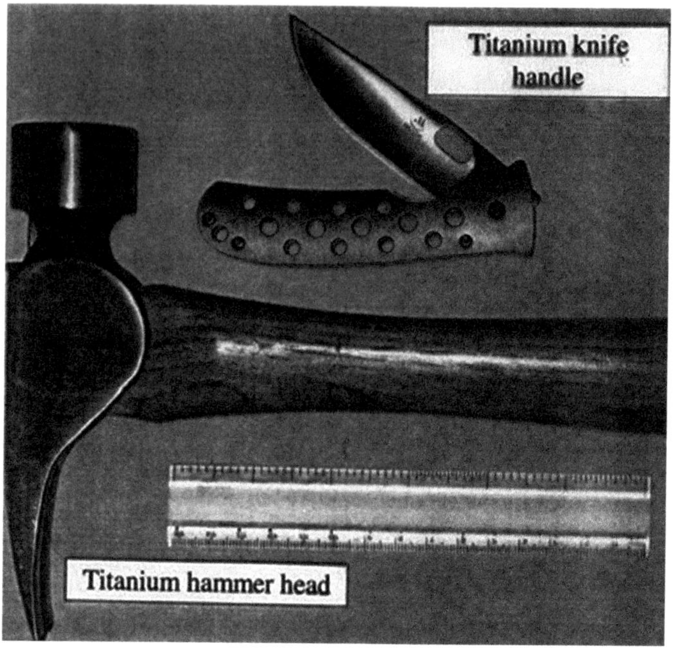

Fig. 8 A knife handle and a hammer head fabricated from titanium.

Fig. 9 Titanium bolts.

variety of bright colors (Fig. 12). With this coloration and chemical inertness, titanium is often used for body piercing.

Titanium alloys are used quite extensively in high-end spectacle frames (Fig. 13) where they feature high durability, low density, and compatibility with human skin, although stiffness may become an issue if the titanium frame is too thin.

Fig. 10 High-end kitchenware made from titanium.

Fig. 11 A titanium ring, colored electrolytically with different thicknesses of oxide.

1.8 Other consumer applications of titanium

Titanium has shown some use in noncirculating coins and medals. Gibraltar released the world's first titanium coin to celebrate the millennium.

Additive manufacturing or 3D printing has quickly become a primary method of manufacturing highly intricate titanium products and/or artworks. As an example, Fig. 14 shows a beautifully intricate 3D printed

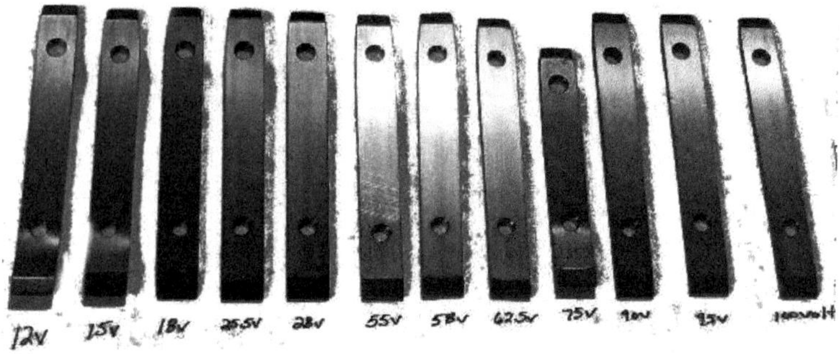

Fig. 12 The variety of colors that can be achieved on the surface of titanium by varying the thickness of the oxide film by electrolysis.

Fig. 13 Oakley sunglasses manufactured from titanium.

Fig. 14 A beautifully intricate titanium (Ti-6Al-4V) ceremonial mace printed using the selective laser melting (SLM) process by the RMIT Centre for Additive Manufacturing. The finished mace is 1.2 m in length and weighs only 1.6 kg.

titanium (Ti-6Al-4V) ceremonial mace for RMIT University, Melbourne, Australia, printed using the selective laser melting (SLM) process by the RMIT Centre for Additive Manufacturing. The finished mace is 1.2 m in length and, despite that impressive size, weighs only 1.6 kg, or 3.5 pounds. This new 3D-printed titanium mace represents the new capabilities from a combination of digital design and metal 3D-printing technology in creating new titanium products or artworks.

Drones also feature titanium.

2 What's in this chapter

In this chapter, we defined consumer applications as automotive use, sporting applications, backpacking, architecture, firearms and laptops, tools and jewelry, as well as watches and drones. The sections in this chapter are based on this definition.

3 Summary

An overview of the history and general characteristics of titanium has been documented. This was followed by a review of the consumer applications of titanium, defined as those areas listed in "What's in this Chapter".

References

[1] A.D. McQuillan, M.K. McQuillan, in: H.M. Finniston (Ed.), Metallurgy of the Rarer Metals, Academic Press, Inc, New York, 1956, pp. 335.
[2] R.A. Wood, The Titanium Industry in the Mid-1970s, Battelle Report MCIC-75-26, Battelle Memorial Institute, Columbus, OH, 1975.
[3] H.B. Bomberger, F.H. (Sam) Froes, P.H. Morten, in: F.H. (Sam) Froes, D. Eylon, H.B. Bomberger (Eds.), Titanium Technology: Present Status and Future Trends, The Titanium Development Association, 1985, pp. 3–18.
[4] M. Qian, F.H. (Sam) Froes (Eds.), Titanium Powder Metallurgy: Science, Technology and Applications, Elsevier, 2015.
[5] S.R. Seagle, J.R. Wood, F.H. (Sam) Froes, T. Khan, Synthesis, Processing and Modelling of Advanced Materials, vol. 77–78, Trans. Tech. Publications, Aedermannsdorf, Switzerland, 1993, pp.91–102.
[6] F.H. Froes, I.L. Caplan, Titanium 92 Science and Technology, TMS, Warrendale, PA, 1993.
[7] P.A. Blenkinsop, W.J. Evans, H.M. Flowers, Titanium 95 Science and Technology, The Institute of Materials, London, UK, 1996.
[8] F.H. (Sam) Froes, Third ASM Int. Paris Conf. on Synthesis, Processing and Modeling of Advanced Materials, ASM International, Materials Park, OH, 1997, pp.3–38.
[9] F.H. (Sam) Froes, P.G. Allen, M. Niiomi (Eds.), Non-Aerospace Applications of Titanium, TMS, Warrendale, PA, 1998.
[10] K.L. Housley, Black Sand: The History of Titanium, Metal Management Aerospace, Hartford, CT, 2007.

[11] A. Imam, F.H. (Sam) Froes, K.L. Housley, Titanium and Titanium Alloys, Kirk Othmer Encyclopedia, 2010, pp.1–41. (Online Only).
[12] M. Molchanova, Phase Diagrams of Titanium Alloys, Israel Program for Scientific Translations, Jerusalem, Israel, 1965.
[13] F.H. (Sam) Froes, D. Eylon, H.B. Bomberger (Eds.), Titanium Technology: Present Status and Future Trends, Titanium Development Association (now International Titanium Association), Boulder, CO, 1985.
[14] R. Boyer, G. Welsch, E. Collings (Eds.), Materials Property Handbook: Titanium Alloys, ASM International Materials Park, OH, 1994.
[15] M.J. Donachie, Titanium, A Technical Guide, ASM International, Metals Park, OH, 1988.
[16] I.J. Polmear, Light Alloys, Metallurgy of the Light Metals, third ed., Edward Arnold, London, 1996.
[17] H. Margolin, H. Neilson, Titanium metallurgy, in: H.H. Hauser (Ed.), Modern Materials, Advances in Development and Applications, vol. 2, Academic Press, New York, 1960, pp. 225–325.
[18] F.H. (Sam) Froes, Titanium powder metallurgy: developments and opportunities in a sector poised for growth, Powder Metall. Rev. 2 (4) (2013) 29–43 (Winter).
[19] B. Dutta, F.H. Froes, Additive Manufacturing of Titanium Alloys, Elsevier, Butterworth-Heinemann, 2016.
[20] N. Ohta, Chem. Eco. Eng. Rev. 13 (1981) 22.
[21] G.E. Haddeland, S. Morikawa, Titanium Dioxide Pigment, Process Economics Program Report No. 117, Stanford Research Institute International, Menlo Park, CA, 1978.
[22] Energy Use Patterns in Metallurgical and Nonmetallic Mineral Processing, Report No. PB-246 357, Battelle Columbus Laboratories, U.S. Department of Commerce, Washington, D.C., 1975.
[23] J.A. Slatnick, et al., Availability of Titanium in Market Economy Countries, IC 9413, Bureau of Mines Information Circular, Washington, D.C., 1994.
[24] W.W. Minkler, E.F. Baroch, The Production of Titanium, Zirconium and Hafnium, 1981.
[25] R.C. Weast (Ed.), CRC Handbook of Chemistry, 62nd ed., CRC Press, Inc., Boca Raton, FL, 1982.
[26] K.S. Weil, Y. Hovanski, C.A. Lavender, Effect of $TiCl_4$ purity on the sinterability of Armstrong-processed Ti powder, J. Alloy Compd. 473 (1–2) (2009).
[27] K.L. Housley, Kroll process may be on the verge of replacement or modification, in: Titanium News, Metal Management Aerospace, Inc., Hartford, CT, 2007. (Summer).
[28] Summary of Emerging Titanium Cost Reduction Technologies, A Study Performed for the U.S. Department of Energy and Oak Ridge National Laboratory, Subcontract 4000023694 by EHK Technologies, Vancouver, WA, 2004.
[29] P. Sun, Z. Fang, et al., A novel method for production of spherical Ti-6Al-4V powder for additive manufacturing, Powder Technol. 301 (2016) 331–335.
[30] Mineral Commodity Summaries, Department of the Interior Geological Survey and Bureau of Minus, Washington D.C., 1996.
[31] Titanium Facts, International Titanium Association, Broomfield, CO, 2007.
[32] F.H. (Sam) Froes, M. Qian (Eds.), Titanium in Medical and Dental Applications, Elsevier Publishing, 2018.

CHAPTER 2

Titanium products for everyday life by *Panzhihua Story*

Huan Yang[a], Shudong Luo[b]
[a]Panzhihua Sunshine Poetry Culture Propagation Co. Ltd., Panzhihua, China
[b]RMIT University, School of Engineering, Centre for Additive Manufacturing, Melbourne, VIC, Australia

Contents

1 Introduction

Titanium (Ti) has attracted increasing interest in the consumer products sector, where its attributes such as light weight, corrosion resistance, biocompatibility, hypoallergenicity, low thermal conductivity, and surface appearance are key factors, yet its mechanical properties are secondary to its marketing advantages. Commercially pure (CP) Ti such as ASTM grade 2 is the material of choice for several good reasons: (1) it is more affordable than alloyed Ti; (2) it offers excellent plasticity (elongation to fracture ≥20%) as well as sufficient strength (yield strength 275–410 MPa) allowing for severe plastic deformation with a low demand for equipment; (3) it exhibits adequate machinability for economic machining through most conventional

Titanium for Consumer Applications
https://doi.org/10.1016/B978-0-12-815820-3.00012-5

processes; (4) it shows good weldability suitable for various welding techniques such as tungsten inert-gas (TIG) welding; and (5) the ease of surface treatments facilitates the realization of distinguished surface features and colors. These advantages give rise to great flexibility in consumer product design and manufacturing with the use of CP Ti.

Titanium consumer products cover a broad spectrum, including automotive parts, architecture, sporting goods, kitchenware, etc. (see other chapters). The focus of this chapter is Ti products used in everyday life. The products presented herein were provided by Panzhihua Sunshine Poetry Culture Propagation Co., Ltd, a China firm specializing in the design and development of tourism products branded as *Panzhihua Story*.

2 Product manufacturing

2.1 Shaping

There readily exist a few forms of CP Ti materials on the market in use for customer product manufacturing, such as sheets, plates, rods, tubing, etc. These forms are produced from ingots primarily through operations such as hot slabbing, forging, rolling, etc. The excellent plasticity of CP Ti along with its moderate strength endows good formability allowing for forming even under ambient conditions. Nevertheless, the forming procedure practically requires special considerations due to some special characteristics of Ti including notch sensitivity, galling, embrittlement due to contamination or hydrogen/oxygen absorption, and high springback [1]. Cold forming is a cost-effective and ecofriendly technique compared to hot forming, and thus cold rolling, drawing, stretching, bending, and other such operations are usually employed in customer product manufacturing with CP Ti. Various tooling materials are available including tool steel, cast zinc, cast bronze, etc. Cold forming is usually assisted with the use of a lubricant such as a graphite suspension in a suitable solvent. For good formability, a slow forming speed is always preferred. Worth mentioning is that overforming and/or hot sizing after cold forming is essentially performed to cope with the high springback of Ti. Annealing treatment appears to be critical in cold forming, which is usually carried out at 480–595°C for 15–240 min in the case of CP Ti. There are multiple reasons for this: reducing stress, improving the part contour, restoring compressive yield strength after tension (Bauschinger effect), and avoiding delayed cracking and stress corrosion.

Unalloyed Ti is readily machinable and suitable for most conventional machining processes including turning, milling, tapping, sawing, boring, grinding, etc. [2]. But the machinability band is narrow relative to steel. This is largely due to the low elastic modulus and poor thermal conductivity intrinsic with Ti. The former can make the vibration of Ti a concern during machining, and accordingly the use of rigid workpiece holders is strictly recommended. The latter causes chips and poor heat dissipation through the workpiece and, as a consequence, significantly traps the heat in the cutting edge and tools. In this regard, the use of coolant is a necessity for Ti machining. There are some vital factors to successful machining of Ti, such as tool selection, coolant, workpiece holders, machining speed, and feeding rate. These factors vary significantly with the specific machining process. To improve the machining result and prolong tool life, there are a few general helpful tips: sharp tools, coolant-fed tools, high coolant concentration, and pressurized coolant to blast away chips.

TIG welding is a commonly practiced method in Ti consumer product manufacturing, which provides more control on heat input and the welding puddle than other welding processes [3]. To guarantee the welding quality, filler metal selection, the use of shielding gas, and cleanness are the most critical technical considerations. The selection of filler metal follows the criterion that it must exactly match the grade of Ti metal being welded. For example, unalloyed Ti rod such as ER Ti-2 (0.08–0.16 wt.%O) is recommended for the welding of ASTM grade 2. Occasionally, filler metal with lower yield strength is likely chosen if a more ductile joint is expected. Because Ti is extremely reactive, especially at high temperatures, the heat-affected zone must be fully covered with inert gas during welding to prevent oxidation and contamination. Commonly used shielding gas includes pure argon and argon/helium (75/25) mixture. The shielding gas must contain no more than 20 ppm free oxygen and have a dew point ideally greater than −50°C [4]. High reactivity of Ti also requires extreme cleanness for the filler metal, the base metal, and the surrounding environment. Oxidized surface at the joint of Ti metal must be removed with a stainless steel brush or die grinder prior to welding. Any contamination from body oil, cutting fluid, lubricant, paint, dust, dirt, etc. can embrittle the joint and lead to weld failure.

2.2 Surface treatment

Surface treatment is essentially applied to Ti consumer products manufacturing for improved wear resistance and/or corrosion resistance [5]. Surface hardening is one such routine surface treatment, which either transforms the vulnerable surface into a hard layer through nitridation, oxidation, boronizing, or ion implantation, or coats the surface with wear-resistant materials such as chromium, tungsten carbide, or alumina. The oxide layer, which naturally forms on Ti surface when exposed to an aerated environment, imparts good corrosion resistance to Ti, but its thickness is usually less than 5 nm. To enhance corrosion resistance, coating Ti with TiN, TiB, or TiBN is a common practice, which can be carried out through physical/chemical vapor deposition, magnetron sputtering, cathodic arc deposition, or laser cladding process [6, 7]. Electropolishing is another useful treatment to Ti consumer products, generally employed to remove sharp edges, burrs, cracks, and microholes for improved corrosion resistance and geometrical appearance.

Titanium consumer products usually require unique surface features for aesthetic or identification purposes, including pattern, texture, picture, logo, symbol, etc. Their fabrication involves various techniques such as screen printing, embossing, engraving, etching, and so on. Such techniques have been well established for other metals, but their application to Ti needs to be careful with respect to the high chemical reactivity of Ti.

2.3 Coloring

The aim of Ti coloring treatment is to form a thin (10–500 nm) oxide (TiO_2) film on Ti surface. The color of Ti results from the thin-film interference of selective wavelengths of light—the transparent TiO_2 film leading to a phase difference between the light reflected from the oxide film and the light reflected from the Ti metal below the film. The color changes with the TiO_2 film thickness. For example, a film thickness of 30, 60, and 120 nm, respectively, corresponds to yellow, light blue, and green on Ti [8].

Mainly for aesthetic purposes, coloring treatment has been intensively applied to Ti consumer products like jewelry for half a century. A variety of coloring techniques have been developed so far, however, anodic oxidation and heat oxidation are the two primary approaches. The former is accomplished through Ti anode absorption of oxygen released from electrolyte whereas the latter relies on the Ti-O reaction at an elevated temperature. Voltage and temperature are their main processing parameters determining the color of Ti products. In addition, TiN surface is pursued for Ti consumer products in some cases. TiN is a gold-yellow ceramic with a decomposition temperature of 2949°C and can be formed by direct nitridation, ion implantation, physical vapor deposition, etc. [9]. It is worth noting that the thick oxide or nitride film also provides Ti consumer products with good resistance to decolorization, wear, and corrosion. Coloring by forming oxide/nitride films rather than traditional painting means a safer and healthier use of Ti consumer products.

3 Representative products

3.1 Bottles/thermos/cups

A range of such products have been manufactured with various shapes and colors, as shown in Fig. 1. The marketing appeals of using Ti as bottles, thermos, and cups mainly include the excellent biocompatibility and hypo-allergenicity of Ti, which ensure a safe long-term contact with the human body. The passive oxide/nitride surface layer avoids metal ion eluting when in contact with water, and its corrosion resistance allows for containing any juices and beverages. The thermal properties of Ti add more attractive features to such consumer products. For instance, Ti possesses low specific heat capacity, and therefore these products do not feel cold in winter; the poor thermal conductivity of Ti excels at retaining heat, ideal for a thermos. An interesting comparison conducted by Japanese HORIE Corporation proved the outstanding heat-retaining ability of double-walled Ti tumblers: ice cubes in a glass cup completely melted whereas half of the ice in a Ti tumbler retained after being kept at 20°C for 3 h [10].

Deep drawing or superplasitic forming is the main manufacturing process for these products. Thread tapping is necessary to process the bottle lip and cap. Shot pinning, electropolishing, or a kiln firing process can be conducted when necessary, which largely contributes to the elegant surface textures. Finely tuned colors by anodic oxidation renders Ti products unique and charming in terms of aesthetics.

3.2 Eating utensils

Good biocompatibility and hypoallergenicity of Ti are also motivators for the manufacturing of eating utensils with CP Ti. Fig. 2 shows some such products. To prevent any defect formation during the frequent contact with other utensils/kitchenwares made of metals or porcelains, surface treatment such as anodic oxidation is essential, which offers good scratch resistance and durability. Apart from an aesthetic purpose, good surface finish, which is generally realized by electropolishing, is also expected from the washing perspective. Electropolishing treatment is also necessary for the removal of sharp edges and burrs.

3.3 Tea sets

Tea set is another application of biocompatibility and nontoxicity of Ti in everyday life consumer products. The dense oxide surface can effectively protect the set against any invasion of chemicals released from the tea such

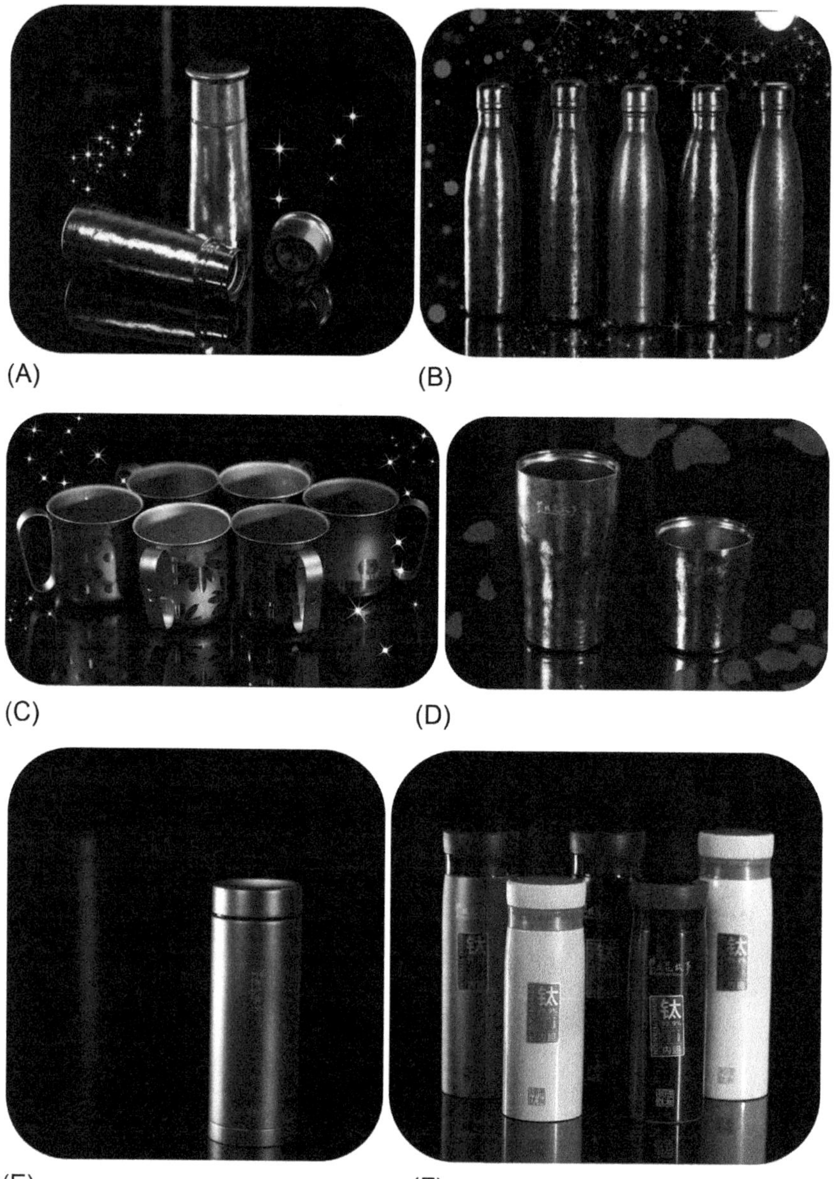

Fig. 1 (A, B) Bottles, (C, D) cups/tumblers, and (E, F) thermos made from CP Ti. Except cups shown in (C), all other products are double-walled and evacuated between the walls. Thermos shown in (F) are coated with paint on the exterior surface.

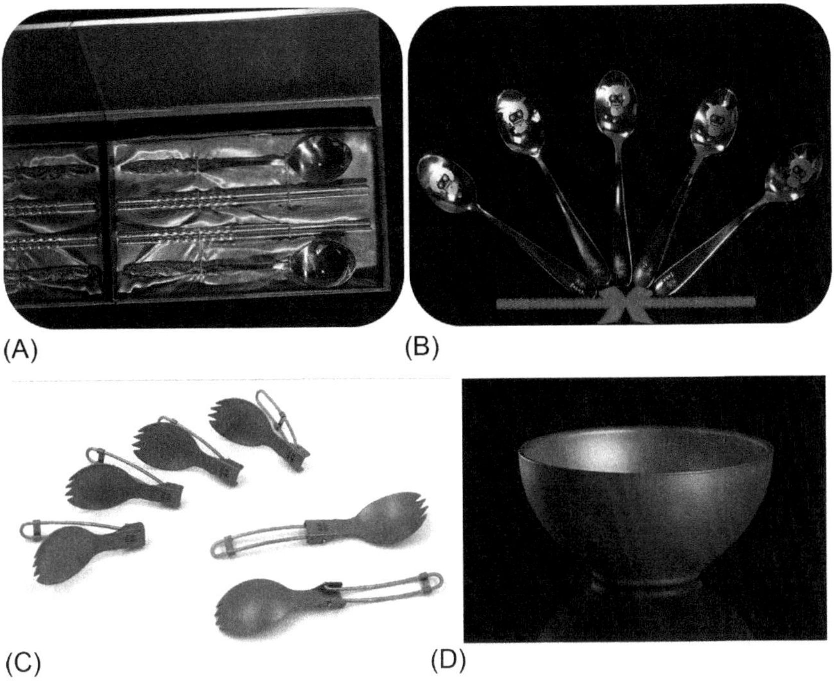

Fig. 2 Titanium eating utensils: (A) chopstick, (B) spoon, (C) spork, and (D) bowl. The bowl is double-walled for improved heat preservation.

as caffeine, theobromine, pigment, etc. Also, the surface plays a key role in avoiding metal ions eluting when the tea set is in contact with boiling water. The light weight of Ti makes the tea set portable. China has had a tea culture for thousands of years, in which the tea set is as important as tea itself. Tea drinking provides a refreshing, perfuming, and delightful experience, but the tea set is more associated with a deep spiritual experience such as harmony, serenity, and peace of mind. To some extent, the tea set made from Ti satisfies these expectations in terms of its exquisite appearance, texture, and colors. Such tea sets as shown in Fig. 3 display charming quality, elegance, grace, and solemnness.

3.4 Jewelry

Titanium jewelry has a history for half a century. Clearly its marketing advantage is not due to rareness in the Earth's crust unlike other precious jewelry materials such as silver, gold, platinum, diamond, etc. Key factors for the success of Ti jewelry include: (1) biocompatibility, hypoallergenicity, and durability of Ti makes it an ideal source metal relative to other engineering

Fig. 3 Titanium tea set: (A, B) teapot, cup, and container, (C) tool kit, and (D) portable tea set.

metals in terms of comfort and wearability; (2) insignificant material cost makes Ti jewelry much more affordable than their counterparts; (3) low material cost alongside a low density (4.51 g/cm^3) allows designers to focus on the aesthetics and artistic aspects of jewelry instead of concerns on the mass and dimensions; and (4) good forming, machining, welding, and coloring abilities of Ti offers tremendous flexibility in design and manufacturing.

The value and preciousness of a Ti jewelry item are completely tied to its creative and unique expression and design. Fig. 4 presents a few items including rings, bracelets, and pendants, all showing attractive appearance, fine texture, and glamorous color.

3.5 Sporting goods

The application of Ti to sporting goods is largely inspired by the high specific strength and corrosion resistance intrinsic with Ti. Bikes and golf heads are such classic products (Fig. 5A). Golf heads made of Ti have a reduced weight, which allows the design of a larger head for a more accurate and longer distance shot.

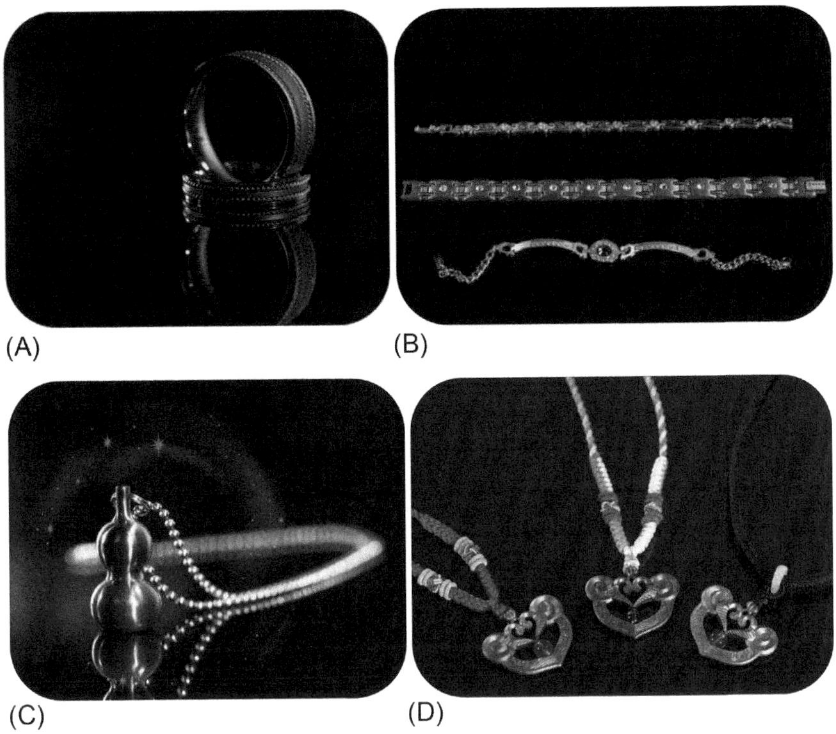

Fig. 4 Titanium jewelry items: (A) rings, (B) bracelets, and (C, D) pendants.

Fig. 5 shows three other sporting goods made from Ti: multifunctional shovel, fishing rod, and Chinese chess. Compared to their counterparts made of other materials, these Ti products exhibit advantageous features such as light weight, affordability, corrosion resistance, and nondecolorization. Take Chinese chess as an example: the passive oxide surface can effectively protect the chess pieces against corrosion by sweat and body oil, and moreover the thin layer makes the chess set washable with a degreasing agent. In addition, adequate impact toughness of Ti allows the chess set to be used roughly like slapping with no concern of the fracture or cracking.

3.6 Decorative items

Similar to jewelry, the easy coloring ability of Ti has inspired craftsmen to create a range of brilliant decorative products such as paintings, sculptures, wall hangings, etc. Fig. 6 shows two items made from Ti, which are impressive in terms of sophistication and vivid colors. Coloring through anodic or heat oxidization treatment instead of the use of pigment, dye, or paint makes

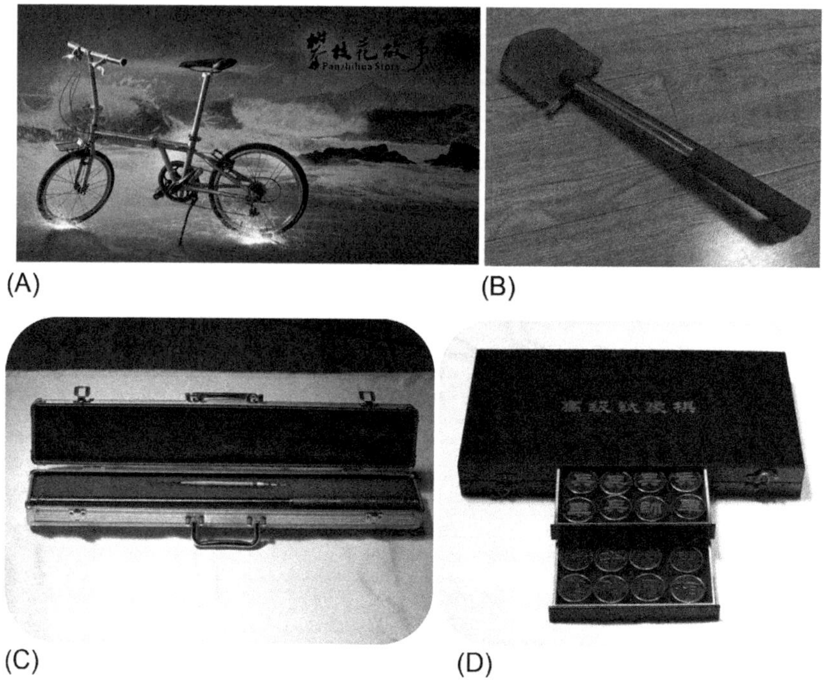

Fig. 5 Titanium sporting goods: (A) bike, (B) multifunctional shovel, (C) fishing rod, and (D) Chinese chess.

Fig. 6 Titanium artworks: (A) decorative plate and (B) engraved print.

such products extremely durable (no fading over time). The manufacturing of the decorative items in Fig. 6 is essentially a pattern/picture transfer involving engraving, acid etching, and coloring techniques. These techniques are well established and are applied to other metals, but in the case of Ti, more attention is needed considering the highly chemical reactivity of Ti.

3.7 Hair comb and hairpin

Titanium is unique among the candidate materials for hair combs such as wood, plastics, horn, or other metals such as stainless steel and aluminium. Clearly, Ti hair combs exhibit better mechanical performance than these made of wood, plastics, and horn. They do not crack, distort, or fade over time and are washable with tap water, offering better durability and easier maintenance. Compared to their counterparts made from stainless steel and aluminum, Ti hair combs have a lighter weight or more choices of colors provided by well-developed coloring techniques. Fig. 7A shows two types of Ti hair combs, both with gradient colors.

The titanium hairpin as shown in Fig. 7B is another interesting decoration, which can vary significantly in shape and surface features. Such products are advantageous largely due to their durability and easy maintenance.

3.8 Other products

Stationary is currently a small-class Ti consumer products. A limited number of products have been developed, such as ballpoint pen shell (Fig. 8), paper knife, and card case. The marketing appeal of a Ti pen shell is largely due to the durability of Ti and, more importantly, the manufacturing service in accordance to individual customer preference.

(A) (B)

Fig. 7 Hairbrush set made of Ti: (A) hair comb and (B) hairpin.

Fig. 8 Ti ballpoint pen shell.

4 Concluding remarks

Titanium consumer products for everyday life use are an emerging sector in the Ti industry with a limited number of manufacturing firms such as Seven Seven Co., Ltd.; Baoji Baoti Metal Products Co., Ltd.; HORIE Corporation; Keith Titanium Co., Ltd.; etc. To some degree, these products are accepted as artworks or light luxury items, and therefore people pursuing health, fashion, and life quality are the main customers. Largely due to the affordability and accessibility, the market scale of such products is limited relative to other sectors such as architecture and automotive parts. Nevertheless, the intrinsic attributes of Ti such as biocompatibility, nontoxicity, hypoallergenicity, and corrosion resistance provide these Ti products with overwhelming marketing advantages over similar products made from plastic, stainless steel, and porcelain. If the cost can be reduced and more manufacturers are involved, the market of Ti consumer products for everyday life use is expected to flourish in the near future.

Acknowledgments

The authors wish to thank Distinguished Professor Ma Qian of RMIT University for his encouragement, suggestions, and comments on the manuscript. Dr. Shudong Luo acknowledges the financial support from the Australian Research Council (ARC) through LP LP140100608.

References

[1] J.D. Beal, R. Boyer, D. Sanders, Forming of titanium and titanium alloys, in: S.L. Semiatin (Ed.), ASM Handbook, Volume 14B: Metalworking: Sheet Forming, 2006, pp. 656–669.
[2] E.O. Ezugwu, Z.M. Wang, Titanium alloys and their machinability—a review, J. Mater. Process. Technol. 68 (1997) 262–274.
[3] A.B. Short, Gas tungsten arc welding of alpha plus beta titanium alloys: a review, Mater. Sci. Technol. 25 (2009) 309–324.
[4] Weld integrity and performance: a source book adapted from ASM International Handbooks, in: S. Lampman (Ed.), Conference Proceedings, and Technical Books, ASM International, 1997, pp. 317.
[5] A. Zhecheva, W. Sha, S. Malinov, A. Long, Enhancing the microstructure and properties of titanium alloys through nitriding and other surface engineering methods, Surf. Coat. Technol. 200 (2005) 2192–2207.
[6] V. Vijay, K. Arjun, H. Arun Kumar, A. Vinoth Kumar, P. Chandrasekar, V. Balusamy, Review of various surface treatment techniques on titanium alloys and their protective effects against corrosion, J. Surf. Sci. Technol. 23 (2007) 49–58.
[7] M.T. Mohammed, Z.A. Khan, A.N. Siddiquee, Surface modifications of titanium materials for developing corrosion behaviour in human body environment: a review, Procedia Mater. Sci. 6 (2014) 1610–1618.
[8] E. Gaul, Coloring titanium and related metals by electrochemical oxidation, J. Chem. Educ. 70 (1993) 176–178.
[9] R.P. van Hove, I.N. Sierevelt, B.J. van Royen, P.A. Note, Titanium-nitride coating of orthopaedic implants: a review of the literature. BioMed Res. Int. (2015) https://doi.org/10.1155/2015/485975.
[10] http://www.horieusa.com/tableware/kamazukuri.htm.

CHAPTER 3

Porous titanium materials and applications

K. Pałka[a], R. Pokrowiecki[b], M. Krzywicka[c]
[a]Department of Materials Engineering, Lublin University of Technology, Lublin, Poland
[b]Department of Head and Neck Surgery-Maxillofacial Surgery, Otolaryngology and Ophtalmology, Olsztyn, Poland
[c]Department of Fundamentals of Technology, University of Life Sciences in Lublin, Lublin, Poland

Contents

1 Introduction

The development of industry in the early 1960s, especially aviation and astronautics, forced the development of new inventions in the field of innovative material assets with unprecedented properties. One of them were

https://doi.org/10.1016/B978-0-12-815820-3.00013-7

exceptionally lightweight metal foams that combine within themselves a number of attractive advantages, such as low thermal conductivity, impact energy absorption, air and water permeability, unusual acoustic properties, and good electrical insulating properties.

Porous titanium has been known since 1910 as a titanium sponge produced in the Hunter process (invented in 1910). However, it was only after World War II that titanium foam gained industrial significance as a result of the development of commercial production of titanium sponge using the Kroll process (in 1948).

The interest and demand of the industry for metal foams can be illustrated by the number of published scientific papers. A detailed literature review made by Tang et al. [1] included articles and patents since 1910, showing a significant increase in the number of publications since 1990, which indicates the intensive development of research in this area and the constantly growing interest in science and industry. Analysis of the scientific activity on Web of Science showed exponential growth of the number of publications (Fig. 1). Almost 900 articles were published in 2017, and in mid-2018 this number reached 550. It points that the limit of 1000 publications should be exceeded this year.

Increased interest in porous titanium in this time period is also associated with the development and evolution of manufacturing processes, including powder metallurgy and additive technologies [2–13]. They allow

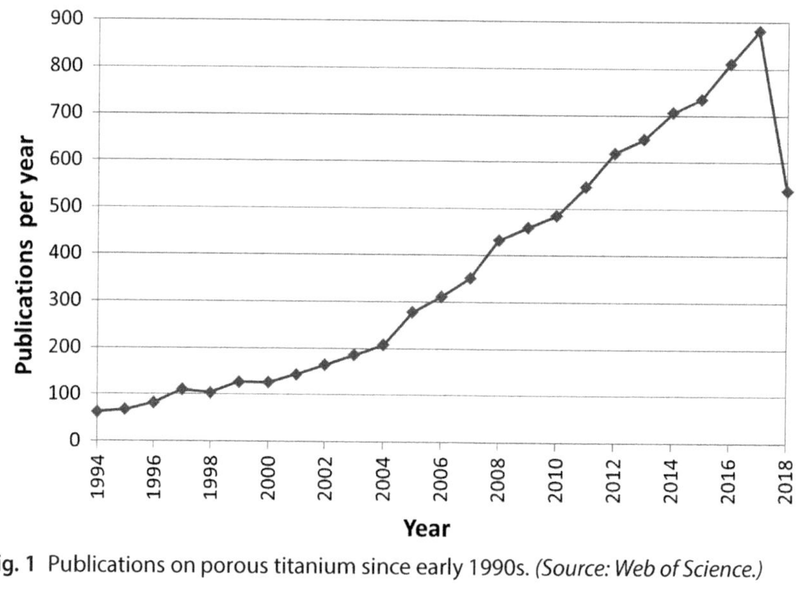

Fig. 1 Publications on porous titanium since early 1990s. *(Source: Web of Science.)*

us to obtain complicated shapes of elements and the desired microstructure to achieve required properties. Great progress has also been made in the field of design and simulation of porous structures [14–19]. Mathematical models describing the relationship between porosity structure parameters and mechanical, hydraulic, etc., properties were developed and used in FEA simulations [20–24].

Porous titanium still finds new applications, including shock and impact absorbers, mostly in sandwich structures, sound and thermal insulators, fluid filters, oxygenators, porous electrodes, bulb filaments, high-temperature gaskets and self-adjusting seals, flame arresters, catalyst supporters, etc.

The main purpose of this chapter is the systematic description of the materials, methods of design, and manufacture, as well as properties and applications of titanium foams. The first part concerns titanium and titanium alloys in solid and porous form, as well as its properties related to the chemical composition and the structure that results from the manufacturing processes. The second part describes the design methodology and manufacturing processes of porous titanium structures. The third part presents examples of commercial applications of porous titanium and titanium-related components.

2 Titanium and porous titanium properties

Commercially pure titanium shows insufficient mechanical properties in most applications. However, its corrosion resistance is significantly higher than titanium alloys [25]. Currently, many works focus on obtaining good mechanical properties of titanium alloys as well as corrosion resistance (which is responsible, among others, for biocompatibility). The works concern the optimization of chemical and phase composition [26–29].

Titanium alloys are created using three types of alloying elements:

(1) α phase stabilizers: aluminum (Al), oxygen (O), carbon (C), and nitrogen (N)

(2) β phase stabilizers:

- **(a)** isomorphous elements: vanadium (V), molybdenum (Mo), niobum (Nb), and tantalum (Ta)
- **(b)** elements that cause eutectoidal transformation: iron (Fe), manganese (Mn), chromium (Cr), nickel (Ni), copper (Cu), silicon (Si), and hydrogen (H)

(3) structurally neutral elements: zirconium (Zr) and tin (Sn).

Commercially pure titanium and α-titanium alloys are characterized by hexagonal lattice structure. They usually contain substitutional elements (as Al or Sn) and/or interstitial ones (O, N, or C) that are soluble in the α phase (Fig. 2A). What's important, fundamentally all α-structured titanium

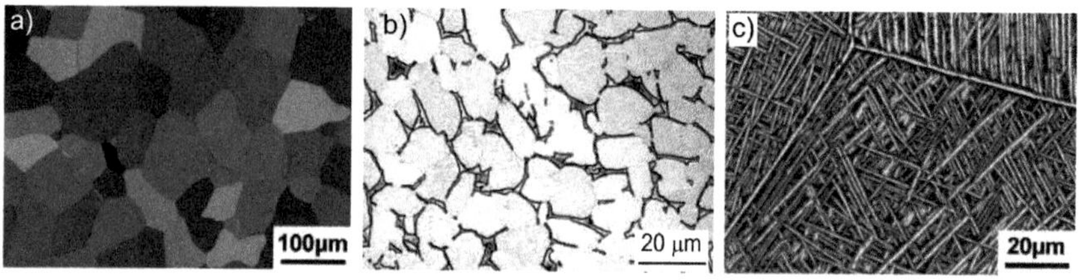

Fig. 2 Microstructures of titanium alloys: (A) CP-Ti (α alloy), (B) $\alpha + \beta$ alloy, and (C) annealed β alloy [30].

alloys contain a small amount (a few percent) of β phase, which prevents the growth of grains in such alloys [30].

Alpha-structured titanium alloys are actually not subjected to heat treatment due to a small amount of alloying additions. Strengthening of α phase alloys is possible through solid solution, by the grain size, by texture, and by precipitation hardening (formation of $\alpha_2 = Ti_3Al$ phase).

The formability of such alloys is limited by elements dissolved in solid solution and the small grains. The same applies to the twinning—the main deformation mechanism. Strengthening mechanism by texture is also achievable, however it shows an extremely directional nature with strong influence on plasticity. Precipitation hardening causes strain localization, which can considerably reduce the susceptibility to plastic deformation [30].

The two-phase titanium alloys ($\alpha + \beta$ structure) contain approximately 20 vol% of β phase (Fig. 2B), whereas when the content of β phase is less than about 10 vol% they are called "near α" alloys. These alloys can be strengthened by martensitic transformation by quenching from the temperature range of the β phase presence to the ambient temperature.

The elements, like V, Mo, and Nb, are the most frequently used β stabilizers in titanium alloys. These additives lower the range of beta phase stability to the ambient temperature. Tantalum (Ta) and rhenium (Re) are less often used because of their high densities. Zirconium and tin behave mostly neutrally; in lower concentrations, they cause a slight decrease in the temperature of α/β transformation, otherwise they increase it [30].

Dual phase titanium alloys ($\alpha + \beta$) exhibit the best mechanical properties (Table 1). Their density varies slightly depending on the alloying elements used and provides the best value of strength-to-density ratio in comparison to other metals and alloys.

Titanium alloys with a β structure (Fig. 2C) show a low value of elastic moduli [38] and are actually metastable β alloys. These alloys do not exhibit martensitic transformation through quenching from the temperature of the β phase field [30].

Table 1 Mechanical properties of bulk and porous titanium in relation to other metals and alloys [31–37].

Material	Density (g/cm^3)	Young's modulus (GPa)	Yield stress (MPa)	Tensile strength (MPa)	Elongation (%)
Ti (Grade 2)	4.51–4.52	100–115	170–483	240–550	15–24
Ti, 70% porosity	~1.35	~5	25[a]	~53	n.a.
Ti6Al4V (α + β)	4.41–4.45	110	860	930	10–15
Ti13Nb13Zr (near β)	4.65–5.10	79–84	863–908	973–1037	10–16
Ti6Ta4Sn (α + β)	~4.65	113–124	870–885	1030–1040	6–8
Ti6Ta4Sn 75% porosity	~1.16	~4.6	65[a]	n.a.	n.a.
Co-based alloys	9.15–9.25	220–230	450–1500	655–1900	5–30
316L stainless steel	7.87–8.07	193	172–690	485–860	12–40
Tantalum	16.65	188–190	138–345	205–517	1–30
Tantalum ~75% porosity	~4.16	2.5–3.9	5–13[b]	~35	n.a.
Magnesium alloy	1.75–1.85	38–65	70–140	190–250	2–11

[a] *σplateau*, compression mode.
[b] *σplateau*, compression mode, calculated according to Eq. (2).

Medical applications of titanium alloys, besides mechanical properties, require biocompatibility. Aluminum and vanadium, which are the main alloying elements, were found to be toxic or irritating [39]. Therefore, it led to the development of alloys containing replacements of vanadium, such as niobium, iron, and molybdenum, whereas in the case of aluminum replacements, elements were tantalum, hafnium, and zirconium [40]. The most popular titanium alloys in medical applications are Ti15Mo5Zr3Al, Ti12Mo6Zr2Fe, Ti30Nb, Ti30Ta, Ti6Al7Nb ($\alpha + \beta$ alloy), and Ti13Nb13Zr [41].

2.1 Porosity

Porous materials show the presence of empty spaces in their structure with different pore size, pore size distribution, and pore morphology. It is possible to produce titanium porous structures with porosity up to 98% (by using hollow titanium spheres) [42]. Porosity percent and mean pore size depend on the application. For example, the optimal porosity of medical implants efficiently stimulating bone ingrowth is in the range of 20%–50% [43] with a pore size of 100–400 μm [44] wherein it is desired that the porosity is of the open type. Open porosity provides permeability, which allows, for example, vascularization (in medical applications) or filtration or catalytic action (in industrial flow systems). On the other hand, closed porosity allows reduction of the mass, vibration dumping, and sound insulation, as well as absorbing the impact energy. Porous titanium structures may exhibit:

- uniform size and shape distribution—pore size is in the narrow range and the shape is the same in whole volume of the element
- bimodal structure—contains two groups of pores that are considerably different in size (macroporosity, with the pore size in range of 500–1000 μm and microporosity, with pore size less than 100 μm), observed especially in elements produced by powder metallurgy or additive technologies (Fig. 3)
- gradient structure—pore size varies along a given axis, in accordance with the expected mechanical and surface properties
- honeycomb structure—regular shape of pores and repeatable structure of millimeter-sized pores [1]

Microporosity plays an important role in medicine in the case of osseointegration by increasing the implant surface energy that stimulates apatite formation de novo [45].

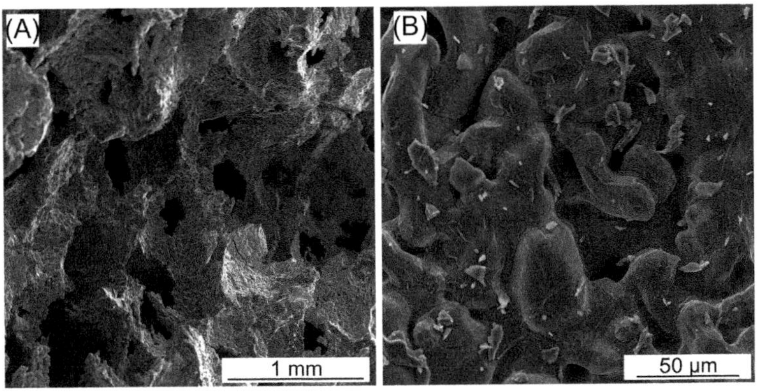

Fig. 3 SEM images of porous titanium developed by space holder technique: (A) macroporosity and (B) microporosity in wall substructure. *(Courtesy of Jarosław Jakubowicz.)*

2.2 Mechanical properties of porous structures

Porous structures exhibit different operational and mechanical properties compared to solid materials. The following will describe the influence of porosity on modulus of elasticity, static and fatigue strength, corrosion resistance, and permeability.

2.2.1 Young's modulus

A mathematical model developed by Gibson and Ashby [46] describes the relation between the elastic moduli of porous and bulk material using the relative density (RD), which is defined as the ratio of the density of porous material ρ to the density of solid metal ρ_s. Porous structures commonly have an RD value of less than about 0.3 and as low as 0.003. However, in the case of porous titanium used in medicine, this value can be in range of 0.3–0.85 [7]. The RD value can be modified during the manufacturing process. For an open cell structure, the relation between elastic moduli and RD can be described as:

$$E / E_s = C_1 (\rho / \rho_s) \tag{1}$$

where E and E_s are the elastic moduli for porous and solid materials, respectively, and $C_1 \approx 1$ includes all of the geometric constants of proportionality [46].

The previous equation indicates approximately the linear relationship between Young's modulus for porous structures with respect to porosity. Literature data indicate an exponential relationship (Fig. 4), These data, however, come from studies of various structures, with different pore sizes

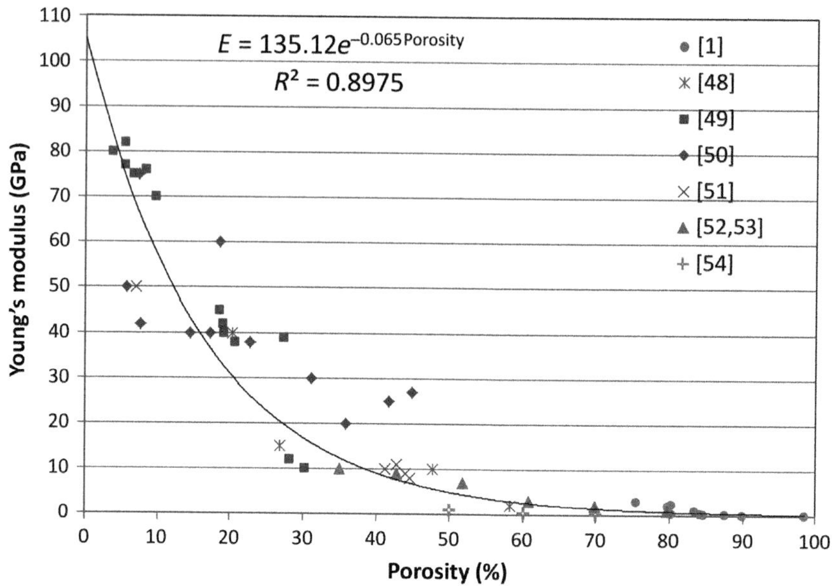

Fig. 4 Young's modulus versus porosity [1, 47–53].

and their morphology [1, 47–52]. Nevertheless this indicates a strong dependence of the elasticity modulus on the structure, unlike solid materials, which basically eliminates linear dependence.

In the case of small strains of porous materials, Young's modulus depends on the type of pores—open or closed. Open-cell structures deforms by wall bending for low RD value. Increasing the wall thickness (RD value more than 0.1) leads to an increase in the proportion of pure tensile or compression of cell walls. The cell edges in closed pore material can bend whereas the cell walls stretch resulting in increased stiffness. The presence of gas locked in the cells also increases their stiffness until the cell is unbroken [46].

In the case of higher strains, the edge can buckle. It leads to a decreasing of the elasticity modulus of the porous structure, whereas in tension mode it increases its value [46].

2.2.2 Compression strength

The strength of porous materials is significantly reduced compared to solid materials due to the reduced cross-section and the stress localization in thin walls. For this reason, such structures are mainly used in compression mode load-bearing applications.

The behavior of porous materials under compression load is described in detail by Gibson and Ashby [46]. Such materials show the presence of a plateau region on the stress-strain curve. In this region, the stress is nearly constant in a wide range of strain. Some materials with low plasticity, including titanium alloys, exhibit stress changes due to wall cracking and material densification in the plateau region. This allows the use of such materials in cases of energy absorption applications (impact energy), where large deformation arises as a result of the constant low-stress action.

Increasing of the porosity, expressed as the inverted RD value, decreases the plateau stress of the material. Thin cell walls plastically collapse (σ_{pl}) with simultaneous material densification. Most porous titanium structures collapse by the brittle crushing σ_{cr} [53, 54]. The value of σ_{pl} (or σ_{cr}) can be estimated as the function of yielding stress σ_{ys} including the RD [46]:

$$\sigma_{pl} = C \cdot (\rho / \rho_s)^n \cdot \sigma_{ys} \tag{2}$$

where $C = 0.3$ and $n = 3/2$ for a stochastic open cellular metal foam.

The compressive strength of porous pure Ti with a porosity of 50%–70% may reach 25–5 MPa, respectively [53]. Increasing the porosity of titanium and its alloys enhances potential suitability, for example, for medical applications. Unfortunately, it also decreases material strength and fatigue resistance [55].

Fig. 5 presents compressive strength in relation to porosity formulated basis on the literature data [1, 47–53, 56–58]. Similarly to the Young's

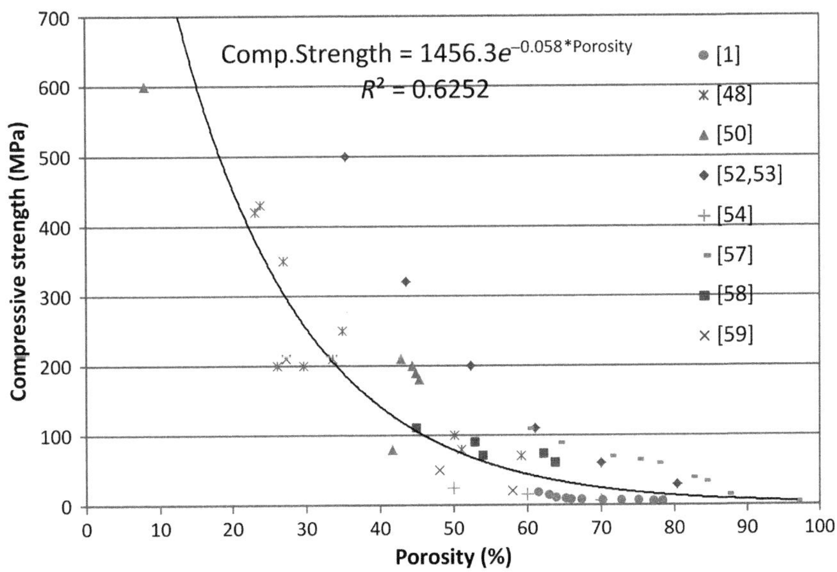

Fig. 5 Compressive strength of titanium foam in relation to porosity [1, 47–53, 56–58].

modulus, the exponential dependence of compressive strength on porosity is observed. However, it should be noted that the pore size, shape, and morphology influence the compressive strength [53]. It is noteworthy that porous titanium alloy has a several times higher strength in relation to CP Ti.

There are also some disadvantages that can limit the potential applications of porous titanium structures. The first is susceptibility of titanium to crack propagation, especially as a combined effect of stress and corrosion at a preferential location along the grain boundaries. It may limit the period of Ti implant's use to only 10–15 years [59]. Strain localization in CP-Ti or Ti-α alloys leads to shear cracking of varying intensity. In the case of α + β alloys, the preferential plastic deformation in the α regions involves concomitant premature crack nucleation in these areas. In β-phase alloys, crack nucleation appears as an effect of the local stress concentration at the grain boundary [30].

2.2.3 Fatigue strength

The fatigue behavior of porous materials can be described in three stages. In the first stage, a slow change of strain is observed. In the second stage, the minimum strain accumulation occurs. The final stage ends with sample fail within a limited number of cycles, preceded by a rapid increase of strain [60]. Structures of higher porosity have a significantly lower fatigue life than low porous materials, and the normalized endurance limits of the porous structures are significantly lower than that of solid titanium alloy [60]. None of the tested porous selected laser-melted samples of Ti6Al4V alloy could withstand 10^6 compression cycles even when the load was as low as $0.2\sigma_{ys}$. The fatigue performance was more than 75% lower than wrought material due to surface finish, porosity, and residual stresses [61].

Fatigue strength of porous Ti-6Al-4 V can be also evaluated using Eq. (2) assuming $n = 2.7$, which is a higher value than for aluminum and nickel foams [62].

Materials that combine porosity with strength higher than traditional porous materials are porosity-graded structures. Such materials, in addition to high load capacity, also show, for example, good osseointegration or permeability [13]. Further reduction of porosity and "sliding" it toward the surface leads to obtaining solid materials with more favorable tribological properties and acceptable osseointegration [63, 64].

2.3 Permeability

Permeability is the property by which we can know the ability of a material to transmit fluid/gases under the pressure difference. Darcy's law describes the permeability coefficient, K:

$$K = \frac{Ql\mu}{A \cdot \Delta p} \tag{3}$$

where Q is fluid flow rate, μ is viscosity of the fluid, l is sample thickness, A is a cross-section area of the sample, and Δp is pressure difference.

The permeability of porous titanium depends not only on the fluid but also on the porosity structure [65]. Work of Yang et al. [66] showed it's possible to control the permeability by varying the size of initial powders. In their study, Ti48Al6Nb porous alloys with various pore sizes were tested (Fig. 6). They developed a formula that describes the relationship between permeability and pore size:

$$K = \varepsilon \frac{D^2}{32\gamma^2} \tag{4}$$

where ε is porosity, D is diameter of pore, and γ is geometric parameter describing pore shape and tortuosity.

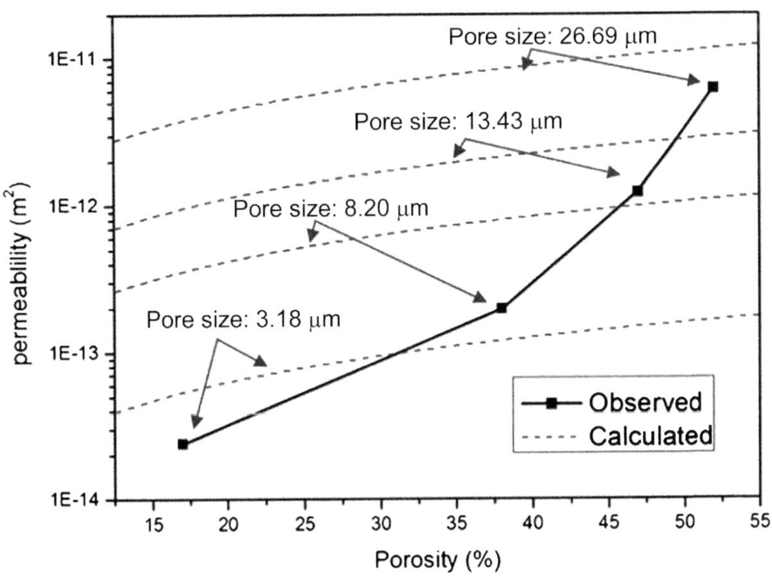

Fig. 6 Permeability as function of porosity for TiAl alloy/N_2 system [66].

Similar results were obtained by Furumoto et al. [67] who described permeability of porous metal structures made of 42CrMo4-Cu_3P-Ni alloy.

It should be emphasized here that the fluid transport through a porous structure is highly dependent on pore size and surface wettability [68]. According to the Laplace equation, fluid transport across two neighboring porous layers is increased when liquid is transported from a hydrophobic layer into a hydrophilic layer, otherwise transport is reduced. Pore size and porosity difference between the two adjacent layers can also affect the fluid transport rate [69].

2.4 Corrosion resistance

The corrosion resistance of titanium and its alloys, as well as other metals, depends on the chemical composition of the material. Suitable alloying elements play a major role in the corrosion resistance of titanium alloys. Addition of Zr, Nb, Ta, and a small amount of Pd decreases the value of critical current density for passivation I_c. The opposite effect on I_c is observed when an addition of Sn is used instead of Zr in Ti alloys [25].

An increase in the porosity of metals leads to a lower corrosion potential value (E_{corr}) value, which results in increased susceptibility of porous materials to local corrosion. It is assumed that the relatively small pores present in the cell walls favor electrolyte placement and oxygen depletion, which is important in the stability and preservation of the oxide layer on titanium. Elements with higher porosity, with open and interconnected pores allow easier electrolyte flow, which complements the oxygen supply during the passivation process [70].

Porous titanium show higher values of E_{corr} when compared to solid counterparts, most probably due to the effect of sinters surface oxidation [71]. Dabrowski et al. [72] came to similar results. They studied the influence of porosity of titanium ranged from 45% to 75% on corrosion resistance and showed that material with higher porosity exhibited less susceptibility to corrosion than those of 45% porosity. However, both elements exhibited lower corrosion resistance than the solid Ti [72]. Chen et al. also confirmed this [73]. Furthermore, they point that the electrolyte flow can also decrease the corrosion rates of both solid and porous materials.

Porous structures show certain hydrophobicity due to the trapping of gas bubbles. This may increase the corrosion resistance due to the limitation of Cl^- active ion transport. Lin et al. described hierarchical 3D porous structures with high capability to trap air bubbles [74]. As shown earlier, there is no unambiguity in assessing the corrosion resistance of porous elements made of titanium or its alloys.

In industrial applications, the elements come into contact with each other often in a corrosive environment. However, the results of tribocorrosion tests are promising. Tribological pairs immersed in corrosive environment presented higher open circuit potential E_{ocp} values, indicating a low susceptibility to corrosion. However, changing the way of contact on sliding diminished values of E_{ocp}. As an effect of sliding, the less mechanical damage on the functionalized surfaces there was observed [75].

The corrosion process can change the porous structure of titanium and weaken the material mechanically. It is therefore desirable to improve the resistance of porous materials to corrosion for example by nitriding or surface alloying with palladium [1].

3 Design and manufacturing of porous titanium structures

3.1 The porous structure design

The porosity of elements is a compromise between mechanical strength and adequate pore size to obtain certain operating properties. For example, porosity in implants provides the space for cell migration as well as for vascularization and new tissue formation. In addition to pore size, the shape of the pores is also an important feature.

Design requirements for titanium porous structures in many cases are contradictory. Especially in medicine, high diffusivity and permeability are required in many tissue engineering applications, whereas limited mass transport is essential in regeneration of cartilage due to the low metabolic rate. Furthermore, mechanical stiffness requirement is in contradiction to the mass transport. The optimization of the element's architecture is still a constant challenge in porous titanium engineering [14].

Functions performed by porous elements force creation of their optimal microstructure, which guarantees obtaining the required mechanical properties, permeability, or energy absorption. However, current production methods and virtual models from which the elements are made have some limitations resulting from the design methodology. Therefore there is a necessity to simplify the model, which match its inner and outer characteristics to specific applications and maximizes its functionality. To do so, two design approaches—irregular (random) and regular structures—are preferable.

3.1.1 Irregular structures

Irregular structures were developed as the mimetic structures of a number of natural materials, for example, bone or wood. Precise imitation of trabecular bone structure in biomimetic scaffolds was possible due to the use of tomography [15]. However, modeling and structural optimization of random pore architecture is still very difficult due to the varying of pore size and shape. Such porous materials are generally manufactured by powder metallurgy [16] or gel-casting [17]. A slightly different manufacturing process was presented by Singh et al. [18], where authors used two processes: production of a porous oxide precursor by sol gel method, followed by electrochemical reduction to a metallic foam. The potential advantage of this method is an ability to tailor the porosity at several scales independently. The newest techniques of manufacturing such structures are based on rapid manufacturing processes where CAD-based models are implemented [19].

3.1.2 Regular structures

Regular structures of porous titanium are a very convenient form both for designing, as well as modifications or simulations, and are most often based on simplified CAD models.

CAD-based design is established on solid or surface modeling systems. In the case of the solid system, models are the composition of standard primitives combined by Boolean operators. This method allows us to create regular solids not including dangling edges. Design using boundary representation is realized by describing solids by their boundaries, including sets of vertices, edges, and loops, however, without regard to the relationship between them [14].

CAD-based FEA simulations can be useful in an optimization process [76]. The cellular architectures mimicking the cuttlebone applied in medicine were described by Chen et al. [77]. CAD-design procedures are often associated with planning medical procedures and implants manufacturing. These are based on computed tomography and surface optical scanning to reveal a CAD model [78], and it was combined with image-based design, subsequently described.

Image-based design was developed by Hollister et al. for the purpose of design and manufacturing of scaffolds intended for a specific patient and location, and having a required, specific structure [79, 80]. The method is a combination of imaging, image processing, and design process connected with manufacturing techniques (Fig. 7). The manufacturing of scaffolds using image-based design consists of utilizing the virtual intersection between 3D images: the first image representing the defect to be reconstructed, and the second one consisting of a binary unit cell stacking.

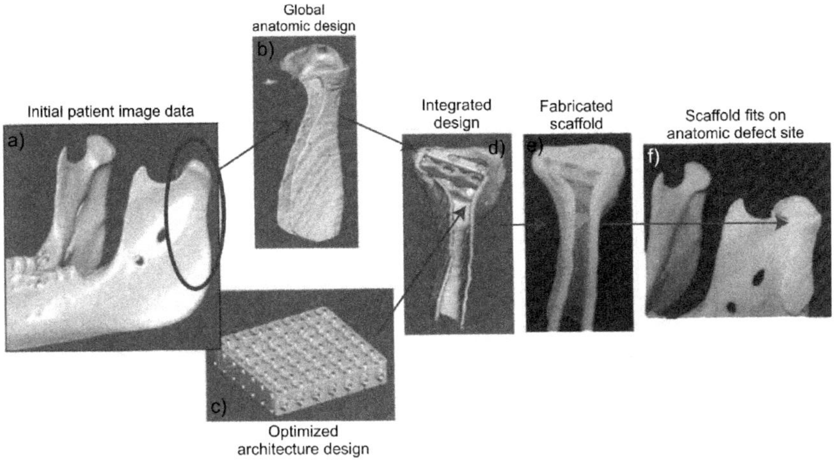

Fig. 7 Image-based design in integration of designed scaffold with anatomical shape: (A) a CT or MRI scan of defected area; (B) modeling the surface geometry; (C) design of internal architecture; (D) integration of (B) and (C) designs by Boolean image techniques; (E) fabrication by additive manufacturing; and (F) placing the scaffold [80].

Implicit surfaces modeling describes the architecture using a single mathematical equation including such boundary conditions as arbitrary pore shapes and pore size gradient. This is a highly flexible method developed by Gabbrielli et al. [81] and Pasko et al. [82]. Examples of such structures are presented in Fig. 8, the trigonometric functions describing the shapes can be found elsewhere [14].

In addition to the design methods previously mentioned, in the case of titanium and its alloys, *the self-organizing technique* [83] and *space-filling curves design* [84, 85] are also possible.

3.2 Optimization of design

The assessment of porous structure design is usually performed using FEA simulations. The results of such assessment are feedback to the design process, as a result of which an optimal structure can be obtained. The software currently present on the market (for example, Abaqus) offers optimization of geometry to obtain the desired utility properties.

The optimization process is realized for a specific purpose. Most authors describe the design of porous elements with subject-specified external shapes and complex structures as well as an evaluation of its mechanical properties [62, 66]. Such results can be considered as a reference point for

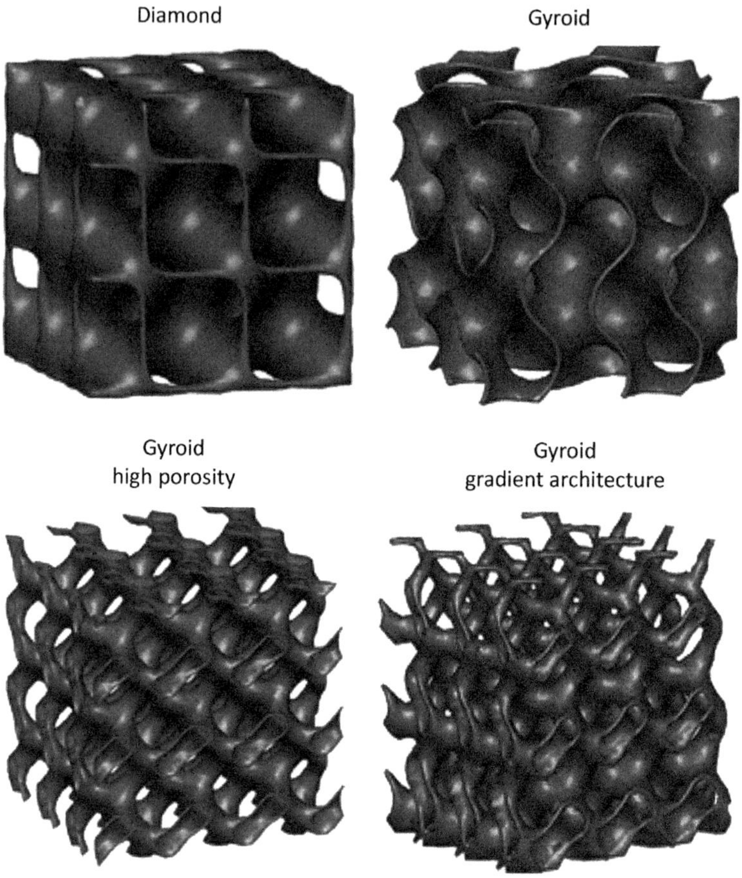

Fig. 8 Porous structures design using implicit surface modeling [14].

the objective of design [44]. In cases where mechanical properties are not so important, works are concerned on the effect of porosity on diffusion process [86] or permeability [87]. However, final verification takes place under real test conditions, because the properties of porous materials can change under operating environments [88].

Iterative design [89] includes the design, optimization, manufacturing, and validation of porous elements. The requirements of the porous element (e.g., structural, mechanical) and the production capability are the initial parameters in the design process [14]. According to this, required material can be selected, and its properties can be applied in appropriate software together with the designed model. The following stage of the FEA analysis is performed in terms of mechanical, hydraulic flow, thermal, etc. behavior of the designed porous element. Each stage of iterative design ends with an evaluation when a positive result starts the next stage. A negative result of middle-stage evaluation requires the return to the early stage of design after improving the initial settings [14, 89].

3.3 Functionally graded porosity

In cases where mechanical properties as well as and good surface permeability are important, the elements with functionally graded porosity should be used (Fig. 9). Functionally graded scaffolds were presented by Khoda et al. including improved permanence and interconnections in areas of dissimilar porosity [90]. The graded structure created by a set of fractal space-filling curves based on CT data was developed by Pandithevan et al. [91]. CAD-based models characterized by high design flexibility are also used to develop the elements of graded porosity. In this case, the collections of unit cells' provided by libraries can be easily used [92]. Moreover, to simplify the CAD design of graded porosity elements, comprehensive databases were established that include the architecture, compressive stiffness, and permeability of many different unit cells [93].

A different case of functionally graded structures are sandwich panels—the light and stiff constructions, consisting of two skins adhesively bonded onto a core material made of aluminum, titanium paper, or other materials.

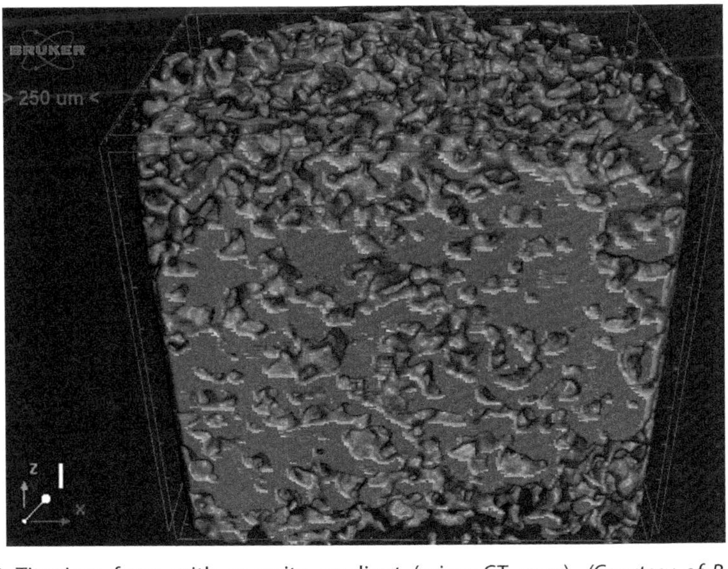

Fig. 9 Titanium foam with porosity gradient (micro-CT scan). *(Courtesy of Barbara Szaraniec (AGH University of Science and Technology) and Krzysztof Pałka.)*

In classic panel constructions, the lightweight materials (mostly consisted of polymeric microsized porous structure) allowing for high stiffness are inserted between the metal sheets of complicated shape. Such composite structures are preferred in high performance applications for filters, catalyst electrodes, heat shields, medical implant, and so on [94]. It is promising that the use of porous metal will allow obtaining high Young's modulus and high functionality. One of the production methods that allows obtaining such structures is metal injection molding (MIM). This method uses a combination of powder metallurgy and plastic injection molding. The particles of space holder are coated with a metal powder using a binder agent. The use of appropriate gradation of particles in subsequent phases of creation enables obtaining a graded porosity. In the next stage, the binder and the space holder particles are removed, then the green compact is sintered [95]. Such microporous material can be applied in many constructions.

3.4 Porous surfaces

Porous surfaces are extremely important in the case of medical devices, especially due to the use of cementless implants [96]. Such surface aims to improve implant performance concerning implant fixation, wear, and corrosion properties, considering it affects tissue ingrowth. The preparation of titanium surface can be made by a number of mechanical, thermal, chemical, electrochemical, and vacuum-based treatments, individually or in combination [97–99]. Usually, in the next stage, a porous, biocompatible material (for example, hydroxyapatite, HA) is applied to the prepared surface. Currently, the most popular technique is plasma spraying. This method produces highly porous surfaces with open and interconnected pores, which can significantly increase bone ingrowth [97, 100, 101]. Furthermore, using plasma spraying it is possible to tailor the elastic moduli of the porous surface to match that of cancellous bone [102]. Other methods for producing porous biocompatible layers are chemical methods [103], laser [104], and sputtering methods [101].

A separate group of surface layers on implants are those made of a basic material as a result of processing or by creating a covering made of different metal. Such surfaces can be obtained by, for example, plasma [105], laser treatment [104], electron beam melting (EBM) [56], and many others.

Surface porosity, and in particular porosity with a trabecular structure, favors obtaining both initial and long-term stabilization of the implant. The friction coefficient of such surfaces is significantly higher than that obtained for laser- or plasma-treated surfaces or elements manufactured in AM technology [106]. This enables direct bone apposition by creating scratches that increase initial implant stability. Moreover, it has a higher bone interface shear strength than sintered beads [107]. Surface trabecular structures, however, are mainly made of tantalum, not titanium.

Changing the tribological properties of the titanium surface can also be obtained by laser surface texturing (LST). Titanium alloys have a poor tribological behavior, including high friction coefficients, severe adhesive wear, and sensitivity to fretting wear, mainly due to poor shear strength and low hardness. The development of surface morphologies could improve the tribological characteristics by different mechanisms, for example, by debris trapping inside the texture, by increasing in load-carrying capacity through a hydrodynamic effect, or by creating micrometric reservoirs for lubricants [108, 109]. The wear debris is trapped in the dimples, thus reducing abrasive wear at the sliding interfaces. Even though various designs and optimization of LST patterns have been investigated, the process design methods are still dominated by "trial and error" approaches. There are wide variations in "optimum" designs obtained by various researchers. In unlubricated condition, the wear debris and material stick contribute the main friction forces, and the large dimple diameter with high-textured area proportion could be more effective in friction reduction by trapping debris and reducing contact area [110, 111].

3.5 Manufacturing processes

Shaping of solid titanium or its alloys is a very difficult process due to different deformation mechanisms of these materials under machining conditions in comparison to commonly used steels or aluminum alloys. The low thermal conductivity of titanium reduces the removal of heat, resulting from the cutting process, through the metal and transferred to the fixture or to the air. The temperature increase caused by this effect causes dimension incorrectness and quick wear of the cutting tools. The spring action involved by elasticity and high ratio of yield stress to tensile strength (value about 0.9) creates vibration and elevation of friction and can further raise the temperature [112].

The melting temperature of titanium-based alloys reaches 1610–1670°C. Such conditions and the high reactivity of titanium at elevated temperatures impedes the use of technological processes, such as casting or powder metallurgy [2].

Due to complex shape and structure of products and the costs of production, several manufacturing processes cannot be applied for industrial or medical devices. There are many techniques for producing titanium porous structures, which are described in detail by Qian and Froes [3]. Only the most important and the widely used techniques will be discussed in this chapter.

Powder metallurgy is a very flexible and relatively simple technique for the manufacturing of porous elements made of titanium or its alloys. It is a low-cost method and allows reduction of material losses [6]. Elements with complex geometry that meet high mechanical requirements can be produced using this method.

Production of elements by powder metallurgy can be done by hollow spheres sintering, thermal decomposition, and sintering of powders, as well as compressing and sintering of titanium beads or fibers [7, 8]. Mixture with a fixed composition of titanium powder and removable space holder, which allows obtaining a porosity with a specific structure, were described in the literature [3, 9–11]. Materials used as space holders are saccharose [53], and salts, like sodium fluoride, sodium chloride [12], ammonium hydrogen carbonate [13], as well as polymer granules [4]. Attempts to use magnesium as a porophor to manufacture porous implants have also been described [5]. Nevertheless, the residues of certain porogens may induce the unfavorable creation of inclusions within the foam [5, 12].

The additive manufacturing methods (AMM) usually utilize CAD models to create 3D structures [77]. These technologies enable production of porous elements with the highest accuracy [113] and complex geometry [114], having specific unit cells [85] and are not very time consuming [91]. In short, the AMM are the most progressive and promising techniques.

Manufacturing of porous titanium elements by AMM can be classified in two categories: powder bed fusion (PBF) and directed energy deposition (DED), which were presented in detail in Table 1. Each AMM is characterized by process parameters: energy density, scanning speed, and layer thickness, all of which affect the resulting properties [91].

The PBF consists of spreading the thin layer of metal powder (as thin as the powder diameter) on the work surface (bed) followed by bonding regions (defined in CAD design) using an energy beam (a laser or electron beam). After this, the powder bed is lowered by a height corresponding to the accuracy of manufacture, and the next quantity of powder is applied onto the working area. This allows the elements to be made with accuracy equal to the powder diameter, which is a great advantage of this method. However, the dimensions of working area limit the size of the manufactured part.

The mostly known techniques in the PBF category are selective laser sintering (SLS) [115], selective laser melting (SLM) [116, 117], and EBM [118, 119] as well as direct metal laser sintering (DMLS) and laser melting (LM) or laser cushing [120, 121]. The EBM requires a vacuum environment as opposed to laser techniques using protective inert gas. Vacuum environment, despite the higher costs, shows benefits of lower residual stresses compared to laser-based systems due to slower cooling of the created element.

The *direct energy deposition* methods use a direct metal powder feed into the model building area. The source of energy, like in the PBF methods, is a laser or electron beam (Table 2). These methods allow simultaneous use of various metals during the manufacturing of the porous element. The disadvantage of these methods is the low resolution, reaching only 0.2 mm. The DED methods are characterized by a high rate of model creation and have no limitations as to the size of manufactured elements, in contrast to PBF methods [123–125].

Sintering or melting of titanium powder in both categories of manufacturing affects the microstructure and properties of the obtained elements. These microstructures presented in Fig. 10 show differences in relation to the equilibrium microstructures shown in Fig. 2 due to fast cooling of deposits.

The formation of α'-martensite are observed in elements made by AMM (Fig. 10A and B), which results in higher strength and lower ductility [122]. In the case of electron beam processing, the slower cooling in the vacuum environment causes the formation of $\alpha + \beta$ microstructures (Fig. 10C), which results in lower strength and higher ductility [122]. A large amount of heat added into the deposit in the wire and arc additive manufacturing (WAAM) method creates a harsher microstructure including the presence of the Widmanstätten structure (Fig. 10D) [126]. Such an unfavorable microstructure, however, does not weaken the element; products manufactured by WAAM of Ti6Al4V alloy have only 10% lower strength than elements produced by classical methods [126].

4 Applications of porous titanium materials

One of the criteria for the use of porous titanium applications in industry and medicine is the type of porosity (Fig. 11). Some of them will be discussed in the next part of the chapter.

Table 2 Industrial deployment of the additive manufacturing methods for processing titanium and titanium alloys [122].

Category	Technology	Company	Description
Direct energy deposition (DED)	Direct metal deposition (DMD)	DM3D Technology LLC	Laser/metal powder—melting and depositing, patented close loop process
	Laser engineered net shaping (LENS)	Optomec, Inc	Laser/metal powder—melting and depositing
	Direct manufacturing (DM)	Sciaky Inc.	Electron beam/metal wire—melting and depositing
	Shaped metal deposition or wire and arc additive manufacturing (WAAM)	Not commercialized yet (patented by Rolls Royce Plc.)	Electric arc/metal wire—melting and depositing
Powder bed fusion (PBF)	Selective laser sintering (SLS)	3D Systems Corp	Laser/metal powder—sintering and bonding
	Direct metal laser sintering (DMLS)	EOS GmbH	Laser/metal powder—sintering, melting and bonding
	Laser melting (LM)	Renishaw Inc.	Laser/metal powder—melting and bonding
	Selective laser melting (SLM)	SLM Solutions GmbH	Laser/metal powder—melting and bonding
	LaserCUSING	Concept Laser GmbH	Laser/metal powder—melting and bonding
	Electron beam melting (EBM)	Arcam AB	Electron beam/metal powder—melting and bonding

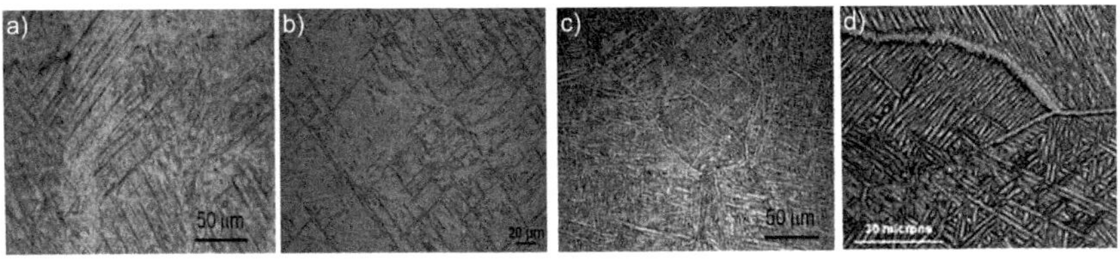

Fig. 10 Typical microstructures of Ti6Al4V obtained by various AMM: (A) DLMS, (B) DMD, (C) EBM, and (D) WAAM [122].

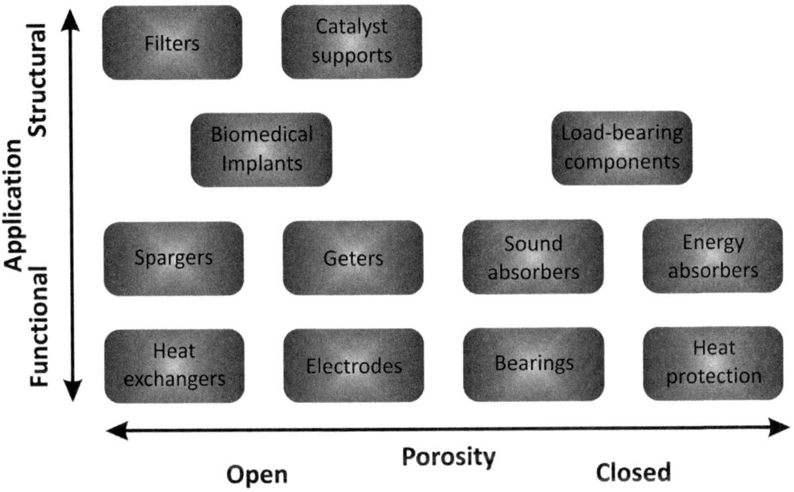

Fig. 11 Applications of porous titanium as the function of porosity [1, 127].

4.1 Medical applications

The advantages of titanium, especially when applied in medicine, are its biocompatibility and excellent corrosion resistance. Commercially available dental implants are usually made of solid titanium. Its surface has been investigated ever since Brannemark described the phenomena of osseointegration, which is a "direct connection of living bone with an implant surface" [128]. The fundamental issue in the use of porous titanium for implants is the porosity of the surface, which should be similar to the tissue porosity (e.g., bone). Such porosity ensures primary osseointegration and vascularization, and also adjusts Young's modulus to a value close to that of bone, which eliminates the effects of stress shielding. Titanium of commercial purity (Grade 2) is the most preferred material when it comes to biocompatibility, but it has relatively low mechanical strength. Therefore, titanium alloys are being developed, but several of their components can be harmful or even toxic, as it was mentioned in Section 2.

The overriding requirement for medical implants is that they have a porous surface to ensure bone ingrowth and thereby improvement of mechanical stabilization [129]. It can be obtained by creating a total porous structure of the implant or by a graded porosity (as explained in Section 3.3) with a solid (or with very low porosity) core and porous covering. The graded structures can be manufactured mainly by plasma spraying, including modifying its chemical composition by using different chemical compounds: titanium oxide, calcium phosphate, or hydroxyapatite [130]. Other methods of producing porous shells are sand blasting with stiff particles [131], a laser micro-machining technique [64, 132], anodization, including nanotubes [133], electron-discharge compaction [134], and the microwave processing method [135]. Additionally, change in roughness also affects the improvement of tribological properties, including in friction pairs, for example, in the hip joint. Obtaining the desired properties is possible due to LST (Fig. 12), as its main parameters are the depth and the diameter of pockets and the area ratio [64].

The 3D printing technology provided the possibility of production of individual metallic biomaterials for reconstruction of craniofacial defects and anomalies. Such may be due to congenital birth defects (clefts, deformations), postablative surgeries (tumors, cancer), and as a result of trauma [136]. Tissue engineering based on digital imaging software platforms and 3D

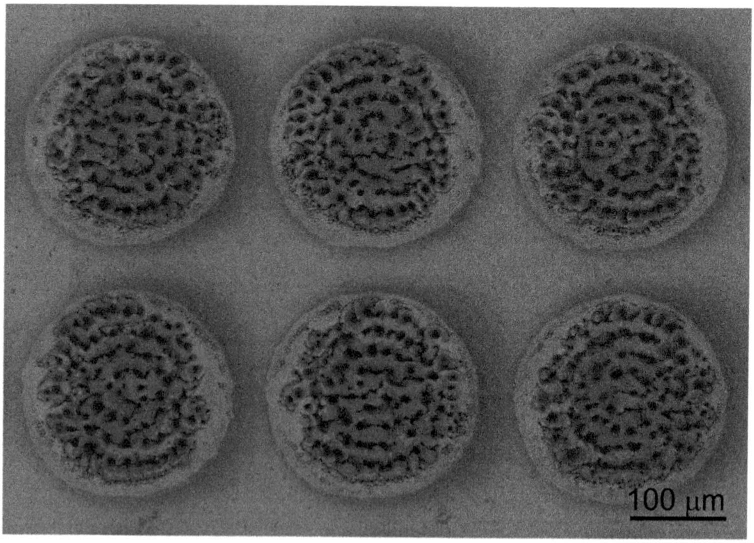

Fig. 12 Laser-textured surface of titanium alloy improving tribological properties. *(Courtesy of Monika Krzywicka and Krzysztof Pałka.)*

printing provides better tools in the treatment of large defects of hard and soft tissues of the head and neck. By using porous implants, it is now possible to print individual bone stabilization plates that are digitally curved to the desired angulation. Such plates help in positioning of the bone fragments and precise fixation, for example, in corrective and orthognathic surgery. Porous titanium sinters are lighter than their solid counterparts. Therefore, they may be used for manufacturing of large custom-made implants for the reconstruction of craniofacial defects, for example, temporo-mandibular-joint prosthesis, cranial vault, orbital floor, or any anatomical site of the head and neck. Also, digitally designed titanium sinters are used as frame support implants for large autologous bone grafts used in multidimensional reconstruction of postcancer-related orofacial defects (Fig. 13) [137]. Such an approach provides more precise surgical planning, time efficient procedures, and reduction of postoperational morbidity of the patients.

A material with excellent biocompatibility and super elasticity, and whose mechanical properties can be modified to meet medical demands, is a porous titanium-nickel alloy, which also exhibits a shape memory effect. Duan et al. presented a gel-casting technique of porous alloy producing with low cost and a wide range of applicability, particularly for producing elements of bigger size and complicated shape showing the transition temperature in the range −50°C to 40°C [138].

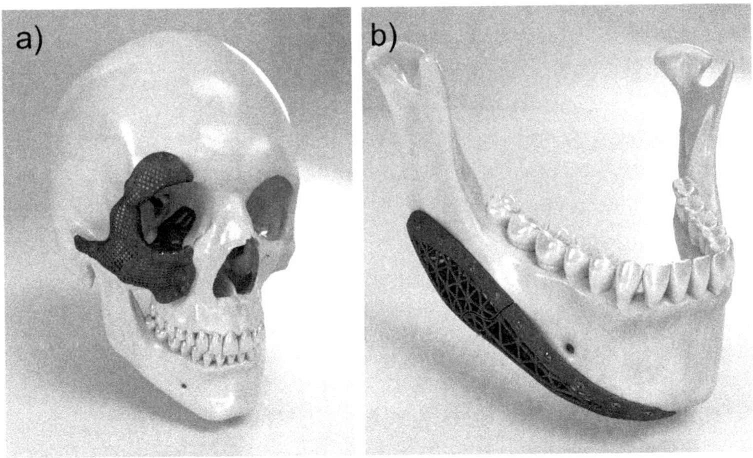

Fig. 13 Examples of individually prepared titanium plates based on 3D skull scan for the reconstruction of the midface (A) and individual, multidimensional plate for stabilization of bone graft reconstruction of postablative mandible defect (B). *(Courtesy of KLS Martin, Gebruder, DE.)*

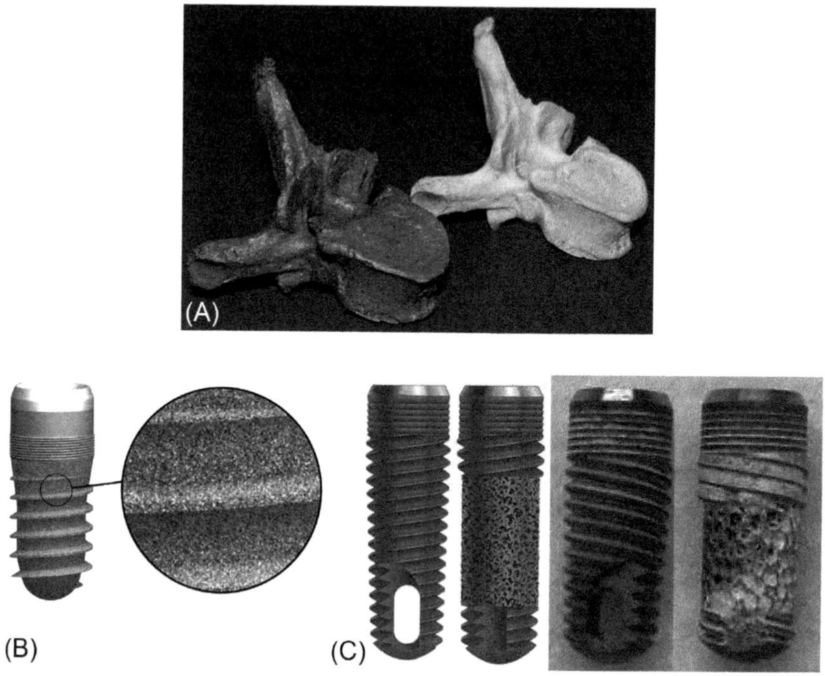

Fig. 14 Examples of porous titanium applications in medicine: (A) CT-based part of a complex human vertebra [141], (B) titanium dental implant with the surface of 2–200 μm-sized pores, and (C) dental implants before and after implantation with visible bone debris on the contact surface [142]. *(Part B: Courtesy of Leader Italia.)*

A dynamic increase in interest in medical devices made of porous titanium has been observed since the early 1970s [139]. These devices demonstrate favorable conditions for osseointegration with the possibility of mechanical property tailoring [7, 44, 47]. The possibilities of their production arose at the moment of the appearance of additional manufacturing techniques and surface treatment techniques using a laser or electron beam. Examples of porous titanium implants are shown in Fig. 14. The range of porous titanium products includes implants for dentistry [7], hip [63] and knee joint prostheses [143], plates for osteosynthesis [144, 145], implants replacing craniofacial bones [140], spinal fusion elements [146], and many others.

4.2 Titanium foam in aerospace industry

The main reasons for using titanium in the aerospace industry are: weight reduction (principally as a steel substitute), space limitation (replacement for Al alloys), maximum service temperature (replacement for Al, Ni, and steel alloys), excellent corrosion resistance (substitute for Al and low alloy steels), and finally composite compatibility (replacement for Al alloys). One of the first applications of titanium in civil aviation was the landing gear beams on the Boeing 747 and 757, which are some of the largest titanium forgings produced. Titanium alloys were also widely used in military aircrafts; a special case is the SR-71 "Blackbird" reconnaissance airplane, where titanium was about 95% of the structural weight. [147]. The largest group of titanium alloys used in aerospace industry are the two-phase structure $\alpha + \beta$ alloys. Their participation in the total titanium production reaches about 60%. The Ti6Al4V alloy accounts for 80% of all titanium alloys used in airframes. Alloys in this group, due to the heat treatment susceptibility, have the best strength, fatigue, and fracture properties. Their compatibility with composites allowed for its use, among others, in the fuselage, nacelles, landing gear, wing, and empennage, moreover fasteners of each type, static and rotating components of gas turbines, fan and compressor disks, and blades [147].

Porous materials have a high capacity to absorb impact energy due to plastic deformation of the walls and struts of the structure. The impact of a large bird on bearing surfaces or the fuselage of a plane can seriously damage a plane flying at high speed, even to a catastrophic point and make flight continuation impossible. In addition, the presence of pores influences the reduction of thermal conductivity and the improvement of sound attenuation; these aspects will be described in the next section. Titanium foams, showing a relatively high value of strength-weight ratio, are therefore a promising material [148]. Twin-skinned sandwich panels for aerospace structures using titanium foam as the cellular core can be found in fuselage, wing, and other components of the aircrafts [149]. The work of Mines et al. showed that Ti6Al4V BCC microlattice structures can replace aluminum honeycomb in the case of impacts in aerospace sandwich panels [150]. The AMM techniques also show new possibilities for engineers to design lightweight and shape-optimized parts for aircrafts [151]. Tancogne-Dejean et al. developed the architecture of microlattice materials for high specific energy absorption under static and dynamic loading. They proved that the specific energy absorption of lattice materials uniformly increases in dependence to RD. The strength of such materials increased by about 30% when increasing the strain rate from 10^{-3}/s to 10^{3}/s. The strain rate sensitivity of the basic material was strongly affected by this effect [152]. Even better results were obtained by Xiao et al. who tested the functionally graded structures made of Ti6Al4V by SLM. The specific strength and specific energy absorption of graded lattice structures of RD in the range 0.139–0.224 was about 28% higher than those of uniform structures [153].

An advantageous method, called the low density core (LDC) process, primarily was developed by Boeing based on an entrapped gas technique, and it was described in detail by Schwartz et al. [154]. The result is the material contains about 20–40 vol% of largely unconnected pores of diameter in range 10–100 μm (Fig. 15). Similar elements can be produced by a novel method of manufacturing titanium porous sheets proposed by Rak et al. The titanium foils are shaped using titanium powder and organic binder

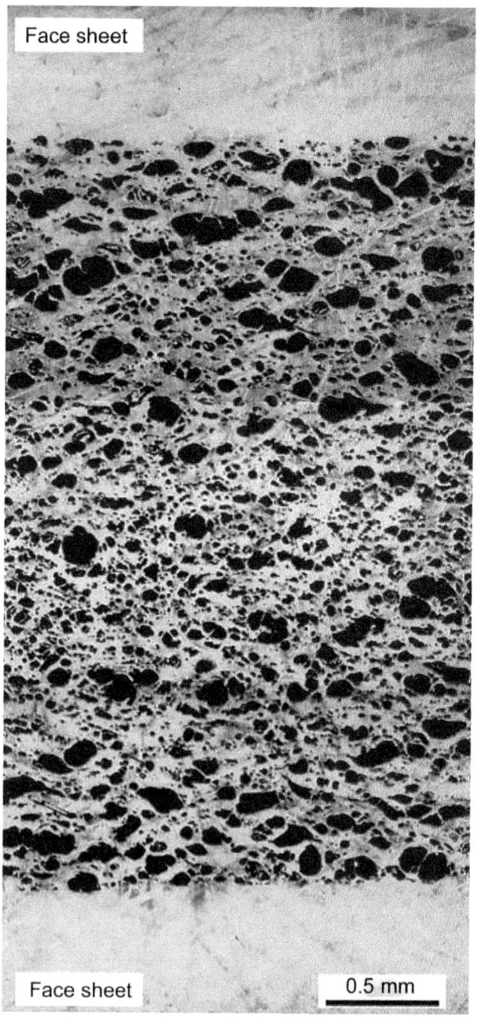

Fig. 15 Cross-section of LDC Ti6Al4V sheet [154].

and then sintered in a vacuum [155]. A significant benefit of such sandwiches is that they can be produced in any shape and curvature also in 3D space in contrary to usually flat honeycomb structures.

A separate issue is the method of combining titanium foam with the skin of a sandwich. Woodwart invented a method of bonding titanium and titanium alloy honeycomb sandwich panel structure by liquid interface diffusion [156]. Commercial brazing pastes and foils are also available [157]. However, the autoclave technique seems to be the most advantageous in this case [158].

Structural parts in turbines are further applications that combine stiffness and damping. Several stages of the engine require seals between them, which are also made of porous metals. These are made by cutting into the cellular material by the turbine blade during its first operation; this achieves almost complete tightness.

The selection of sandwich components depends on the placement of application and the requirements resulting therefrom. External structures (fuselage) undergo large temperature fluctuations and are loaded with aerodrome pressure. Shields and fairings are exposed to impacts due to bird strike, hail, lightning strikes, and abrasion caused by rain and dust whereas the lower part of the aircraft is exposed for impact of runway remains. Floor panels in the passenger compartment are loaded by passenger presence as well as by point pressure caused by, for example, shoe heels. These examples indicate the variety of loading and working conditions on sandwich structures. It is important to consider these requirements at an early stage of the design process to choose optimal components.

4.3 Metallic foam in automotive industry

The light-weight constructional aspect of porous metals is very similar in the aerospace and automotive sector. Increasing demands for vehicle safety enlarge their mass, which interferes with the overriding requirement of low fuel consumption. So, there is a strong need to reduce the weight of the car while maintaining safety. Elements made of titanium can significantly reduce the vibration management system needed for reciprocating weight, furthermore, they had less rotating mass than steel. In addition, the European and Japanese markets require cars with reduced length, which complicates their construction. This creates conflicting requirements for passenger comfort (cabin size), engine design with heat dissipation, as well as maintaining the right crush zone size. The last important requirement is a reduction of acoustic emissions, which requires the use of efficient sound-absorbent materials. The aspects previously raised are depicted in Fig. 16.

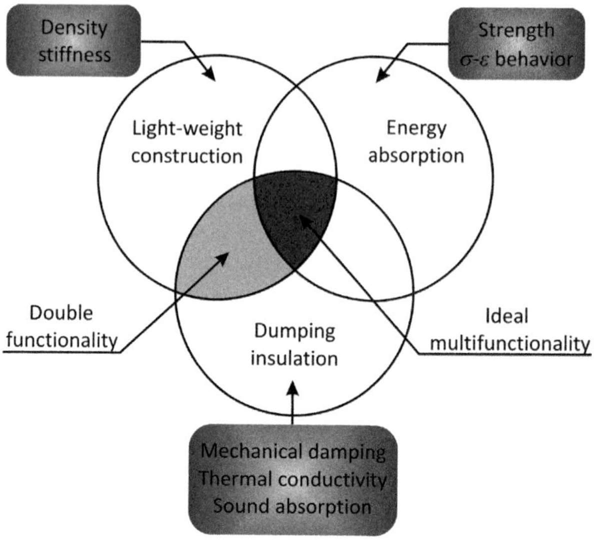

Fig. 16 Structural metal foams in automotive applications [127].

Due to the high costs of titanium, it is not widely used in automotive constructions, especially in the form of foams. An acceptable level of costs with satisfactory efficiency is achieved using aluminum or polymer foams or honeycomb structures [127].

The good susceptibility to forging of the innovative titanium alloys and MMCs in combination with the powder metallurgy technique creates an opportunity for the manufacture of prototype connecting rods for high-performance automotive engines. The use of composites reduces the imperfections of classical titanium alloys, which have a low wear resistance stiffness and creep resistance for optimal performance [159].

4.4 Multifunctional porous titanium foams

Advanced industry sectors, such as aviation, spaceflight, and military applications, put rigorous multitasking requirements on the materials used due to difficult working conditions: high temperature and aggressive environment, as well as mechanical load. Furthermore, simultaneous compliance with these claims are expected.

A functional material that has a higher service temperature than aluminum foams may be required when it needs to be used as a lightweight and denoising structure for aerospace science and technology applications. Furthermore, an engineering material that is more resistant to corrosion than aluminum foams may be needed for lightweight and silencing structures in aggressive environments. Liu et al. presented the good sound absorption performance achieved from highly porous titanium foams, with total porosities as high as 86–90% vol. and main pores that were spherical and millimeter-scaled [160].

The dual objective in multifunctional thermal barrier coatings utilized in sandwich panels was discussed by Nguyen et al. They presented results obtained for a typical aerospace panel mimicking thermal loading at a supersonic high-altitude flight that reduced heat transfer and showed better thermomechanical behavior, including higher strength and stiffness. Furthermore, the objectives were realized with the overriding premise of obtaining a minimum weight [161].

Multifunctional highly porous (about 88% vol.) titanium foam has also been developed by Liu et al. (Fig. 17). This foam of three-dimensional reticular structure characterized by open interconnected cells was made by impregnating a polymer foam in slurry and then sintering. It exhibits electromagnetic shielding performance together with a relatively low thermal conductivity. The electromagnetic shielding was measured in the frequency range of 0.3–3000 MHz (which covers that of radio waves), and it pointed to an evident effectiveness especially at low frequencies; in addition, the tendency of decrease of shielding efficiency was observed with the increase of the frequency. This effectiveness could be significantly increased for the higher thickness of the sample with the same pore parameter. The thermal conductivity of titanium foam had the value in range 0.4–0.8 W/(m·K) at room temperature for the sample with the porosity of 87%–89%. Significant reduction of thermal conductivity was also observed along with increased porosity [162].

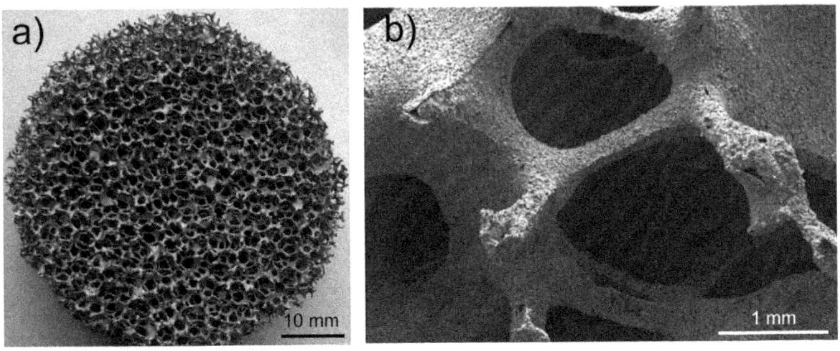

Fig. 17 (A) The reticular structure of titanium foam sample and (B) the pore structure in higher magnification [162].

A sound absorption high porosity titanium foam (86%–90%) with spherical pores of millimeter scale was achieved by Liu et al. by improved foaming method of melting the metallic powder. In a frequency range of 3150–6300 Hz, its sound absorption coefficient reached 0.6 and even exceeded 0.9 at the resonance frequency. Authors proposed two mechanisms of sound absorption: when frequency is lower than 4250 Hz, this should be interference silencing due to the surface reflection; above this frequency range, the viscous dissipation occurs [163].

Porous structures have many favorable properties and usually perform many functions in constructions at the same time. The most common case is combining the energy absorption function and the thermal barrier and/or sound insulation. It is worth noting that metal foams, especially titanium foams, due to much higher strength and stiffness than aluminum foams with an analogous porosity, have a higher energy absorption capacity and a much lower thermal conductivity coefficient.

4.5 Porous titanium in flow systems

4.5.1 Filters and gas sensing devices

Excellent corrosion resistance of titanium is the main reason for its use as filters in the chemical processing industry in applications involving heat, gases, aggressive chemicals, cryogenics, or polymers. Such filters can be made in various forms and shapes with a fixed pore structure mainly by using powder metallurgy. Porous titanium filters are produced in a range of configurations and micron ratings (Fig. 18A) to perform in a variety of

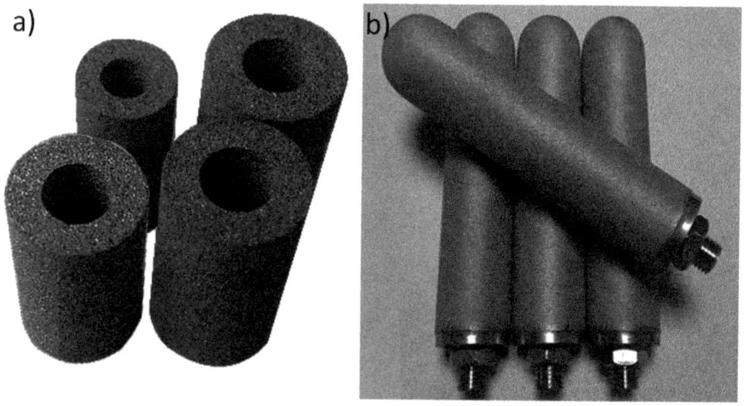

Fig. 18 Sintered porous titanium: (A) tubes and pipes and (B) filter cartridges. *(Courtesy of Qingan Fort Industrial Park [164].)*

liquid and gas applications such as molten sulfur filtration, amino-naphthol sulfonic acid solution filtration, dairy industry, and so on [1, 165].

Porous titanium filters are made of titanium powder of high purity using high temperature sintering (Fig. 18B). In recent years, they successfully replaced filters made of polymers, ceramic, stainless steel, nickel, and so on. They are widely used in liquid separation and purification, including pharmaceutical industry, water treatment industry, food industry, bioengineering, chemical and petrochemical industry, drug manufacturing, medical equipment, metallurgical industry, gas purification, and many other fields. Titanium filter precision available on the market are in range 0.22–100 μm [164, 166]

Increasing the efficiency of internal combustion engines contributes to reducing fuel consumption but at the same time requires emission control of harmful substances. Gas sensors are of great importance here, because the information flowing from them is the input data of the engine's control system. Sensors are based on two types of materials: zeolites and metal-organic frameworks (MOFs), which can detect the presence of water vapor, oxygen, NOx, carbon monoxide and carbon dioxide, hydrocarbons and volatile organic compounds, ammonia, hydrogen sulfide, sulfur dioxide, hydrogen, and many others. The porous MOF sensors do not contain pure titanium but only its compounds as chromium titanium oxide or titanium dioxide. Chemical composition of such sensors affects hydrophilicity or hydrophobicity, crystal size, and orientation, thus enabling detection and differentiation between various gases and vapors [167].

4.5.2 Gas getters

The first use of getters was recorded in 1894 when Malignani applied a suspension of red phosphorous onto the inside of the exhaust tube of an incandescent lamp to improve the vacuum [168]. Materials currently used as getters include aluminum, magnesium, barium, thorium, zirconium, uranium, alloys of rare-earth elements, and titanium. Gas absorbing by titanium depends on its temperature. Above 700°C, titanium will getter oxygen, nitrogen, and carbon dioxide, meanwhile hydrogen is absorbed in the temperature range of 25–400°C. Releasing hydrogen from titanium is feasible by heating. In both elevated and reduced temperatures, titanium can absorb water vapor and methane. Titanium can absorb very large amounts of gases—from 10 to 90 atm%. When the saturation is reached, titanium becomes brittle [169].

4.5.3 Spargers and aerators

Gas spargers are required in processes where small gas bubbles can be injected into a liquid or another gas. These elements must be resistant to corrosion in the environment of this liquid or gas, and must also have high resistance to mechanical loads and thermal shocks because of their complex hydrodynamic behavior [170]. Porous metal spargers (Fig. 19B) are widely used in many industrial processes, for example, water purification, CO_2 sparging for carbonated beverages and beers, and chlorine sparging for bleaching pulp in paper manufacturing. Due to excellent titanium corrosion resistance to ozone, such porous spargers are recommended to spray high-concentrated ozone for disinfecting and deodorizing air or waste treatment. In comparison to 316L stainless steel, titanium has two to three times higher permissible concentration of ozone [1].

Flat discs of porous titanium with a thickness of 0.6–10 mm and pore size of 0.2–50 μm (Fig. 19) can also be used as the main element of artificial cardiopulmonary oxygenators [166] and can easily compete with commonly used polymers. When an extracorporeal circulation is needed during a surgery, an artificial lung (oxygenator) is used that supplies oxygen to the body and removes carbon dioxide from it. Titanium has excellent biocompatibility, and when using compounds like TiO_2, SiC, or TiN as surface layers, reduction or elimination of blood clotting as well as increased endothelial adhesion and proliferation (e.g., in case of implants) are obtained [171, 172].

Microporous titanium aerators are made of pure titanium powder (purity over 99.67%) by high temperature vacuum sintering, and their pore sizes are in range of 0.22–100 μm. Such aerators are characterized by different shapes and efficiency, and allow for uniform bubbles distribution. They have excellent corrosion resistance, decreased 40% than conventional

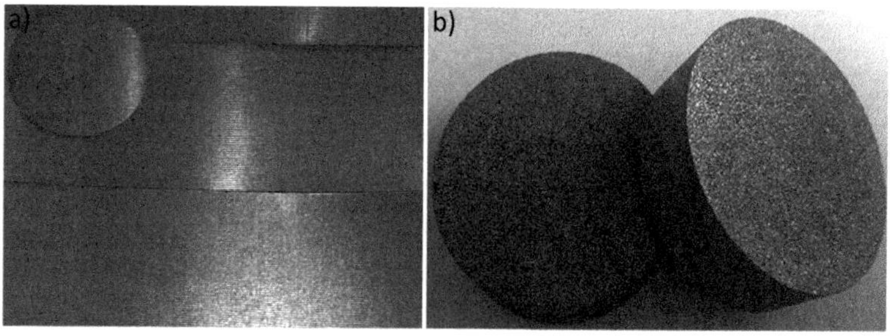

Fig. 19 (A) Porous filter plates and (B) discs. *(Courtesy of Qingan Fort Industrial Park [164].)*

aeration gas consumption and low hydraulic resistance. These are especially applicable to the city sewage, new expansion of large water plant, and transformation of the old aeration tanks [166].

4.5.4 Catalyst carriers

Attractive properties of porous titanium as a support for catalyst that may be used in the electrochemical processes (Ti-based electrodes), photoelectrochemical reactors (support for titania photo-catalyst), organically modified solar cells (current collector), chemical processes, etc., make their use constantly increasing [173, 174]. The techniques of manufacturing thin, porous titanium panels developed in recent years enable their use in such applications [155, 175].

Porous titanium substrate was also applied by Zhang et al. as a three-dimensional lead dioxide electrode (denoted as 3D-Ti/PbO_2 electrode with porosity of about 54%) for electrocatalytic activity for phenol degradation. It showed more active sites than the lead dioxide electrode electrodeposited on two-dimensional titanium substrate. The spatial structure increases the stability and service life (up to 350 h) of the electrodes obtained in this way [176].

A porous semiconductor material made of a porous titanium-niobium oxide was also described as a photocatalyst for the photolysis of organic compounds in water. Such catalyst was compared to a commercial non-porous TiO_2 catalytic powder. Results showed that photocatalytic activity of developed porous material was slightly higher to the commercial TiO_2 powder when pH = 2.1, whereas at higher pH values, the results were similar for both materials. Furthermore, the porous titanium-niobium oxide favors sedimentation better than the TiO_2 powder, resulting in a possibility to improve the quality of filtration and catalyst recovery after the removal of micropollutants from water [177].

The high mechanical strength of titanium in combination with corrosion and temperature resistance makes it possible to compete titanium foams with classic catalysts made of brittle ceramics.

4.6 Electrochemical applications

Recently, interest in hydrogen has increased as a renewable energy source. With the increase of power demand, currently exceeding the gigawatt range, it is also necessary to use renewable energy sources of the same size. Ensuring a sustainable solution for the production of hydrogen is provided by polymer electrolyte membrane (PEM) electrolysis. One of the main elements of PEM water electrolysis cells are current collectors. These elements should have a complex porosity structure to provide water flow through the large pores and to carry the product gas away from the electrodes through fine pores. Due to the corrosive condition, carbon materials (such as carbon paper, cloth, and fleece) cannot be applied in PEM electrolysis. Therefore, current collectors are mainly used as porous titanium plates typically prepared by thermal sintering of spherically shaped titanium powder. These collectors are more efficient and more durable than that made of stainless steel [178, 179].

Insoluble electrodes for electrochemical processes can also be made of porous titanium. Such electrodes can be used in the electroextraction of metals such as copper, nickel, zinc, and electrolytic manganese dioxide production. The use of lead protects porous titanium from passivation whereas the latter stabilizes the lead from spalling, which is a major problem for conventional lead anodes [1, 180].

4.7 Fuel cells

A fuel cell is an electrochemical device that converts chemical energy contained in fuel into electricity using electrochemical reactions, usually of hydrogen with an oxidant (Fig. 20). The cells consist of an oxidizer electrode, a fuel electrode (Fig. 19A) and an electrolyte. If gaseous products are formed as a result of the reaction, it is necessary to use porous electrodes, usually containing a catalyst. Wet-proofing of the electrodes is necessary to establish and maintain the three phase interface between gaseous reactant, liquid electrolyte, and solid catalyst. Furthermore, electrode materials should withstand elevated temperatures [181, 182].

In applications for fuel cell electrodes, a porous titanium of about 40%–75% porosity is preferred. The optimum pore size is determined in the

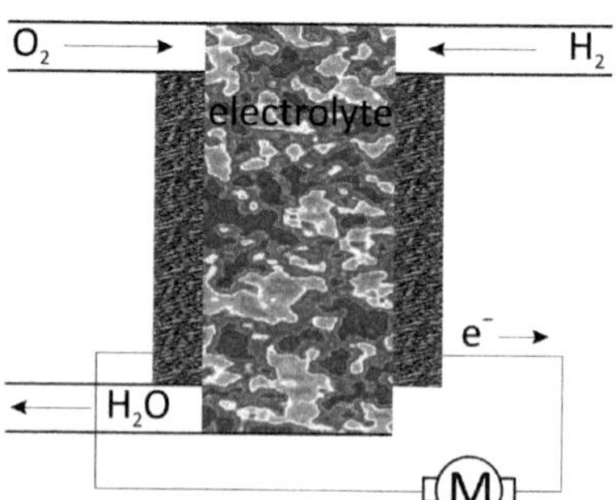

Fig. 20 Schematic diagram of oxygen-hydrogen fuel cell.

range of 8 μm, whereas the catalytic layer usually made of the metal from the platinum group. It is recommended that the titanium porosity is not lower than 40%, as in this case a high gas pressure (fuel and oxidant) is required. Due to such a high pressure, it is possible to fracture and damage the electrodes, especially when their thickness is low [182].

In hydrogen fuel cells applied commercially, porous titanium discs made of Grade 2 CPTi are used. Their diameter is 230 mm with the thickness 1.5 mm and porosity at least 40%. For methanol-reforming hydrogen fuel cells, an anode is also made of porous titanium plate of thickness 0.6–10 mm [166].

4.8 Other applications

An evolutional example of the use of porous titanium is the development of lightweight and durable materials that can be used, for example, in transport vehicles. An example of this are Ti-Mg pseudo-alloys made by infiltration of porous titanium by magnesium and its alloys [183].

A product from a completely different technical area are flash lamps for photographic dedications as well as scientific, medical, and industrial applications (e.g., lasers, strobe lamps). They emit very strong and very short flashes of white light to illuminate the subject or to achieve a strobe effect. Flashtubes are usually made in the form of closed tubes with electrodes on both ends and are filled with gas, which, under the influence of high voltage, ionizes and conducts a current pulse producing intense light. The electrodes are commonly made of tungsten, however, some works are aimed at using other high-melting metals for this purpose, including titanium, as well as porous titanium. This allows for a 20-time reduction in the cost per unit of volume of electrode material [37], reduced weight, and increased surface of the electrode, which may result in improving the efficiency [184].

5 Summary and future considerations

Porous titanium, including foams and rod structures, has become an important group of metallic material with a favorable combination of mechanical and functional properties. They have use in many applications, starting with medicine, through flow systems such as filters, to solutions in aviation and aeronautics. The main advantages of these materials are resistance to corrosion, low weight, and relatively high mechanical strength. Importantly, these properties can be adjusted through the use of appropriate pore structure and morphology. The pore architecture can be uniform, bimodal, gradient, or honeycomb, and the pores can be open or closed, which determines the application.

Both mechanical properties and corrosion resistance can be modified by choosing the proper chemical composition of the titanium alloy. The interstitial impurities (O, N, and C) in porous titanium should be strictly controlled due to their negative influence on most of the properties. However, medical applications limit the use of certain alloying elements due to their harmfulness or potential adverse effects on the human body. These elements are, for example, aluminum and vanadium, the most popular alloy additions in high-strength titanium alloys.

A significant limitation of the use of titanium or titanium alloys as porous materials is their price, resulting from the costs of the raw material and of manufacturing the products. However, to maintain a significant position in the industry, future efforts must be directed toward reducing the costs of titanium porous structures. Whereas powder metallurgy is already a well-established technology, obtaining a controlled morphology of porosity is still difficult and has some randomness. The promising ability to produce complex porous structures with strictly controlled architecture, while providing freedom of design, are additive technologies. These methods also enable the production of elements displaying a concentration gradient of elements by in situ alloying of metal powders of different composition.

Although the manufacturing of medical implants and elements of flow systems made of porous titanium has already been implemented, the use of such materials in other areas, such as aerospace or vehicles, faces significant cost constraints. Despite favorable properties, such as high strength-to-weight ratio, excellent corrosion resistance, and compatibility with composite structures, porous titanium structures are still not widely used in the afore-mentioned areas. Global market forecasts for 2013–32 indicate that, in the field of civil aviation, approximately 5% per year of passenger transport increase is expected, which will translate into a demand for new aircraft [151]. The expansion of porous titanium products into these industries allows an increase in the efficiency and safety of vehicles, as well as meeting the demands to reduce fuel consumption. Compliance with restrictive regulations on pollutant emissions, such as CO_2 or NO_x, will also benefit the aviation community by providing a more stable production base, which should offer greater price stability [147]. This approach is a challenge for constructors and producers of structural parts and engines for aircrafts and vehicles.

Acknowledgments

The authors would like to acknowledge the financial support of statutory research S2/M/2018 (KP), TKT/MN/4 (MK), and AOCMF, Project No AOCMFS-18-14P (RP).

References

[1] H.P. Tang, J. Wang, M. Qian, Porous titanium structures and applications. in: Titanium Powder Metallurgy: Science, Technology and Applications, Butterworth-Heinemann, 2015, pp. 533–554, https://doi.org/10.1016/B978-0-12-800054-0.00028-9.

[2] M.T. Andani, N. Shayesteh Moghaddam, C. Haberland, D. Dean, M.J. Miller, M. Elahinia, Metals for bone implants. Part 1. Powder metallurgy and implant rendering. Acta Biomater. 10 (2014) 4058–4070, https://doi.org/10.1016/j.actbio.2014.06.025.

[3] M. Qian, F.H. Froes, Titanium Powder Metallurgy, Butterworth-Heinemann, 2015.

[4] B. Arifvianto, J. Zhou, Fabrication of metallic biomedical Scaffolds with the space holder method: a review. Mater. Met. 7 (2014) 3588–3622, https://doi.org/10.3390/ma7053588.

[5] S.W. Kim, H. Do Jung, M.H. Kang, H.E. Kim, Y.H. Koh, Y. Estrin, Fabrication of porous titanium scaffold with controlled porous structure and net-shape using magnesium as spacer. Mater. Sci. Eng. C 33 (2013) 2808–2815, https://doi.org/10.1016/j.msec.2013.03.011.

[6] A. Nouri, P.D. Hodgson, C. We, Biomimetic porous titanium scaffolds for orthopedic and dental applications. in: Biomimetics Learning From Nature, InTech, 2010, https://doi.org/10.5772/8787.

[7] Z. Wally, W. van Grunsven, F. Claeyssens, R. Goodall, G. Reilly, Porous titanium for dental implant applications. Mater. Met. 5 (2015) 1902–1920, https://doi.org/10.3390/met5041902.

[8] M. Mour, D. Das, T. Winkler, E. Hoenig, G. Mielke, M.M. Morlock, A.F. Schilling, Advances in porous biomaterials for dental and orthopaedic applications. Mater. Met. 3 (2010) 2947–2974, https://doi.org/10.3390/ma3052947.

[9] A. Amherd Hidalgo, R. Frykholm, T. Ebel, F. Pyczak, Powder metallurgy strategies to improve properties and processing of titanium alloys: a review. Adv. Eng. Mater. 19 (2017) 1–14, https://doi.org/10.1002/adem.201600743.

[10] Z.Z. Fang, J.D. Paramore, P. Sun, K.S.R. Chandran, Y. Zhang, Y. Xia, F. Cao, M. Koopman, M. Free, Powder metallurgy of titanium—past, present, and future. Int. Mater. Rev. (2017) 1–53, https://doi.org/10.1080/09506608.2017.1366003.

[11] K. Pałka, R. Pokrowiecki, Porous titanium implants: a review. Adv. Eng. Mater. (2018) 1700648, https://doi.org/10.1002/adem.201700648.

[12] B. Lee, T. Lee, Y. Lee, D. Jun, J. Jeong, J. Yuh, S. Ho, H. Seop, C. Soo, Space-holder effect on designing pore structure and determining mechanical properties in porous titanium. Mater. Des. 57 (2014) 712–718, https://doi.org/10.1016/j.matdes.2013.12.078.

[13] B. Szaraniec, K. Pałka, Microstructural analysis of porosity-graded titanium sinters using microtomography, Eng. Biomater. 15 (2012) 26–29. http://www.academia.edu/21401391/Microstructural_analysis_of_porosity-graded_titanium_sinters_using_microtomography, (Accessed August 16, 2017).

[14] S.M. Giannitelli, D. Accoto, M. Trombetta, A. Rainer, Current trends in the design of scaffolds for computer-aided tissue engineering. Acta Biomater. 10 (2014) 580–594, https://doi.org/10.1016/j.actbio.2013.10.024.

[15] Z. Chen, Z. Su, S. Ma, X. Wu, Z. Luo, Biomimetic modeling and three-dimension reconstruction of the artificial bone. Comput. Methods Prog. Biomed. 88 (2007) 123–130, https://doi.org/10.1016/j.cmpb.2007.08.001.
[16] B. Dabrowski, W. Swieszkowski, D. Godlinski, K.J. Kurzydlowski, Highly porous titanium scaffolds for orthopaedic applications. J. Biomed. Mater. Res. (2010) https://doi.org/10.1002/jbm.b.31682.
[17] F. Maes, P. Van Ransbeeck, H. Van Oosterwyck, P. Verdonck, Modeling fluid flow through irregular scaffolds for perfusion bioreactors. Biotechnol. Bioeng. 103 (2009) 621–630, https://doi.org/10.1002/bit.22277.
[18] R. Singh, P.D. Lee, J.R. Jones, G. Poologasundarampillai, T. Post, T.C. Lindley, R.J. Dashwood, Hierarchically structured titanium foams for tissue scaffold applications. Acta Biomater. 6 (2010) 4596–4604, https://doi.org/10.1016/j.actbio.2010.06.027.
[19] Y. Wang, Y. Shen, Z. Wang, J. Yang, N. Liu, W. Huang, Development of highly porous titanium scaffolds by selective laser melting. Mater. Lett. 64 (2010) 674–676, https://doi.org/10.1016/j.matlet.2009.12.035.
[20] L.J. Gibson, M.F. Ashby, The Structure of Cellular Solids. Cambridge University Press, 1999, https://doi.org/10.1017/CBO9781139878326.004.
[21] L.J. Gibson, M.F. Ashby, The Mechanics of Foams: Basics Results, Cambridge University Press, 1997.
[22] S.Y. Min, Y. Kim, T.K. Lee, S. Seo, H.S. Jung, D. Kim, J. Lee, B.J. Lee, The effect of porosity on the elasticity of pure titanium: an atomistic simulation. Met. Mater. Int. 16 (2010) 421–425, https://doi.org/10.1007/s12540-010-0612-8.
[23] A. Zargarian, M. Esfahanian, J. Kadkhodapour, S. Ziaei-Rad, Numerical simulation of the fatigue behavior of additive manufactured titanium porous lattice structures. Mater. Sci. Eng. C 60 (2016) 339–347, https://doi.org/10.1016/j.msec.2015.11.054.
[24] J. Parthasarathy, B. Starly, S. Raman, A design for the additive manufacture of functionally graded porous structures with tailored mechanical properties for biomedical applications. J. Manuf. Process. 13 (2011) 160–170, https://doi.org/10.1016/j.jmapro.2011.01.004.
[25] Y. Okazaki, A. Ito, T. Tateishi, Y. Ito, Effects of alloying elements on anodic polarization properties of titanium alloys in acid solutions, Mater. Trans. JIM 35 (1994) 58–66.
[26] G. Lütjering, J.C. Williams, Titanium, Springer-Verlag, Berlin Heidelberg, 2010.
[27] V.S. De Viteri, E. Fuentes, Titanium and titanium alloys as biomaterials. in: J. Gegner (Ed.), Tribology—Fundamentals and Advancements, InTech, 2013, pp. 155–182, https://doi.org/10.5772/3427.
[28] Z. Tarzimoghadam, S. Sandlöbes, K.G. Pradeep, D. Raabe, Microstructure design and mechanical properties in a near-α Ti-4Mo alloy. Acta Mater. 97 (2015) 291–304, https://doi.org/10.1016/j.actamat.2015.06.043.
[29] Y. Kirmanidou, M. Sidira, M.E. Drosou, V. Bennani, A. Bakopoulou, A. Tsouknidas, N. Michailidis, K. Michalakis, New Ti-alloys and surface modifications to improve the mechanical properties and the biological response to orthopedic and dental implants: a review, Biomed. Res. Int. 2016 (2016) https://doi.org/10.1155/2016/2908570.
[30] G. Lütjering, J.C. Williams, Titanium, Springer-Verlag, Berlin Heidelberg, 2007.
[31] J.A. Helsen, J.H. Breme, Metals as Biomaterials, Wiley, New York, 1998.
[32] R. Narayan, Biomedical Materials. Springer US, New York, 2013, https://doi.org/10.1017/CBO9781107415324.004.
[33] J.Y. Wong, J.D. Bronzino, Biomaterials, CRC Press, New York, 2007.
[34] R. Fazel-Rezai, N. Root, A. Rabbi, D. Lee, W. Ahmad, Biomedical signal transceivers. in: Biomedical Engineering, InTech, Rijeka, Croatia, 2011, pp. 63–65, https://doi.org/10.5772/22513.

[35] N. Li, Y. Zheng, Novel magnesium alloys developed for biomedical application: a review. J. Mater. Sci. Technol. 29 (2013) 489–502, https://doi.org/10.1016/j.jmst.2013.02.005.

[36] S. Wu, X. Liu, K.W.K. Yeung, C. Liu, X. Yang, Biomimetic porous scaffolds for bone tissue engineering. Mater. Sci. Eng. R 80 (2014) 1–36, https://doi.org/10.1016/j.mser.2014.04.001.

[37] Granta Design Limited, CES EduPack. http://www.grantadesign.com/education/edupack/, 2017.

[38] M. González, E. Salvagni, J.C. Rodríguez-Cabello, E. Rupérez, F.J. Gil, J. Peña, J.M. Manero, A low elastic modulus Ti-Nb-Hf alloy bioactivated with an elastin-like protein-based polymer enhances osteoblast cell adhesion and spreading. J. Biomed. Mater. Res. A 101A (2013) 819–826, https://doi.org/10.1002/jbm.a.34388.

[39] B. Świeczko-Żurek, The allergic and irritating reactions to metallic implants with trauma—orthopaedic patients, Int. J. New Technol. Res. 2 (2016) 50–56. https://www.ijntr.org/download_data/IJNTR02020019.pdf.

[40] L. Evrard, Titanium: a new allergen. in: Implant Dentistry—A Rapidly Evolving Practice, InTech, 2011, https://doi.org/10.5772/19931.

[41] E. Eisenbarth, D. Velten, M. Müller, R. Thull, J. Breme, Biocompatibility of β-stabilizing elements of titanium alloys. Biomaterials 25 (2004) 5705–5713, https://doi.org/10.1016/j.biomaterials.2004.01.021.

[42] P. Yu, G. Stephani, S.D. Luo, H. Goehler, M. Qian, Microwave-assisted fabrication of titanium hollow spheres with tailored shell structures for various potential applications. Mater. Lett. 86 (2012) 84–87, https://doi.org/10.1016/j.matlet.2012.07.022.

[43] L.M.R. de Vasconcellos, D.O. Leite, F.N. de Oliveira, Y.R. Carvalho, C.A.A. Cairo, Evaluation of bone ingrowth into porous titanium implant: histomorphometric analysis in rabbits, Implantol. Braz. Oral Res. 24 (2010) 399–405.

[44] S.J. Hollister, Scaffold design and manufacturing: from concept to clinic. Adv. Mater. 21 (2009) 3330–3342, https://doi.org/10.1002/adma.200802977.

[45] X.B. Chen, Y.C. Li, P.D. Hodgson, C. Wen, The importance of particle size in porous titanium and nonporous counterparts for surface energy and its impact on apatite formation. Acta Biomater. 5 (2009) 2290–2302, https://doi.org/10.1016/j.actbio.2009.02.027.

[46] L. Gibson, M.F. Ashby, Cellular Solids: Structure and Properties, Cambridge University Press, 1999. https://books.google.pl/books/about/Cellular_Solids.html?id=I-ySUr5sn4N8C&redir_esc=y, (Accessed August 16, 2017).

[47] B.V. Krishna, S. Bose, A. Bandyopadhyay, Low stiffness porous Ti structures for load-bearing implants. Acta Biomater. 3 (2007) 997–1006, https://doi.org/10.1016/j.actbio.2007.03.008.

[48] I.H. Oh, N. Nomura, N. Masahashi, S. Hanada, Mechanical properties of porous titanium compacts prepared by powder sintering. Scr. Mater. 49 (2003) 1197–1202, https://doi.org/10.1016/j.scriptamat.2003.08.018.

[49] Y. Torres, J.J. Pavón, I. Nieto, J.A. Rodríguez, Conventional powder metallurgy process and characterization of porous titanium for biomedical applications. Metall. Mater. Trans. B Process Metall. Mater. Process. Sci. 42 (2011) 891–900, https://doi.org/10.1007/s11663-011-9521-6.

[50] Y. Torres, J.A. Rodríguez, S. Arias, M. Echeverry, S. Robledo, V. Amigo, J.J. Pavón, Processing, characterization and biological testing of porous titanium obtained by space-holder technique. J. Mater. Sci. 47 (2012) 6565–6576, https://doi.org/10.1007/s10853-012-6586-9.

[51] C.E. Wen, M. Mabuchi, Y. Yamada, K. Shimojima, Y. Chino, T. Asahina, Processing of biocompatible porous Ti and Mg. Scr. Mater. 45 (2001) 1147–1153, https://doi.org/10.1016/S1359-6462(01)01132-0.

[52] C.E. Wen, Y. Yamada, K. Shimojima, Y. Chino, H. Hosokawa, M. Mabuchi, Novel titanium foam for bone tissue engineering. J. Mater. Res. 17 (2002) 2633–2639, https://doi.org/10.1557/JMR.2002.0382.

[53] J. Jakubowicz, G. Adamek, K. Pałka, D. Andrzejewski, Micro-CT analysis and mechanical properties of Ti spherical and polyhedral void composites made with saccharose as a space holder material. Mater. Charact. 100 (2015) 13–20, https://doi.org/10.1016/j.matchar.2014.12.006.

[54] K. Pałka, G. Adamek, J. Jakubowicz, Compression behavior of Ti foams with spherical and polyhedral pores. Adv. Eng. Mater. 18 (2016) 1511–1518, https://doi.org/10.1002/adem.201600169.

[55] V. Chastand, A. Tezenas, Y. Cadoret, P. Quaegebeur, W. Maia, E. Charkaluk, Fatigue characterization of titanium Ti-6Al-4V samples produced by additive manufacturing. Procedia Struct. Integr. 2 (2016) 3168–3176, https://doi.org/10.1016/j.prostr.2016.06.395.

[56] P. Heinl, L. Müller, C. Körner, R.F. Singer, F.A. Müller, Cellular Ti-6Al-4V structures with interconnected macro porosity for bone implants fabricated by selective electron beam melting. Acta Biomater. 4 (2008) 1536–1544, https://doi.org/10.1016/j.actbio.2008.03.013.

[57] M. Sharma, G.K. Gupta, O.P. Modi, B.K. Prasad, A.K. Gupta, Titanium foam through powder metallurgy route using acicular urea particles as space holder. Mater. Lett. 65 (2011) 3199–3201, https://doi.org/10.1016/j.matlet.2011.07.004.

[58] W. Xue, B.V. Krishna, A. Bandyopadhyay, S. Bose, Processing and biocompatibility evaluation of laser processed porous titanium. Acta Biomater. 3 (2007) 1007–1018, https://doi.org/10.1016/j.actbio.2007.05.009.

[59] A. Choubey, B. Basu, R. Balasubramaniam, Tribological behaviour of Ti-based alloys in simulated body fluid solution at fretting contacts. Mater. Sci. Eng. A 379 (2004) 234–239, https://doi.org/10.1016/j.msea.2004.02.027.

[60] S.A. Yavari, R. Wauthle, J. Van Der Stok, A.C. Riemslag, M. Janssen, M. Mulier, J.P. Kruth, J. Schrooten, H. Weinans, A.A. Zadpoor, Fatigue behavior of porous biomaterials manufactured using selective laser melting. Mater. Sci. Eng. C 33 (2013) 4849–4858, https://doi.org/10.1016/j.msec.2013.08.006.

[61] P. Edwards, M. Ramulu, M. Science, A. Engineering, Fatigue performance evaluation of selective laser melted Ti-6Al-4V. Mater. Sci. Eng. A 598 (2014) 327–337, https://doi.org/10.1016/j.msea.2014.01.041.

[62] S.J. Li, L.E. Murr, X.Y. Cheng, Z.B. Zhang, Y.L. Hao, R. Yang, F. Medina, R.B. Wicker, Compression fatigue behavior of Ti-6Al-4V mesh arrays fabricated by electron beam melting. Acta Mater. 60 (2012) 793–802, https://doi.org/10.1016/j.actamat.2011.10.051.

[63] S. Ghosh, S. Abanteriba, Status of surface modification techniques for artificial hip implants. Sci. Technol. Adv. Mater. 17 (2016) 715–735, https://doi.org/10.1080/14686996.2016.1240575.

[64] B. Antoszewski, M. Krzywicka, S. Tofil, Laser surface texturing of titanium alloys for biomedical applications. in: Technical Transactions, 2016, pp. 3–8, https://doi.org/10.4467/2353737XCT.16.113.5724. 3–M.

[65] W.R. Wise, A new insight on pore structure and permeability. Water Resour. Res. 28 (1992) 189–198, https://doi.org/10.1029/91WR02522.

[66] F. Yang, L. Zhang, J. Lin, Y. Liang, Y. He, S. Shang, Z.K. Liu, Pore structure and gas permeability of high Nb-containing TiAl porous alloys by elemental powder metallurgy for microfiltration application. Intermetallics 33 (2013) 2–7, https://doi.org/10.1016/j.intermet.2012.07.022.

[67] T. Furumoto, A. Koizumi, M.R. Alkahari, R. Anayama, A. Hosokawa, R. Tanaka, T. Ueda, Permeability and strength of a porous metal structure fabricated by additive manufacturing. J. Mater. Process. Technol. 219 (2015) 10–16, https://doi.org/10.1016/j.jmatprotec.2014.11.043.

[68] M.J. Blunt, Effects of heterogeneity and wetting on relative permeability using pore level modeling. SPE J. 2 (1997) 70–87, https://doi.org/10.2118/36762-PA.
[69] A. Jo, S. Ahn, K. Oh, W. Kim, H. Ju, Effects of metal foam properties on flow and water distribution in polymer electrolyte fuel cells (PEFCs). Int. J. Hydrog. Energy 43 (2018) 14034–14046, https://doi.org/10.1016/j.ijhydene.2018.01.134.
[70] K.H.W. Seah, R. Thampuran, S.H. Teoh, The influence of pore morphology on corrosion. Corros. Sci. 40 (1998) 547–556, https://doi.org/10.1016/S0010-938X(97)00152-2.
[71] G. Adamek, K. Pałka, J. Jakubowicz, Corrosion properties of Ti scaffolds prepared with sucrose as a space holder. Solid State Phenom. 227 (2015) 519–522, https://doi.org/10.4028/www.scientific.net/SSP.227.519.
[72] B. Dabrowski, J. Kaminski, W. Swieszkowski, K.J. Kurzydlowski, Porous titanium scaffolds for biomedical applications: corrosion resistance and structure investigation. Mater. Sci. Forum 674 (2011) 41–46, https://doi.org/10.4028/www.scientific.net/MSF.674.41.
[73] X. Chen, Q. Fu, Y. Jin, M. Li, R. Yang, X. Cui, M. Gong, In vitro studying corrosion behavior of porous titanium coating in dynamic electrolyte. Mater. Sci. Eng. C 70 (2017) 1071–1075, https://doi.org/10.1016/j.msec.2016.03.044.
[74] Y. Lin, Y. Shen, A. Liu, Y. Zhu, S. Liu, H. Jiang, Bio-inspiredly fabricating the hierarchical 3D porous structure superhydrophobic surfaces for corrosion prevention. Mater. Des. 103 (2016) 300–307, https://doi.org/10.1016/j.matdes.2016.04.083.
[75] F. Toptan, A.C. Alves, A.M.P. Pinto, P. Ponthiaux, Tribocorrosion behavior of bio-functionalized highly porous titanium. J. Mech. Behav. Biomed. Mater. 69 (2017) 144–152, https://doi.org/10.1016/j.jmbbm.2017.01.006.
[76] X. Luo, B. Yang, L. Sheng, J. Chen, H. Li, L. Xie, G. Chen, M. Yu, W. Guo, W. Tian, CAD based design sensitivity analysis and shape optimization of scaffolds for bio-root regeneration in swine. Biomaterials 57 (2015) 59–72, https://doi.org/10.1016/j.biomaterials.2015.03.062.
[77] Y.H. Chen, J. Cadman, S.W. Zhou, Q. Li, Computer-aided design and fabrication of bio-mimetic materials and scaffold micro-structures. Adv. Mater. Res. 213 (2011) 628–632, https://doi.org/10.4028/www.scientific.net/AMR.213.628.
[78] S. Barone, M. Casinelli, M. Frascaria, A. Paoli, A.V. Razionale, Interactive design of dental implant placements through CAD-CAM technologies: from 3D imaging to additive manufacturing. Int. J. Interact. Des. Manuf. 10 (2016) 105–117, https://doi.org/10.1007/s12008-014-0229-0.
[79] S.J. Hollister, R.A. Levy, T.-M. Chu, J.W. Halloran, S.E. Feinberg, An image-based approach for designing and manufacturing craniofacial scaffolds. Int. J. Oral Maxillofac. Surg. 29 (2000) 67–71, https://doi.org/10.1034/j.1399-0020.2000.290115.x.
[80] S.J. Hollister, Porous scaffold design for tissue engineering. Nat. Mater. 4 (2005) 518–524, https://doi.org/10.1038/nmat1421.
[81] R. Gabbrielli, I.G. Turner, C.R. Bowen, Development of modelling methods for materials to be used as bone substitutes. Key Eng. Mater. 361–363 (II) (2008) 901–906, https://doi.org/10.4028/www.scientific.net/KEM.361-363.903.
[82] A. Pasko, O. Fryazinov, T. Vilbrandt, P.A. Fayolle, V. Adzhiev, Procedural function-based modelling of volumetric microstructures. Graph. Model. 73 (2011) 165–181, https://doi.org/10.1016/j.gmod.2011.03.001.
[83] A. Bayu, D. Nandiyanto, N. Hagura, F. Iskandar, K. Okuyama, Design of a highly ordered and uniform porous structure with multisized pores in film and particle forms using a template-driven self-assembly technique. Acta Mater. 58 (2010) 282–289, https://doi.org/10.1016/j.actamat.2009.09.004.
[84] I. Zein, D.W. Hutmacher, K.C. Tan, S.H. Teoh, Fused deposition modeling of novel scaffold architectures for tissue engineering applications. Biomaterials 23 (2002) 1169–1185, https://doi.org/10.1016/S0142-9612(01)00232-0.

[85] S.M. Ahmadi, S.A. Yavari, R. Wauthle, B. Pouran, J. Schrooten, H. Weinans, A.A. Zadpoor, Additively manufactured open-cell porous biomaterials made from six different space-filling unit cells: the mechanical and morphological properties. Mater. Met. 8 (2015) 1871–1896, https://doi.org/10.3390/ma8041871.

[86] J. Woo Jung, H.-G. Yi, T.-Y. Kang, W.-J. Yong, S. Jin, W.-S. Yun, D.-W. Cho, Evaluation of the effective diffusivity of a freeform fabricated scaffold using computational simulation. J. Biomech. Eng. 135 (2013) 084501, https://doi.org/10.1115/1.4024570.

[87] S. Truscello, G. Kerckhofs, S. Van Bael, G. Pyka, J. Schrooten, H. Van Oosterwyck, Prediction of permeability of regular scaffolds for skeletal tissue engineering: a combined computational and experimental study. Acta Biomater. 8 (2012) 1648–1658, https://doi.org/10.1016/j.actbio.2011.12.021.

[88] C.X.F. Lam, M.M. Savalani, S.-H. Teoh, D.W. Hutmacher, Dynamics of *in vitro* polymer degradation of polycaprolactone-based scaffolds: accelerated versus simulated physiological conditions. Biomed. Mater. 3 (2008) 034108, https://doi.org/10.1088/1748-6041/3/3/034108.

[89] M.F. Ashby, H. Shercliff, D. Cebon, Materials: Engineering, Science, Processing and Design, Butterworth-Heinemann, 2010.

[90] A.K.M.B. Khoda, Engineered tissue scaffolds with variational porous architecture. J. Biomech. Eng. 133 (2010) 011001, https://doi.org/10.1115/1.4002933.

[91] P. Pandithevan, G. Saravana Kumar, Personalised bone tissue engineering scaffold with controlled architecture using fractal tool paths in layered manufacturing. Virtual Phys. Prototyp. 4 (2009) 165–180, https://doi.org/10.1080/17452750903055512.

[92] M.A. Wettergreen, B.S. Bucklen, M.A.K. Liebschner, W. Sun, CAD assembly process for bone replacement scaffolds in computer-aided tissue engineering. in: Virtual Prototyping and Bio Manufacturing in Medical Applications, Springer US, Boston, 2008, pp. 87–111, https://doi.org/10.1007/978-0-387-68831-2_5.

[93] N. Sudarmadji, J.Y. Tan, K.F. Leong, C.K. Chua, Y.T. Loh, Investigation of the mechanical properties and porosity relationships in selective laser-sintered polyhedral for functionally graded scaffolds. Acta Biomater. 7 (2011) 530–537, https://doi.org/10.1016/j.actbio.2010.09.024.

[94] J.R. Vinson, Sandwich structures: past, present, and future. in: O.T. Thomsen, E. Bozhevolnaya, A. Lyckegaard (Eds.), Sandwich Structures 7: Advancing With Sandwich Structures and Materials, Springer-Verlag, Berlin/Heidelberg, 2005, pp. 3–12, https://doi.org/10.1007/1-4020-3848-8_1.

[95] K. Nishiyabu, S. Matsuzaki, S. Tanaka, Fabrication and high-functionality of micro porous metals by powder space holder method. Proc. Mater. Process. Conf. 2004 (12) (2017) 83–84, https://doi.org/10.1299/jsmemp.2004.12.83.

[96] E.E. Pakos, T. Xenakis, Titanium porous-coated implant-bone interface in total joint arthroplasty. in: T. Karachalios (Ed.), Bone-Implant Interface in Orthopedic Surgery: Basic Science to Clinical Applications, Springer London, London, 2014, pp. 67–81, https://doi.org/10.1007/978-1-4471-5409-9_6.

[97] C. Larsson, P. Thomsen, B.O. Aronsson, M. Rodahl, J. Lausmaa, B. Kasemo, L.E. Ericson, Bone response to surface-modified titanium implants: studies on the early tissue response to machined and electropolished implants with different oxide thicknesses. Biomaterials 17 (1996) 605–616, https://doi.org/10.1016/0142-9612(96)88711-4.

[98] S. Nishiguchi, T. Nakamura, M. Kobayashi, H.M. Kim, F. Miyaji, T. Kokubo, The effect of heat treatment on bone-bonding ability of alkali-treated titanium. Biomaterials 20 (1999) 491–500, https://doi.org/10.1016/S0142-9612(98)90203-4.

[99] M. Baleani, M. Viceconti, A. Toni, The effect of sandblasting treatment on endurance properties of titanium alloy hip prostheses. Artif. Organs 24 (2000) 296–299, https://doi.org/10.1046/j.1525-1594.2000.06486.x.

[100] C. Zhang, Y. Leng, J. Chen, Elastic and plastic behavior of plasma-sprayed hydroxyapatite coatings on a Ti-6Al-4V substrate. Biomaterials 22 (2001) 1357–1363, https://doi.org/10.1016/S0142-9612(00)00289-1.

[101] C. Massaro, M.A. Baker, F. Cosentino, P.A. Ramires, S. Klose, E. Milella, Surface and biological evaluation of hydroxyapatite-based coatings on titanium deposited by different techniques. J. Biomed. Mater. Res. 58 (2001) 651–657, https://doi.org/10.1002/jbm.1065.

[102] B.J.V. Yoon, F. Xavier, B.R. Walker, S. Grinberg, F.P. Cammisa, C. Abjornson, Optimizing surface characteristics for cell adhesion and proliferation on titanium plasma spray coatings on polyetheretherketone. Spine J. 16 (2016) 1238–1243, https://doi.org/10.1016/J.SPINEE.2016.05.017.

[103] M. Fujio, S. Komasa, H. Nishizaki, T. Sekino, J. Okazaki, Biocompatibility of titanium surface nanostructures following chemical processing and heat treatment. Front. Nanosci. Nanotechnol. 2 (2016) https://doi.org/10.15761/FNN.1000143.

[104] R. Brånemark, L. Emanuelsson, A. Palmquist, P. Thomsen, Bone response to laser-induced micro- and nano-size titanium surface features. Nanomedicine 7 (2011) 220–227, https://doi.org/10.1016/j.nano.2010.10.006.

[105] M. Echeverry-Rendón, O. Galvis, R. Aguirre, S. Robledo, J.G. Castaño, F. Echeverría, Modification of titanium alloys surface properties by plasma electrolytic oxidation (PEO) and influence on biological response. J. Mater. Sci. Mater. Med. 28 (2017) https://doi.org/10.1007/s10856-017-5972-x.

[106] Y. Zhang, P.B. Ahn, D.C. Fitzpatrick, A.D. Heiner, R.A. Poggie, T.D. Brown, Interfacial frictional behavior: cancellous bone, cortical bone, and a novel porous tantalum biomaterial, J. Musculoskelet. Res. 3 (1999) 245–251.

[107] T.L. Karachalios, Bone-Implant Interface in Orthopedic Surgery. 2014, https://doi.org/10.1007/978-1-4471-5409-9.

[108] B. Antoszewski, P. Sęk, Influence of laser beam intensity on geometry parameters of a single surface texture element. Arch. Metall. Mater. 60 (2015) 2215–2219, https://doi.org/10.1515/amm-2015-0367.

[109] J.I. Ahuir-Torres, M.A. Arenas, W. Perrie, J. de Damborenea, Influence of laser parameters in surface texturing ofTi6Al4V and AA2024-T3 alloys. Opt. Lasers Eng. 103 (2018) 100–109, https://doi.org/10.1016/j.optlaseng.2017.12.004.

[110] D. Bhaduri, A. Batal, S.S. Dimov, Z. Zhang, H. Dong, M. Fallqvist, R. M'Saoubi, On design and tribological behaviour of laser textured surfaces. Procedia CIRP 60 (2017) 20–25, https://doi.org/10.1016/j.procir.2017.02.050.

[111] Q. Ding, L. Wang, L. Hu, Tribology optimization by laser surface texturing: from bulk materials to surface coatings. in: J. Lawrence, D.G. Waugh (Eds.), Laser Surface Engineering: Processes and Applications, Woodhead Publishing, 2014, pp. 405–422, https://doi.org/10.1016/B978-1-78242-074-3.00016-7.

[112] A. Pramanik, Problems and solutions in machining of titanium alloys. Int. J. Adv. Manuf. Technol. 70 (2014) 919–928, https://doi.org/10.1007/s00170-013-5326-x.

[113] L. Le Guéhennec, A. Soueidan, P. Layrolle, Y. Amouriq, Surface treatments of titanium dental implants for rapid osseointegration. Dent. Mater. 23 (2007) 844–854, https://doi.org/10.1016/j.dental.2006.06.025.

[114] L.E. Murr, S.M. Gaytan, F. Medina, H. Lopez, E. Martinez, B.I. Machado, D.H. Hernandez, L. Martinez, M.I. Lopez, R.B. Wicker, J. Bracke, Next-generation biomedical implants using additive manufacturing of complex, cellular and functional mesh arrays, Philos. Trans. R. Soc. A Math. Phys. Eng. Sci. 368 (2010) 1999–2032, https://doi.org/10.1098/rsta.2010.0010.

[115] S. Leuders, M. Thöne, A. Riemer, T. Niendorf, T. Tröster, H.A. Richard, H.J. Maier, On the mechanical behaviour of titanium alloy TiAl6V4 manufactured by selective laser melting: fatigue resistance and crack growth performance. Int. J. Fatigue 48 (2013) 300–307, https://doi.org/10.1016/j.ijfatigue.2012.11.011.

[116] D. Gu, Y.C. Hagedorn, W. Meiners, G. Meng, R.J.S. Batista, K. Wissenbach, R. Poprawe, Densification behavior, microstructure evolution, and wear performance of selective laser melting processed commercially pure titanium. Acta Mater. 60 (2012) 3849–3860, https://doi.org/10.1016/j.actamat.2012.04.006.

[117] B. Wysocki, P. Maj, A. Krawczyńska, K. Rożniatowski, J. Zdunek, K.J. Kurzydłowski, W. Święszkowski, Microstructure and mechanical properties investigation of CP titanium processed by selective laser melting (SLM). J. Mater. Process. Technol. 241 (2017) 13–23, https://doi.org/10.1016/j.jmatprotec.2016.10.022.
[118] N. Taniguchi, S. Fujibayashi, M. Takemoto, K. Sasaki, B. Otsuki, Effect of pore size on bone ingrowth into porous titanium implants fabricated by additive manufacturing: an in vivo experiment. Mater. Sci. Eng. C 59 (2016) 690–701, https://doi.org/10.1016/j.msec.2015.10.069.
[119] N. Hrabe, T. Gnäupel-Herold, T. Quinn, Fatigue properties of a titanium alloy (Ti-6Al-4V) fabricated via electron beam melting (EBM): effects of internal defects and residual stress. Int. J. Fatigue 94 (2017) 202–210, https://doi.org/10.1016/j.ijfatigue.2016.04.022.
[120] T.M. Mower, M.J. Long, Mechanical behavior of additive manufactured, powder-bed laser-fused materials. Mater. Sci. Eng. A 651 (2016) 198–213, https://doi.org/10.1016/j.msea.2015.10.068.
[121] R.O. Withers, J.C. Loutfy, S.M.S.M. Pickard, Additive manufacturing to produce standard and custom alloy titanium. in: TMS 2017 146th Annual Meeting and Exhibition Supplemental Proceedings, John Wiley & Sons, Inc., Hoboken, NJ, USA, 2017, pp. 81–89, https://doi.org/10.1007/978-3-319-51493-2_9.
[122] B. Dutta, F.H.(.S.). Froes, The additive manufacturing (AM) of titanium alloys. Met. Powder Rep. 72 (2017) 96–106, https://doi.org/10.1016/j.mprp.2016.12.062.
[123] L. Hirt, A. Reiser, R. Spolenak, T. Zambelli, Additive manufacturing of metal structures at the micrometer scale. Adv. Mater. 29 (2017) https://doi.org/10.1002/adma.201604211.
[124] R. Banerjee, S. Nag, S. Samuel, H.L. Fraser, Laser-deposited Ti-Nb-Zr-Ta orthopedic alloys. J. Biomed. Mater. Res. A 78 (2006) 298–305, https://doi.org/10.1002/jbm.a.30694.
[125] S.L. Sing, J. An, W.Y. Yeong, F.E. Wiria, Laser and electron-beam powder-bed additive manufacturing of metallic implants: a review on processes, materials and designs. J. Orthop. Res. 34 (2016) 369–385, https://doi.org/10.1002/jor.23075.
[126] F. Wang, S. Williams, P. Colegrove, A.A. Antonysamy, Microstructure and mechanical properties of wire and arc additive manufactured Ti-6Al-4V. Metall. Mater. Trans. A Phys. Metall. Mater. Sci. 44 (2013) 968–977, https://doi.org/10.1007/s11661-012-1444-6.
[127] J. Banhart, Manufacture, characterisation and application of cellular metals and metal foams. Prog. Mater. Sci. 46 (2001) 559–632, https://doi.org/10.1016/S0079-6425(00)00002-5.
[128] G.R. Parr, Tissue-Integrated Prostheses: Osseointegration in Clinical Dentistry. Quintessence, 1985, https://doi.org/10.1016/0022-3913(85)90460-3.
[129] J. Wieding, T. Lindner, P. Bergschmidt, R. Bader, Biomechanical stability of novel mechanically adapted open-porous titanium scaffolds in metatarsal bone defects of sheep. Biomaterials 46 (2015) 35–47, https://doi.org/10.1016/j.biomaterials.2014.12.010.
[130] L. Gaviria, J.P. Salcido, T. Guda, J.L. Ong, Current trends in dental implants. J. Korean Assoc. Oral Maxillofac. Surg. 40 (2014) 50, https://doi.org/10.5125/jkaoms.2014.40.2.50.
[131] A.E. Medvedev, H.P. Ng, R. Lapovok, Y. Estrin, T.C. Lowe, V.N. Anumalasetty, Effect of bulk microstructure of commercially pure titanium on surface characteristics and fatigue properties after surface modification by sand blasting and acid-etching. J. Mech. Behav. Biomed. Mater. 57 (2016) 55–68, https://doi.org/10.1016/j.jmbbm.2015.11.035.
[132] W.T. Hsiao, H.C. Chang, A. Nanci, R. Durand, Surface microtexturing of Ti-6Al-4V using an ultraviolet laser system. Mater. Des. 90 (2016) 891–895, https://doi.org/10.1016/j.matdes.2015.11.039.

[133] L. Salou, A. Hoornaert, G. Louarn, P. Layrolle, Enhanced osseointegration of titanium implants with nanostructured surfaces: an experimental study in rabbits. Acta Biomater. 11 (2015) 494–502, https://doi.org/10.1016/j.actbio.2014.10.017.
[134] F. Meyer, J. Wardale, S. Best, R. Cameron, N. Rushton, R. Brooks, Effects of lactic acid and glycolic acid on human osteoblasts: a way to understand PLGA involvement in PLGA/calcium phosphate composite failure. J. Orthop. Res. 30 (2012) 864–871, https://doi.org/10.1002/jor.22019.
[135] S.D. Luo, M. Qian, Microwave processing of titanium and titanium alloys for structural, biomedical and shape memory applications: current status and challenges. Mater. Manuf. Process. (2016) 1–15, https://doi.org/10.1080/10426914.2016.1257800.
[136] C.D. Lopez, L. Witek, A. Torroni, R.L. Flores, D.B. Demissie, S. Young, B.N. Cronstein, P.G. Coelho, The role of 3D printing in treating craniomaxillofacial congenital anomalies. Birth Defects Res. 110 (2018) 1055–1064, https://doi.org/10.1002/bdr2.1345.
[137] M. Kozakiewicz, T. Wach, P. Szymor, R. Zieliński, Two different techniques of manufacturing TMJ replacements—a technical report. J. Cranio-Maxillofac. Surg. 45 (2017) 1432–1437, https://doi.org/10.1016/j.jcms.2017.06.003.
[138] B.H. Duan, H.X. Hong, D.Z. Wang, H.J. Liu, X.J. Dong, D.D. Liang, Porous nickel-titanium alloy prepared by gel-casting. Rare Metals 33 (2014) 394–399, https://doi.org/10.1007/s12598-013-0195-x.
[139] F.A. Young, Porous titanium dental implants. J. Biomed. Mater. Res. 8 (1974) 401–407, https://doi.org/10.1002/jbm.820080420.
[140] J. Parthasarathy, 3D modeling, custom implants and its future perspectives in craniofacial surgery. Ann. Maxillofac. Surg. 4 (2014) 9, https://doi.org/10.4103/2231-0746.133065.
[141] A.T. Sidambe, Biocompatibility of advanced manufactured titanium implants—a review. Materials 7 (2014) 8168–8188, https://doi.org/10.3390/ma7128168.
[142] D.G. Kim, Y.H. Jeong, H.H. Chien, A.M. Agnew, J.W. Lee, H.B. Wen, Immediate mechanical stability of threaded and porous implant systems. Clin. Biomech. 48 (2017) 110–117, https://doi.org/10.1016/j.clinbiomech.2017.08.001.
[143] H.M. Schrøder, A. Berthelsen, G. Hassani, E.B. Hansen, S. Solgaard, Cementless porous-coated total knee arthroplasty: 10-year results in a consecutive series. J. Arthroplast. 16 (2001) 559–567, https://doi.org/10.1054/arth.2001.23565.
[144] M. Takemoto, S. Fujibayashi, M. Neo, J. Suzuki, T. Matsushita, T. Kokubo, T. Nakamura, Osteoinductive porous titanium implants: effect of sodium removal by dilute HCl treatment. Biomaterials 27 (2006) 2682–2691, https://doi.org/10.1016/j.biomaterials.2005.12.014.
[145] C. Zhao, X. Zhu, K. Liang, J. Ding, Z. Xiang, H. Fan, X. Zhang, Osteoinduction of porous titanium: a comparative study between acid-alkali and chemical-thermal treatments. J Biomed Mater Res B Appl Biomater 95B (2010) 387–396, https://doi.org/10.1002/jbm.b.31728.
[146] M. Takemoto, S. Fujibayashi, M. Neo, K. So, N. Akiyama, T. Matsushita, T. Kokubo, T. Nakamura, A porous bioactive titanium implant for spinal interbody fusion: an experimental study using a canine model. J. Neurosurg. Spine 7 (2007) 435–443, https://doi.org/10.3171/SPI-07/10/435.
[147] R.R. Boyer, An overview on the use of titanium in the aerospace industry. Mater. Sci. Eng. A 213 (1996) 103–114, https://doi.org/10.1016/0921-5093(96)10233-1.
[148] A. Salimon, Y. Bréchet, M.F. Ashby, A.L. Greer, Potential applications for steel and titanium metal foams. J. Mater. Sci. 40 (2005) 5793–5799, https://doi.org/10.1007/s10853-005-4993-x.
[149] A.S. Herrmann, P.C. Zahlen, I. Zuardy, Sandwich structures technology in commerical aviation: present applications and future trends. Sandw. Struct. Adv. Sandw. Struct. Mater. 7 (2005) 13–26, https://doi.org/10.1007/1-4020-3848-8_2.

[150] R.A.W. Mines, S. Tsopanos, Y. Shen, R. Hasan, S.T. McKown, Drop weight impact behaviour of sandwich panels with metallic micro lattice cores. Int. J. Impact Eng. 60 (2013) 120–132, https://doi.org/10.1016/j.ijimpeng.2013.04.007.
[151] E. Uhlmann, R. Kersting, T.B. Klein, M.F. Cruz, A.V. Borille, Additive manufacturing of titanium alloy for aircraft components. Procedia CIRP 35 (2015) 55–60, https://doi.org/10.1016/j.procir.2015.08.061.
[152] T. Tancogne-Dejean, A.B. Spierings, D. Mohr, Additively-manufactured metallic micro-lattice materials for high specific energy absorption under static and dynamic loading. Acta Mater. 116 (2016) 14–28, https://doi.org/10.1016/j.actamat.2016.05.054.
[153] L. Xiao, W. Song, Additively-manufactured functionally graded Ti-6Al-4V lattice structures with high strength under static and dynamic loading: experiments. Int. J. Impact Eng. 111 (2018) 255–272, https://doi.org/10.1016/j.ijimpeng.2017.09.018.
[154] D.S. Schwartz, D.S. Shih, R.J. Lederich, R.L. Martin, D.A. Deuser, Development and scale-up of the low density core process for Ti-64, in: D.S. Schwartz, D.S. Shih, A.G. Evans, W.N. Haydn (Eds.), Porous and Cellular Materials for Structural Applications, Materials Research Society, San Francisco, 1998, pp. 225–230.
[155] Z.S. Rak, J. Walter, Porous titanium foil by tape casting technique. J. Mater. Process. Technol. 175 (2006) 358–363, https://doi.org/10.1016/j.jmatprotec.2005.04.066.
[156] J. Woodward, Liquid Interface Diffusion Method of Bonding Titanium and Titanium Alloy Honeycomb Sandwich Panel Structure, US3769101A 1971.
[157] M. Singh, G.N. Morscher, T.P. Shpargel, R. Asthana, Active metal brazing of titanium to high-conductivity carbon-based sandwich structures. Mater. Sci. Eng. A 498 (2008) 31–36, https://doi.org/10.1016/j.msea.2007.11.151.
[158] P. Jakubczak, J. Bieniaś, R.J. Mania, K. Majerski, Forming of thin-walled profiles made of FML in autoclave process, Aircr. Eng. Aerosp. Technol. 90 (2018) 506–514.
[159] S. Abkowitz, S. Abkowitz, H. Fisher, Titanium alloy components manufacture from blended elemental powder and the qualification process. in: Titanium Powder Metallurgy: Science, Technology and Applications, Butterworth-Heinemann, 2015, pp. 299–312, https://doi.org/10.1016/B978-0-12-800054-0.00017-4.
[160] X. Jiang, Z. Yang, Z. Wang, F. Zhang, F. You, C. Yao, Preparation and sound absorption properties of a barium titanate/nitrile butadiene rubber-polyurethane foam composite with multilayered structure. Materials (Basel) 11 (2018), https://doi.org/10.3390/ma11040474.
[161] C.H. Nguyen, K. Chandrashekhara, V. Birman, Multifunctional thermal barrier coating in aerospace sandwich panels. Mech. Res. Commun. 39 (2012) 35–43, https://doi.org/10.1016/j.mechrescom.2011.10.003.
[162] P.S. Liu, H.B. Qing, H.L. Hou, Y.Q. Wang, Y.L. Zhang, EMI shielding and thermal conductivity of a high porosity reticular titanium foam. Mater. Des. 92 (2016) 823–828, https://doi.org/10.1016/j.matdes.2015.12.105.
[163] P.S. Liu, H.B. Qing, H.L. Hou, Primary investigation on sound absorption performance of highly porous titanium foams. Mater. Des. 85 (2015) 275–281, https://doi.org/10.1016/j.matdes.2015.06.118.
[164] Qingan Fort Industrial Park, Sintered Titanium Filters, 2016, http://sagafilter.com/products/sintered-titanium-filters/ (Accessed 30 August 2018).
[165] Graver Technologies, Filtration/Separation/Purification TPE ™ Series Filter Cartridges, Graver Technologies, Glasgow, 2016. www.gravertech.com.
[166] Baoji Fitow Metal Co. Ltd, Sintering Titanium Powder Filter, 2016, http://porousfilter.com/index.php/list/index/id/6.html (Accessed 30 August 2018).
[167] D.J. Wales, J. Grand, V.P. Ting, R.D. Burke, K.J. Edler, C.R. Bowen, S. Mintova, A.D. Burrows, Gas sensing using porous materials for automotive applications. Chem. Soc. Rev. 44 (2015) 4290–4321, https://doi.org/10.1039/c5cs00040h.
[168] P.A. Redhead, Vacuum Science and Technology: Pioneers of the 20th Century, AIP Press for the American Vacuum Society, 1994.

[169] V.L. Stout, M.D. Gibbons, Gettering of gas by titanium. J. Appl. Phys. 26 (1955) 1488–1492, https://doi.org/10.1063/1.1721936.

[170] A.B. Pandit, J.B. Joshi, Mixing in mechanically agitated gas-liquid contactors, bubble columns and modified bubble columns. Chem. Eng. Sci. 38 (1983) 1189–1215, https://doi.org/10.1016/0009-2509(83)80040-2.

[171] J.Y. Chen, Y.X. Leng, X.B. Tian, L.P. Wang, N. Huang, P.K. Chu, P. Yang, Antithrombogenic investigation of surface energy and optical bandgap and hemocompatibility mechanism of Ti(Ta(+5))O2 thin films. Biomaterials 23 (2002) 2545–2552, https://doi.org/10.1016/S0142-9612(01)00389-1.

[172] T. Wierzchoń, E. Czarnowska, A. Sowińska, T. Borowski, M. Ossowski, I. Pomnik, Z. Dziecka, Procesy azotowania i tlenoazotowania jarzeniowego stopu tytanu Ti6Al4V w aspekcie zastosowania na implanty kardiologiczne, Publishing House of Warsaw University of Technology, Warsaw, 2004.

[173] S. Ichikawa, R. Doi, Hydrogen production from water and conversion of carbon dioxide to useful chemicals by room temperature photoelectrocatalysis. Catal. Today 27 (1996) 271–277, https://doi.org/10.1016/0920-5861(95)00198-0.

[174] F. Fathollahi, M. Javanbakht, M.R. Ganjali, P. Nourozi, Electrochemical investigation of $Ru_{0.3}Ti_{0.7}O_2$ and $Ir_{0.3}Ti_{0.7}O_2$ coated titanium anodes for oxygen production processes, Anal. Bioanal. Electrochem. 5 (2013) 689–697.

[175] S. Oppenheimer, D.C. Dunand, Solid-state foaming of Ti-6A1-4V by creep or superplastic expansion of argon-filled pores. Acta Mater. 58 (2010) 4387–4397, https://doi.org/10.1016/j.actamat.2010.04.034.

[176] W. Zhang, H. Kong, H. Lin, H. Lu, W. Huang, J. Yin, Z. Lin, J. Bao, Fabrication, characterization and electrocatalytic application of a lead dioxide electrode with porous titanium substrate. J. Alloys Compd. 650 (2015) 705–711, https://doi.org/10.1016/j.jallcom.2015.07.222.

[177] G.B. Saupe, Y. Zhao, J. Bang, N.R. Yesu, G.A. Carballo, R. Ordonez, T. Bubphamala, Evaluation of a new porous titanium-niobium mixed oxide for photocatalytic water decontamination. Microchem. J. 81 (2005) 156–162, https://doi.org/10.1016/j.microc.2005.01.002.

[178] M. Carmo, D.L. Fritz, J. Mergel, D. Stolten, A comprehensive review on PEM water electrolysis. Int. J. Hydrog. Energy 38 (2013) 4901–4934, https://doi.org/10.1016/j.ijhydene.2013.01.151.

[179] W. Xu, K. Scott, The effects of ionomer content on PEM water electrolyser membrane electrode assembly performance. Int. J. Hydrog. Energy 35 (2010) 12029–12037, https://doi.org/10.1016/j.ijhydene.2010.08.055.

[180] A. Mirza, M. Burr, T. Ellis, D. Evans, D. Kakengela, L. Webb, J. Gagnon, F. Leclercq, A. Johnston, Corrosion of lead anodes in base metals electrowinning. J. S. Afr. Inst. Min. Metall. 116 (2016) 533–538, https://doi.org/10.17159/2411-9717/2016/v116n6a7.

[181] B. Sørensen, Hydrogen and Fuel Cells: Emerging Technologies and Applications, Academic Press, 2012. http://books.google.com/books?hl=en&lr=&id=1YOZuVtTAxYC&oi=fnd&pg=PP1&dq=Hydrogen+and+Fuel+Cells+Emerging+technologies+and+applications&ots=zdKgdl-Sc_&sig=kWWM2Sk3yVggMSqskrQ0I-xNKUQ, (Accessed September 3, 2018).

[182] A.P. Hauel, J. Hill, W. Reilly, Fuel Cell With Titanium-Containing Electrode and Process of Use Thereof, US Patent No US3437525A 1964.

[183] I.N. Frantsevich, D.M. Karpinos, L.I. Tuchinskii, A.B. Sapozhnikova, L.N. Klimenko, Kinetics of infiltration of porous titanium magnesium. Sov. Powder Metall. Met. Ceram. 16 (1977) 534–537, https://doi.org/10.1007/bf00790979.

[184] Q. Zhang, J.Y. Lin, X. Guo, The research of porous titanium cathode, Rare Met. 23 (1999) 72–74.

CHAPTER 4

Automobile applications of titanium

Tadahiko Furuta
Applied Powder Metallurgy and Additive Manufacturing Laboratory, Totoya Central R&D Labs., Inc., Nagakute, Japan

Contents

1 Consumer applications

1.1 Automotive use

1.1.1 Introduction

Titanium is an attractive metal for transportation equipment in the aerospace and automotive industries due to its high specific strength, heat resistance, and high corrosion resistance compared with other lightweight materials, such as aluminum, magnesium alloys, and carbon fiber-reinforced plastics. On the one hand, titanium is identified as a high-versatility material for aerospace use [1]. On the other hand, however, in the automotive industry, steel is better known for its high versatility. This is because automotive parts manufactured with titanium are generally too expensive for the consumer automotive market [2, 3]. The high cost of making titanium parts hinders its expansion into applications for mass-produced parts. Instead, it is identified mainly as a special material, for example, for automobile and motorcycle racing parts, where low weight and high strength and rigidity are critical [4].

Fig. 1 shows the change of fuel economy of mass-produced automobiles as a function of vehicle weight. Fig. 2 shows details of low-fuel-consumption technologies. Lightweighting of vehicles is effective for the reduction of running resistance, and lightweighting of reciprocating parts is more effective for improving the efficiency of single components, from the viewpoint of the reduction of losses due to friction. Above all, the effect of lightweighting reciprocating parts on the improvement of the fuel

Titanium for Consumer Applications
https://doi.org/10.1016/B978-0-12-815820-3.00006-X

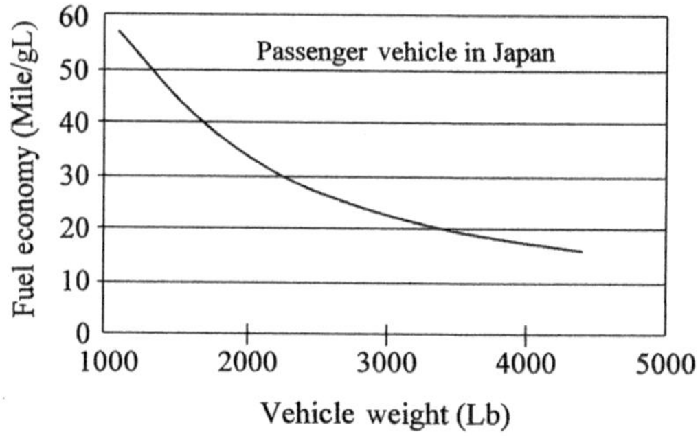

Fig. 1 Change of fuel economy of mass-produced automobiles as a function of vehicle weight [2].

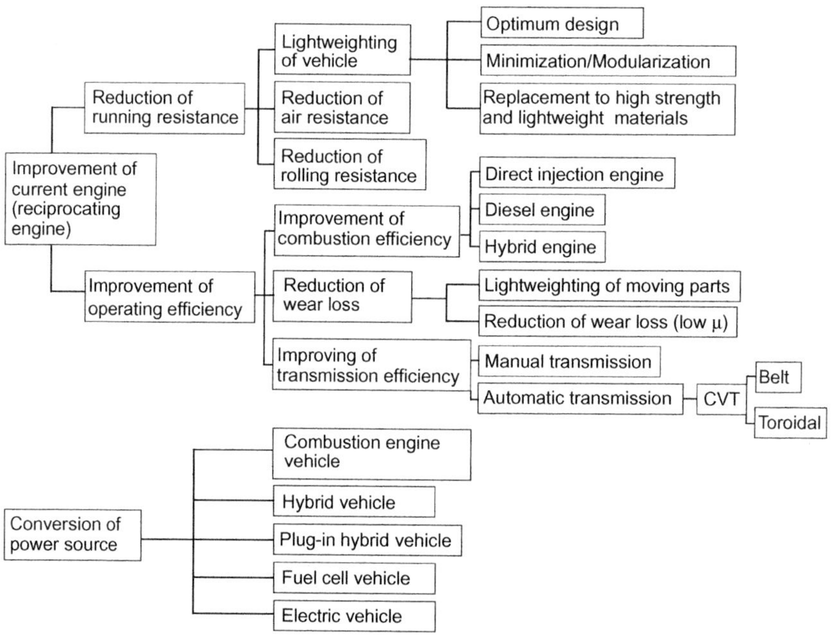

Fig. 2 Details of low-fuel-consumption technologies.

consumption is tremendous at the start of acceleration. In fact, the total energy expenditure at this point can be reduced by half.

The primary reason for applying titanium to automotive parts is to save weight. The engine valve train and connecting rods are typical titanium applications for reducing inertial forces in a reciprocating engine. However, the environment surrounding automotive design has recently changed due to an unprecedented revolution of the powertrain to reduce CO_2 emissions.

Therefore, the trends for titanium use have been changing. Namely, not only is titanium applied to lightweighting, it is also being newly applied to other automotive parts as the powertrain system evolves from a reciprocating engine to new propulsion systems for hybrid vehicles (HVs), plug-in hybrid vehicles (PHVs), fuel cell vehicles (FCVs), and electric vehicles (EVs), as shown in Fig. 2. A typical example of such new applications is the titanium fuel separator in the fuel cell (FC) stack in an FCV [5]. This separator is required to have high specific strength for weight saving and high corrosion resistance due to extremely severe operating conditions. It is reported [5] that the total cost of an FC stack can be reduced significantly because of eliminating high-cost surface treatments by adopting a separator made of titanium instead of stainless steel.

To increase the general use of titanium for automobile components in the future, it is important to consider the merits of practical use of titanium and its alloys in a total life cycle assessment (LCA) or value added as a unit or system only. In the following subsections, the detailed development history of titanium parts is introduced from perspectives of both alloy design and material processing for use in automotive systems. An overview of future prospects is also presented.

1.1.2 Reciprocating engine

Reciprocating parts are the most effective applications for titanium, and they include intake and exhaust valves, valve spring retainers, valve springs, valve lifters, connecting rods, and piston pins as shown in Fig. 3. Among them, intake valves, valve lifters, and connecting rods are usually designed to use typical structural $\alpha+\beta$-titanium alloys, such as Ti-6Al-4V, due to operating environments up to 300°C. This is because the large end of the connecting rod needs to have an optimum shape with sufficient rigidity to suppress strength reduction by fretting fatigue. The weight saving of that with piston pins lead to significantly reduced noise, vibration, and harshness, and improved engine performance including fuel consumption. In contrast, valve spring retainers are designed using β-titanium alloys, such as Ti-22V-4Al (DAT51) [6], which have a cold deformability. Also, Ti-6.8Mo-4.5Fe-1.5Al (or low-cost beta, LCB) alloys [7] are suitable for making valve spring retainers. Specifically, LCB alloy has low Young's modulus, high strength, and low manufacturing cost in comparison with Ti-6Al-4V. The details of spring properties required for automotive use are discussed later. The crankshaft, however, is the only reciprocating component of the engine that is difficult to effectively fabricate from titanium. Because

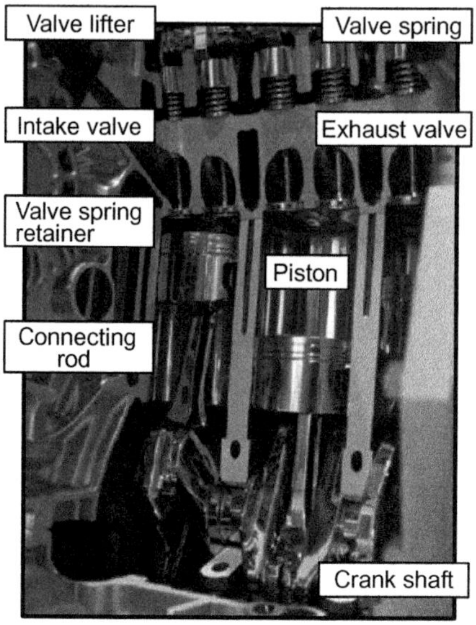

Fig. 3 Reciprocating engine cut model (https://pixta.jp).

the function of the crankshaft is to convert the reciprocating motion of pistons into torque to create the driving force, weight-saving materials lack the rigidity essential for this function.

To consider the application of titanium to exhaust valves from the viewpoint of manufacturing cost, we may say that replacing steel reciprocating engine parts is most effective for parts having expensive alloy compositions due to high-temperature operation, that is, around 800°C. Because iron-based exhaust valves consist mainly of Fe-21Cr-0.4N (21-4N) [8, 9], it has high austenitic heat resistance but is expensive. Therefore, since the 1980s, high-temperature near-α titanium alloys, specifically Ti-5.8Al-4Sn-3.5Zr-0.5Mo-0.7Nb-0.35Si-0.06C (IMI834) [10] and Ti-6Al-2.7Sn-4Zr-0.4Mo-0.45Si (Ti-1100) [11], have been developed by IMI and TIMET. In the 1990s, moreover, a large number of high-temperature α-titanium alloys were developed for exhaust valves as exhaust gas temperatures increased due to the increased power of reciprocating engines and stringent emissions regulations. These trends prompted new titanium technologies, such as unique alloy designs and microstructure control, as shown in Fig. 3. For example, 5 vol% TiB/Ti-6Al-4Sn-4Zr-1Mo-1Nb-0.2Si (TiB MMC) [12, 13], Ti-5.8Al-4Sn-3.5Zr-2.8Mo-0.7Nb-0.35Si-0.06C (DAT54) [14], and Ti-6Al-4Sn-3.5Zr-0.5Mo-1Ta-0.35Si [15] are typical of titanium alloys developed for exhaust valves. In particular, TiB MMC has unique properties, such as ultra-high strength and heat resistance up to around 900°C, high-temperature fatigue strength, and wear resistance. This heat resistance is due to TiB particles in Ti-MMC, but they are only stable in titanium matrix alloys, after in-situ precipitation by a reaction between TiB_2 and Ti during high-temperature sintering. In contrast, DAT54 attains a balance between creep resistance and low-cycle fatigue resistance at high temperatures because adding 2.8% Mo gives it a unique bimodal microstructure. Ti-1100, which has a β-transformed acicular microstructure with a high creep resistance, also has been applied to exhaust valves [16].

Moreover, to give a boost to the increased use of titanium in reciprocating engine parts, it is important to merge the processing conditions needed to elicit the required properties of each engine part. For example, as mentioned before, the blended elemental (BE) powder metallurgy process and high-temperature extrusion methods are successful for producing almost 100% material yields [13, 17, 18]. Fig. 4 shows the trial products of engine parts fabricated by BE powder metallurgy process [18]. Cold forming is also important for fabricating low-cost titanium engine parts. For example, the formability of Ti-22V-4V for valve spring retainer alloy is superior to that of Ti-6Al-V alloy. Surface treatments, such as shot-peening for high cycle fatigue strength, metal plating, thermal spraying, and hardening by interstitial elements, also add high value and are effective for improving reliability. Especially, oxidation processing is applied to engine valves as a typical low-cost treatment [16, 18]. TiB MMC engine valves with low-cost surface treatment meet severe criteria for wear resistance after engine endurance tests, as described in detail later [13].

Fig. 4 Trial products of engine parts fabricated by BE powder metallurgy process.

1.1.3 Exhaust systems

Titanium is mainly used in mufflers in exhaust systems, as shown in Fig. 5. Advantages of replacing iron-based muffler parts with titanium are the weight savings and aesthetics. The colors of titanium alloys are mysterious and beautiful. The fascination with those colors from the viewpoint of the driver is that driving habits, such as acceleration, can create unique patterns and novel beauty. The exhaust temperatures are one factor that determines the color tones changed by the refraction ratio of light rays. The refraction ratio is strongly related to the thickness of the oxide film formed on the muffler. Ordinarily, automotive muffler parts are placed under the car body, unlike motorcycle mufflers, and they are not as suitable for showing off the characteristic titanium colors.

Therefore, the expansion of titanium use in automotive mufflers is difficult to justify based only on aesthetics. However, an effective weight saving is expected. The mainstream material for mufflers has mostly been pure titanium (Cp-Ti), but recently titanium alloys have been developed due to increasing exhaust temperatures. These alloys have included Ti-1Cu-0.5Nb [19], Ti-0.45Si-0.25Fe [20], Ti-1.5Al [21], Ti-0.5Al-0.45Si-0.2Nb (Ti-ASN) [22], and Ti-0.9SA [23]. Higher-performance titanium alloys are needed for mufflers because they are used under severe conditions characteristic to automobile operation, such as high temperatures and salt corrosion [24].

1.1.4 Automotive body and structural parts

Automotive body and structural parts account for most of the weight in an automobile. Therefore, the effectiveness of incorporating titanium into these parts for weight reduction is enormous. For example, the Thrust SSC (supersonic car) [25] is a British jet-propelled titanium car developed by

Fig. 5 Exhaust system for commercially pure titanium [4].

Richard Noble, Glynne Bowsher, Ron Ayers, and Jeremy Bliss. On October 15, 1997, Andy Green drove the Thrust SSC to a world land speed record of 1228 km/h (763 mi/h), and it became the first car to officially break the sound barrier. However, applying such substantial titanium body content to mass-produced cars is impractical, mostly because titanium is too expensive for the general consumer market compared with aluminum- and iron-based structural parts.

However, suspension parts fabricated from titanium alloys have the potential for general consumer markets, especially for after-market racing parts. They hold a very important role for improving the principal driving performance, such as straight-line drivability and steering stability. The coil spring shown in Fig. 6 [2, 26] has a high fatigue strength and low modulus of transverse elasticity. Generally, the spring constant K is calculated using the following relation:

$$K = Gd^4 / 8D^3N \tag{1}$$

where G is the modulus of transverse elasticity, d is the wire diameter, D is the coil diameter, and N is the number of windings. Following this formula, for a constant value of K, making G smaller reduces N. By using titanium alloys instead of iron alloys, the automotive coil spring can be thereby lightweighted. For example, when using Ti-3Al-8V-6Cr-4Mo-4Zr (Beta C) [27] or LCB alloys, N becomes 1/2 and, as a result, the weight of the titanium coil spring is 1/3 the weight of an iron coil spring. To improve the critical stress up to the fatigue strength required for coil springs, they are fabricated by cold drawing and then receive aging and shot-peening treatment [26].

1.1.5 New applications of titanium automotive parts in FCVs and EVs

As discussed in Section 1.1.1, with the large ongoing movement to change the propulsion source in automotive technology, as shown in Fig. 2, the standard condition or appearance points of titanium are going to change drastically from being merely another choice for structural lightweighting. The separator in FC stacks is required to have very high corrosion resistance capable of simultaneously withstanding the strong acid of an electrolyte film and impacts associated with travel vibration. The titanium separators with a newly developed π-conjugated amorphous carbon [28] treatment in the MIRAI (by Toyota Motor Co.) [5] improve reliability with respect to the SUS316L separators that they replaced and can achieve a weight reduction of approximately 40%. In other words, a titanium alloy specifically obtains the needed results.

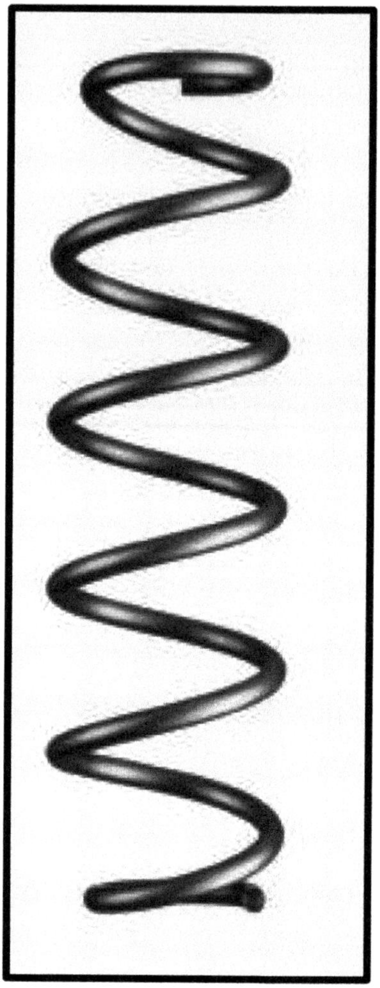

Fig. 6 Titanium coil spring made of LCB [26].

In addition, titanium has unique physical properties, such as being non-magnetic and having a relatively high electrical resistance among metals. As electrical motors become the primary motive force for vehicles, materials with specific electromagnetic properties become more important, and electrical resistance is a very important property for motor materials rotating at high speeds. High electric resistance plays a role in reducing eddy current losses, and it is no exaggeration to say that titanium alloys are indispensable materials for making automobiles that rely on clean energy.

1.1.6 Practical application examples

Fig. 7 shows the development chronologically of typical practical titanium parts, the kinds of titanium and alloys used, and the applied automotive names. Since the 1990s, Ti-6Al-4V, a typical $\alpha+\beta$-titanium alloy, has been used for connecting rods in the reciprocating engines of luxury sports automobiles. NSX (by Honda Motor Co.), with a V-6 cylinder incorporating a connecting rod made of Ti-3Al-2V RE (Rare earth) [29], is first luxury sports automobile. By adding RE, it succeeded in improving the poor machinability of titanium.

Since 2000, not only luxury sports automobiles but also some mass-produced cars have used titanium parts, including intake and exhaust valves, mufflers, turbochargers, and coil springs. Among them, in Oct. 1998, the ALTEZZA (by Toyota) was released into the Japanese market with titanium intake and exhaust valves [30], and in 2000, the LUPO (by Volkswagen AG) [2] was released into the German market with coil springs made of LCB alloy. These were important milestones, as these are advanced markets for consumer goods, especially mass-produced automobiles. The ALTEZZA was the first mass-produced automobile with titanium parts in the world. Fig. 8 shows the appearance of the developed intake valve (left) and exhaust valve (right) in ALTEZZA's high-performance four-cylinder reciprocating engine. The intake valve was made of Ti-6Al-4V alloy manufactured via the cost-effective BE process using TiH_2 powder, which does not require as much heat resistance. By using TiH_2 powder, it overcame a seizure between titanium and a molding die for making a billet and succeeded to construct a mass-production system. In contrast, exhaust valves need resistance to high temperatures and severe operating conditions. The highest head temperatures measured were up to 780°C for 21-4N steel valves and up to 800°C for TiB MMC valves; the titanium valves were a little hotter than the 21-4N steel valves due to the low heat conductivity of iron-based alloys. However, the developed valves passed all of the various actual durability tests and reliability evaluation tests, as shown in Fig. 9.

The developed titanium-based intake and exhaust valves lightened the valve weight by 40% and the valve spring weight by 16%, in comparison with conventional steel valves and springs [30]. As a result, the maximum revolution limit increased by 10%, and the noise in the high-revolution range decreased by 3 dB. Subsequently, these technologies were applied in the latest and toughest racing engines, unlike the normal path of research and development, where racing components are developed first and later passed along to the general retail market. Furthermore, these valves are still

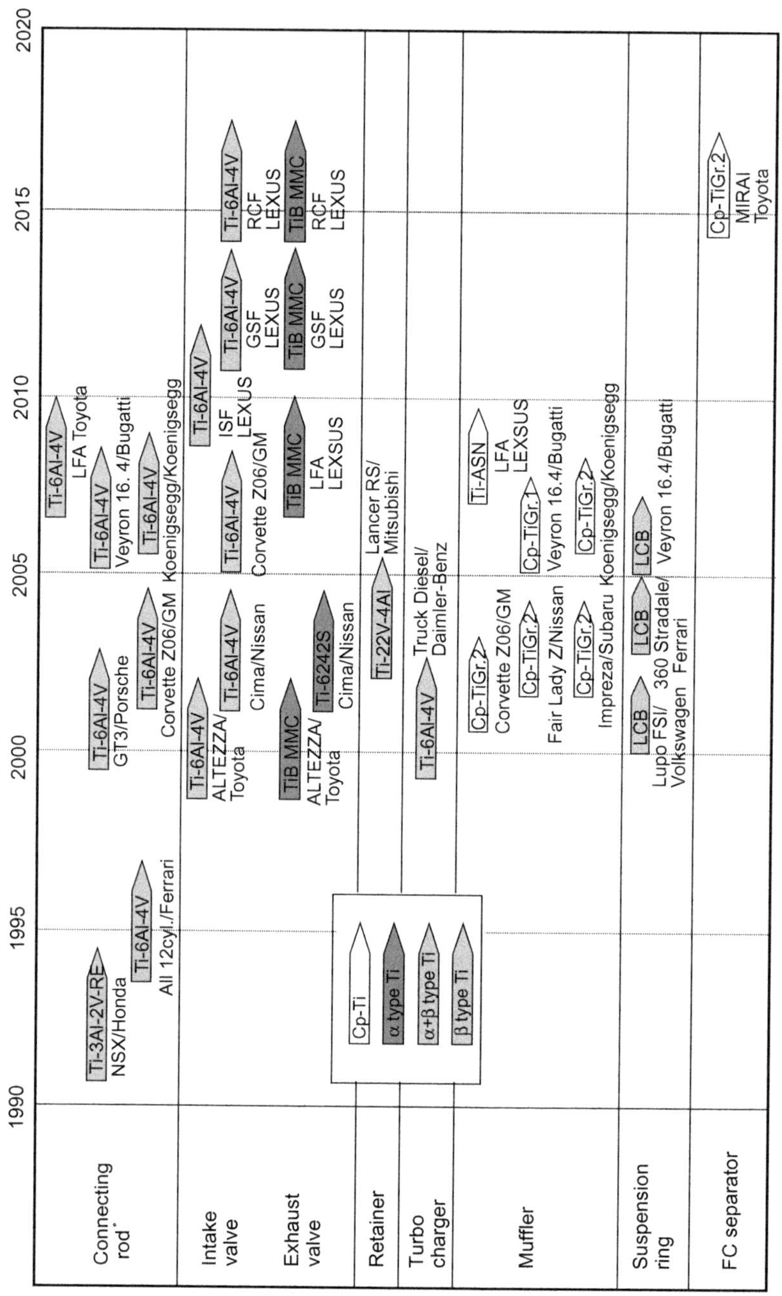

Fig. 7 Development chronologically of typical practical titanium parts.

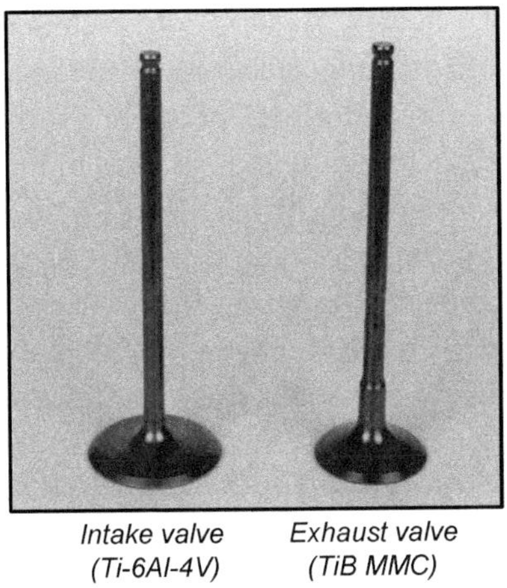

Fig. 8 Appearance of the developed intake valve (left) and exhaust valve (right) [30].

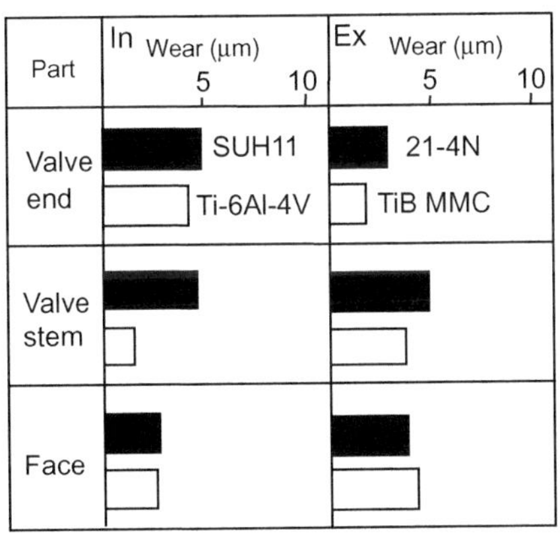

Fig. 9 Comparison of actual durability tests for engine valves [30].

used in the engines of the LFA, GSF, and RCF models (Lexus/Toyota), showing that these valves are core technologies, even 20 years since they entered the commercial market [30].

For the CORVETTE Z06 (by Chevrolet/GM) titanium (Cp-TiGr.2) exhaust system in 2001, the goal was to reduce mass to improve performance. Titanium provided an 11.7 kg dual muffler assembly, achieving a 41% weight reduction and contributing to the best performing CORVETTE ever with a virtually unlimited exhaust life [2]. LFA also applied a titanium muffler made by Ti-ASN with the high temperature salt damage corrosion [24] in 2010.

MIRAI (by Toyota), which has been sold commercially since 2016, also has importance for both titanium and automotive industries. For the titanium industry, the amount of titanium per automobile for MIRAI is remarkably increased in comparison with most engines, as mentioned before, and tens of kilograms of titanium are used in each MIRAI [5], although the commercial production of these automobiles is low. If MIRAI is to have a significant impact on the environment, a larger number of sales is an absolute necessity. However, to gain further market penetration in the future, we need to develop titanium alloys with even more corrosion resistance and durability.

1.1.7 Issues and potential challenges in automotive use

The unavoidable issues for increased automotive use of titanium are, above all else, *cost* and *durability*. Titanium must overcome the hurdle of the current average cost of 1 yen per gram in mass-produced parts. To do so, it is necessary to apply new smelting methods, harmonize the low-cost production processes [31], and improve machining difficulties. However, those steps alone will not sufficiently reduce the cost of titanium. For example, LCA related to the production process is an effective method for solving the cost issue by considering the total cost balance. Furthermore, significant durability improvement can support low titanium automotive parts costs by replacing more iron parts with titanium parts. In that case, we must draw attention to the deterioration of individual titanium components in comparison to their initial properties by evaluating the required criteria for automobile parts used in severe environments. Fig. 10 shows one example

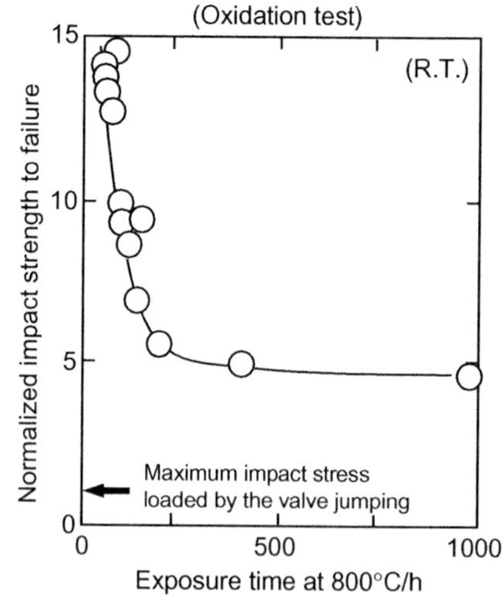

Fig. 10 Results of reliability test for exhaust valves [30].

of reliability test results for exhaust valves [30]. The criterion for these valves was impact strength during anomalous motion under over-speed conditions. It is clear that the impact strength after an assumed 100,000 km of oxidative degradation exceeds by about five times or greater than maximum impact stress loaded by the valve jumping that occurs during anomalous motion on over-speed [29]. This demonstrates the importance of the life cycle cost of automotive parts, not just the initial cost for their manufacture.

To be the third-most versatile material for automotive parts, after iron and aluminum, new uses need to be found for titanium. It is not enough to only replace existing iron and aluminum parts with titanium parts. The key to this is the environment. For example, when FCVs are assessed as the primary "green" propulsion technology for the automotive industry, the number of FCVs dramatically increases, potentially doubling the current volume of wrought titanium materials within 15 years. Titanium is a very attractive metal, and, as discussed earlier, it is important to understand its merits and identify parts and structures that can make the best use of titanium's performance in the context of the environment surrounding the automobile industry and the needs of society.

Reference

[1] F.H. Froes, R.H. Jones, The 14th international Ti applications conference and exhibition, JOM 6 (1999) 40–44.

[2] K. Faller, F.H. Froes, The use of titanium in family automobiles: current trends, JOM 53 (2001) 27–28.

[3] F.H. Froes, Titanium alloys, in: J.K. Wessel (Ed.), Chapter 8 of the Handbook of Advanced Materials, McGraw-Hill Inc., New York, NY, 2001.

[4] F.H. Froes, H. Friedrich, J. Keise, D. Bergoint, Titanium in the family automobile: the cost challenge, JOM 56 (2004) 40–44.

[5] N. Konno, S. Mizuno, N. Hiroki, Y. Ishikawa, SAE Int. J. Alter. Powertrains 4 (1) (2015) 123–129.

[6] A. Fukui, Y. Ohtakara, A. Suzuki, Properties of a beta type new titanium alloy Ti-22V-4Al, Denkiseiko 57 (1986) 303–317.

[7] P.G. Allen, P.J. Bania, A. J. Hutt and Y. Combres: Proceeding on the 8th World Conference on Titanium (1996), 1680–1687.

[8] D. L. Loveless: US Patent 2839391 (1958).

[9] M. Fujita, Forgeability of 21-4N steel, Denkiseiko 42 (1971) 106–112.

[10] D.F. Neal: Proceeding on the 6th World Conference on Titanium (1988), 253–258.

[11] P.J. Bania: Proceeding on the 6th World Conference on Titanium (1988), 825–830.

[12] T. Saito, T. Furuta, Toyota Central Research and Development R&D Review, vol. 29, 1994, pp. 51–56.

[13] T. Furuta, Research of Titanium Based Materials with Controlled Elastic Properties, Doctoral Theses, Tohoku University, 2008, pp. 1–250.

[14] T. Noda, M. Okabe, S. Isobe, Development of high performance heat resistant near-alpha titanium alloy compressor disk, Denkiseiko 67 (1996) 103–107.

[15] T. Matsumoto, Kobe Steel Engineering Reports, vol. 43, 1993, pp. 103–106.

[16] H. Fujii, K. Takahashi, Y. Yamashita, Nippon Steel Technical Report, 88, 2003, pp. 70–75.
[17] S. Abkowitz, S.M. Abkowitz, H. Fisher, P.J. Schwartz, CermeTi^R discontinuously reinforced Ti-matrix composites: manufacturing, properties, and applications, JOM (5) (2004) 37–41.
[18] T. Furuta, T. Yamaguchi, Y. Shibata and T. Saito: Proceeding on the 9th World Conference on Titanium, (1999), 1917–1924.
[19] H. Otsuka, H. Fujii, K. Takahashi and M. Ishii: Proceeding on the 11th World Conference on Titanium, (2007), 1391–1394.
[20] Y. Kosaka, S. P. Fox and K. Faller: Proceeding on the 11th World Conference on Titanium, (2007), 1403–1406.
[21] K. Matsukura, T. Yashiki, Kobe Steel Eng. Rep. 54 (3) (2004) 38–41.
[22] T. Yashiki, K. Ymamoto, Kobe Steel Eng. Rep. 55 (3) (2005) 42–47.
[23] K. Tada, Y. Itsumi, Kobe Steel Eng. Rep. 60 (2) (2005) 42–45.
[24] H. Ohyama, Kobelco Technol. Rev. 35 (2017) 14–77.
[25] http://www.bloodhoundssc.com/node/311/thrustssc.
[26] S.P. Fox, Y. Kosaka, Titanium suspension spring—an update, SAE Technical Paper Series 2002-01-0675, Transaction 109 (2002).
[27] V. R. Jablokov, J. R. Wood and B. G. Drummond: Proceeding on the10th World Conference on Titanium, (2003), 3035–3042.
[28] S. Mizuno, Y. Nakaji, H. Yoshikawa, Toyota Techn. Rev. 61 (2015) 20–25.
[29] S. Nakamura, Development of free-machining pure titanium alloys and free-machining titanium alloys, Denkiseiko 59 (1988) 79–86.
[30] T. Yamaguchi, H. Morishita, S. Iwase, Y. Yamada, T. Furuta, T. Saito, Development of P/M titanium engine valves, SAE 2000 Transactions, vol. 109, 2001, pp. 416–424.
[31] E. H. Kraft: Paper Presented at the ITA Conference, Scottsdale, AZ, September (2005).

CHAPTER 5

Titanium and titanium alloys in drones and other small flying objects

M. Yan[a], T. Ebel[b]

[a]Department of Materials Science and Engineering, and Shenzhen Key Laboratory for Additive Manufacturing of High-performance Materials, Southern University of Science and Technology, Shenzhen, China

[b]Helmholtz-Zentrum Geesthacht, Institute of Materials Research, Geesthacht, Germany

Contents

1 Opening remarks

In this chapter, we introduce applications of Ti and Ti products in drones and other small flying objects, namely micro air vehicles, model aircrafts (aero modeling), paramotors, and in air racing.

General aviation (GA) refers to all civil aviation operations except those scheduled air services and nonscheduled air transport operations for

Titanium for Consumer Applications
https://doi.org/10.1016/B978-0-12-815820-3.00018-6

remuneration or hire [1, 2]. It covers a vast range of aerial activities for example, flying clubs and light aircraft manufacturing and maintenance, and it has many types of aircraft involved, from gliders to powered parachutes. Globally, GA is a trillion-dollar business that is experiencing rapid growth. According to the definition, the five areas to be reviewed by this chapter generally fall into the category of GA.

For GA applications, there are a few important concerns from a material perspective, particularly regarding weight reduction and cost-effectiveness. From this perspective, the relatively high cost of using Ti and Ti alloys have certainly limited their commercial applications to a larger scale. They have to compete harshly with lower cost materials such as carbon fibers, Al alloys, polymers, and composite materials [3, 4].

Nevertheless, Ti and Ti alloys are unique in terms of their great combination of low density and high strength. They can be easily 1000 MPa high in fracture strength, meanwhile only showing ~60% of density of steel (Fig. 1) [5–10]. Apart from this, a large variety of Ti and Ti alloys exist, from the ductile, commercially pure Ti (CP-Ti) to the mostly employed

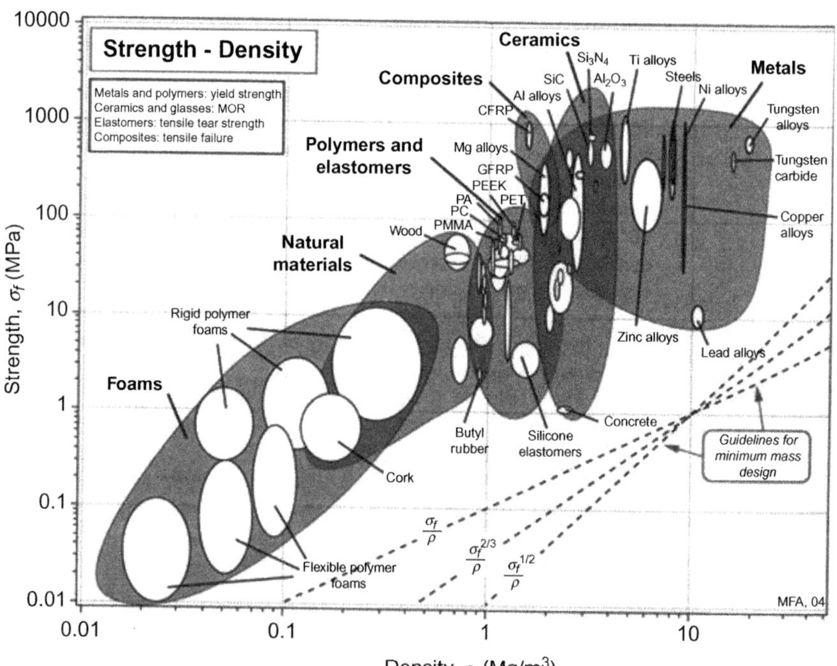

Fig. 1 Comparison of strength and density among a large number of materials including Ti and Ti alloys.

Ti-6Al-4V, to the high-temperature Ti materials (e.g., TiAl intermetallics), ensuring that they can provide service parts in many circumstances, for example, as engine materials whose working temperatures are around 700°C [5–8]. In addition, Ti and Ti alloys also show excellent corrosion resistance, particularly the Ti-Pd alloys, and this gives them another important advantage over other materials for all-weather applications [11].

Most Ti and Ti alloys employed in GA are fabricated by conventional processing means. Recently developed advanced processing techniques, namely additive manufacturing (AM)/3D printing, have emerged as promising approaches toward customized production of Ti and Ti alloys [12–14]. This can be increasingly important because personalized products take a big share of GA. This chapter, therefore, will also briefly introduce the impact brought by AM on the processing of Ti and Ti alloys regarding their GA applications.

2 Ti and Ti alloys in drones

2.1 Brief introduction to drones: Definition, classification, and market

Drones are also known as unmanned aerial vehicles (UAV) [15, 16]. They are becoming increasingly popular and important in recent years for their vast potentials in both military applications and consumer applications. It is reported that there are more than 10 million flying drones so far, ranging from hobby-level ones to million-dollars worth of daunting military tools (Fig. 2) [17].

There are many standards to classify drones [15, 16]. One can roughly separate them into two large categories [15, 16]: One is for military purposes, for example, providing strategic information via long-range, high-altitude surveillance drones, for example, the Global Hawk developed by the Northrop Grumman [17] (Fig. 2). The other is for civilian and consumer usages, covering a large range of applications including picturing and landscape exploration, road surveillance, search and rescue, agriculture, cargo delivery, and policing (Fig. 3) [18].

Drones are becoming an important worldwide business. Sales of civilian and consumer drones are reported to be ~500,000 units per year [19], whereas the overall units of drones are expected to soon reach 70 million a year [20]. The total market of consumer drones reached about U.S. $2 billion in 2018, and a complex annual growth rate (CAGR) of almost 20% is well expected [21].

Drone Diversity

The big aerospace companies that have long led the drone industry offer high-powered, high-priced devices, while a bevy of drone upstarts are pitching lightweight, low-cost drones.

PRICE:	$93 MILLION	$100,000	$10,000–$15,000	$1300
	BIG AEROSPACE COMPANIES		DRONES OF SMALLER STARTUPS	
NAME Manufacturer	**Global Hawk** Northrop Grumman	**ScanEagle** Insitu*	**Lancaster Hawkeye** PrecisionHawk	**Phantom Vision 2+** DJI
PRIMARY USE	Military intelligence	Surveillance	Agriculture	Photography
SIZE	47.6 feet long by 130.9-foot wingspan	5.6 x 10.2 ft. wingspan	4 x 4 ft. wingspan	0.95 x 0.95 ft. quadcopter
WEIGHT	32,250 lb.	48.5 lb.	3 lb.	2.8 lb.
ENDURANCE	28 h	24 h	45 min	25 min
SPEED	357 mph	57 to 69 mph	25 mph	25 mph
OPERATING ALTITUDE	60,000 ft.	19,500 ft.	400 ft.	Less than 400 ft.

Source and photos: the companies *Boeing subsidiary

The Wall Street Journal

Fig. 2 Drone diversity illustration, showing different types of drones, ranging from ~$1000 worthy consumer ones to military drones worth millions of dollars.

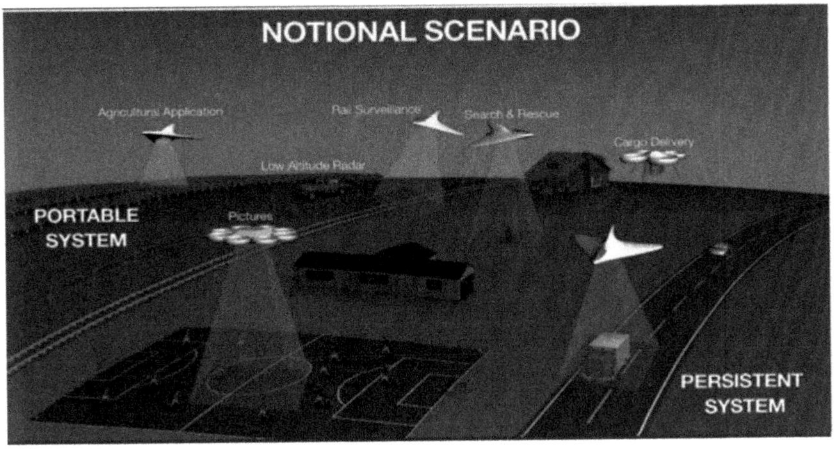

Fig. 3 The applications of commercial drones at a national scenario.

2.2 Ti and Ti alloys in drones: Conventionally processed parts and 3D-printed parts

2.2.1 Structure of a drone

A typical medium-sized drone consists of these essential parts: main body, power supply and propellers, navigation and positioning system, camera system, sensing system, payload system, and data processing system [22]. To make drones fly, there are extra systems and components involved, namely the drones' pilot as well as ground system. Data streams are constantly exchanged between drones and their ground control system.

2.2.2 Typical Ti and Ti alloys parts made by conventional processing methods for drones

A wide survey indicates that parts made of Ti and Ti alloys by conventional processing methods, that is, casting, forging, machining, etc., are employed in drones in the following circumstances, see Fig. 4 and Table 1 for some typical examples [23–26]:

(a) As frame-structure material to support propellers, which is reported by the Niono Drone company [23]. Length of the frame structure varies from 149 to 257 mm, using 57–120 g weight Ti alloy per piece.

(b) As housing materials to hold propellers, which is promoted by the Dynamic UAV for their PHANTOM 4 Prof drones [24].

(c) As accessory parts and fittings such as screws, which are made of Ti-3Al-5Mo-4.5V alloy (TC16) [25].

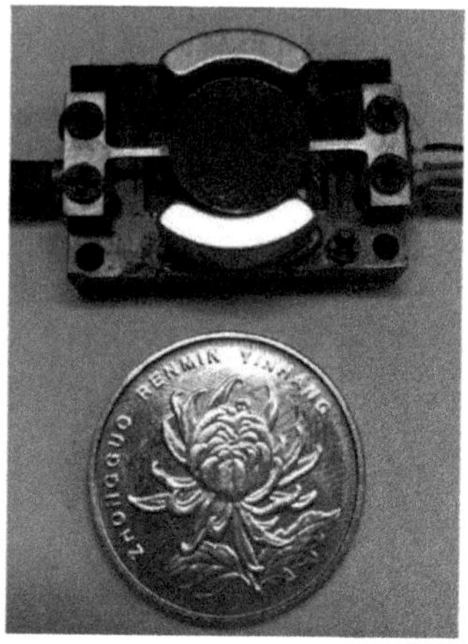

Fig. 4 Applications of Ti and Ti alloys in drones: as substrate material for a micro mirror device.

(d) As substrate material for a micro scanning mirror (micro-electro-mechanical system—MEMS) device of a laser rangefinder equipped on drones (Fig. 4) [26].

2.2.3 Typical Ti and Ti alloys parts processed by 3D printing

One of the mainstream 3D printing techniques, electron beam melting (EBM), has been employed to produce the main body frame structure as well as air foils of a medium-sized drone using Ti alloy, see Fig. 5 [27, 28] and Table 1 for details. The part was produced by CSIROscope Australia, jointly with Melbourne University students, for making bush fire-fighting drones [27, 28]. The Ti alloy used was not specified by the reports, but it was supposed to be Ti-6Al-4V; due to the rapid prototyping capability of AM, parts of the drone were printed out within merely 2 days, which was virtually impossible for any other processing means.

From the reports as well as the authors' understanding, EBM and other AM techniques such as selective laser melting (SLM) can offer the following advantages over conventional processing approaches and are therefore becoming increasingly important processing techniques for Ti and Ti alloys [12–14]:

- Ability to fabricate part of complex geometry and optimal functionality as shown by the example in Fig. 5, which guarantees that most parts can be satisfactorily fabricated by the AM techniques regardless of their processability by conventional means.

Table 1 Applications of Ti and Ti alloys in drones, micro air vehicles, aero modeling, paramotor, and air racing

Application area	Part made by Ti and Ti alloys	Processing method	Ti alloy used	Ref.
Drone	Frame structure	- (Forging?)	Ti-6Al-4V	[23]
	Housing structure	- (Forging?)	Ti-6Al-4V	[24]
	Screw	Cold forming	Ti-3Al-5Mo-4.5V	[25]
	Substrate for micromirror	Electro-sparking	Ti-6Al-4V	[26]
	Main body frame structure as well as air foil	3D printing	–	[27] [28]
Micro air vehicle	Wing frame structure	Chemical etching and patterning	Ti-6Al-4V	[29]
Aero modeling and model aircraft	Turnbuckle for the Al 6061 landing gear kit	- (Rolling?)	Ti-6Al-4V	[30]
	Screw for DIY model aircraft	- (Forging?)	Ti	[31]
	Bolt for DIY model aircraft	- (Forging?)	Ti-6Al-4V	[32]
Paramotor	Exhaust for engine	- (Forging?)	- (Ti-6Al-4V?)	[33]
	Frame structure material	- (Forging?)	- (Ti-6Al-4V?)	[34]
Air racing	Bolt	- (Forging?)	- (Ti-6Al-4V?)	[35]

(A)

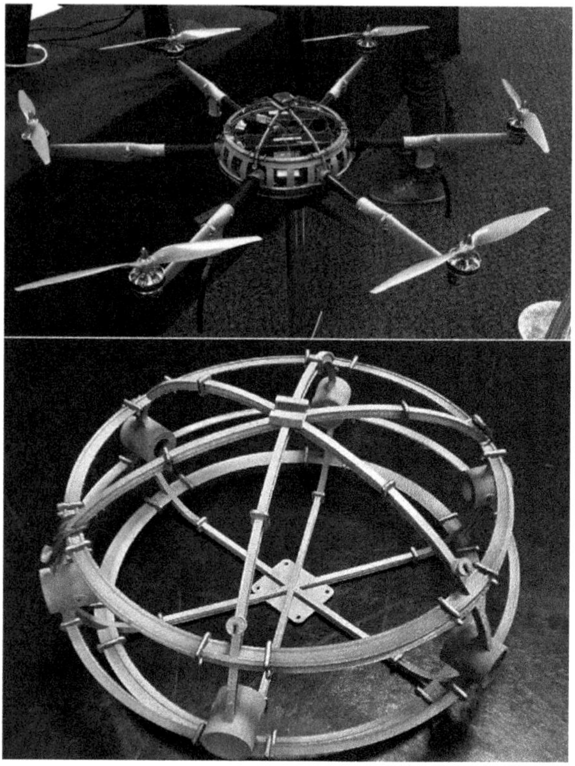

(B)

Fig. 5 (A) A picture of bushfire surveillance drone whose main frame structure and air foils are made of 3D-printed Ti alloy and (B) an enlarged picture of the frame structure.

- Near net-shaping capability and the associated high buy-to-fly ratio, which is valuable for saving costly Ti and Ti raw materials.
- Moldless process and short lead time, which eliminates cost on mold making and provides rapid response to commercial opportunities and research needs.
- Ability to reduce the number of joints as a result of topography optimization and additive design prior to actual AM processing.
- Ability to process virtually all types of Ti alloys, implying that parts can be made by AM as long as one can find the most suitable Ti alloy for meeting individual requirement.
- Ability to satisfy the request on customization and personalization, as virtually all the processing procedures of AM are computer designed and automatically implemented.

Certain drawbacks of AM also exist, and these include [12–14]:

- high cost in raw (powder) materials, equipment, and know-how;

- potential safety hazard in handling raw (powder) materials; and
- relatively slow process rate, for example, most SLM machines currently print at a speed of ~30 cm^3, corresponding to merely ~250 g/h for steel.

3 Ti and Ti alloys in micro air vehicles

3.1 Brief introduction to micro air vehicles: Definition, classification, and market

Micro air vehicle (MAV), also referred to as micro aerial vehicle, belongs to UAVs as well, but they are much smaller in size [36, 37]. Modern MAVs can be as small as 5 cm; see Fig. 6, the robot hummingbird MAV for a typical example [38].

Compared to larger flying objects such as light planes, MAVs are also much lighter in weight, normally less than 1 kg, and their Reynolds numbers generally range between 10^2 and 10^4, suggesting that they are more easily influenced by flying medium (i.e., air) (Fig. 7A) [29].

Despite their small sizes, MAVs can be classified into three main categories like other normal aircraft, namely the flapping wings, rotorcrafts, and fixed wings (Fig. 7B). Their flying times are mostly within hours, often powered by various types of batteries [39].

The currently available MAVs are made for hobby purposes such as aerial contests and aerial photography, research purposes, or military applications. The commercial market for MAVs is not fully developed yet; the size of the global market is hard to estimate, but it is reported that in Canada

Fig. 6 Picture of a hummingbird micro air vehicle.

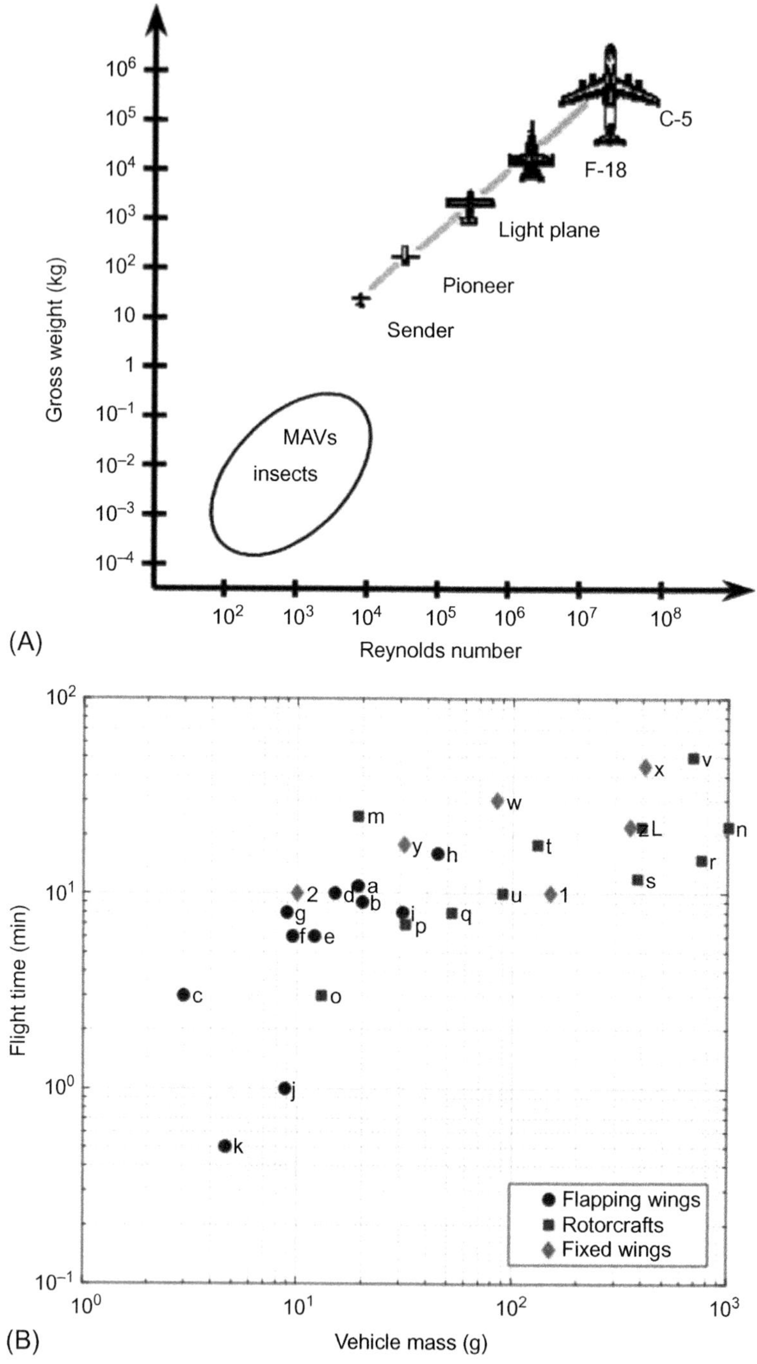

Fig. 7 (A) Classification of air vehicles in terms of their weight and Reynolds number and (B) typical micro air vehicles and their corresponding flight time and weight. English characters and numbers (i.e., 1 and 2) in the figure represent the following types of micro air vehicles: a: Nano hummingbird, b: DelFly Explorer, c: Delfly Micro, d: H2 Bird, e: MicroBat, f: Bionic Bird, g: Avitron V2.0, h: 36-cm Ornithopter, i: 28-cm Ornithopter, j: 15-cm Ornithopter, k: 10-cm Ornithopter, L: Parrot Bebop drone, m: PD-100 Black Hornet PRS, n: DJI Phantom 2, o: Seiko-Epson uFR-II, p: Ladybird V2, q: Mini X6, r: 350. QX2, s: AR. Drone 2.0, t: QR Y100, u: QR W100S, v: eBee, w: Black widow, x: Wasp III, y: Univ. Florida MAV, z: H301S, 1: Diamond 600 EP, 2: EPFL MC2.

alone the sale of MAVs is about $20 million in 2018, and a ~10% annual increase is well expected for the years to come [40].

3.2 Ti and Ti alloys in micro air vehicles

3.2.1 Structure of an MAV

Because their sizes mostly fall in the same range of insects, many MAVs are biomimetically designed, and their structure is mainly composed of the following key components: main body and wings, power supply, power and control electronics, and various types of sensors and positioning system, see Fig. 8 for a typical example [41].

It has been reported that a 250-μm thick Ti alloy (Ti-6Al-4V) foil was used to make a wing frame structure for an MAV [29], as shown in Fig. 9 and Table 1. The corresponding membrane for the wing is Parylene. The two materials are bonded together by a chemical etching and deposition process. A higher lift coefficient, as large as five times that of fixed-wing aircraft, has been achieved by using this Ti alloy-involved new design.

4 Ti and Ti alloys in model aircraft

4.1 Brief introduction to model aircraft: Definition, classification, and market

Model aircraft, also known as aero modeling, are normally referred to as replicas of existing or past aircraft, except that they are in much scaled-down sizes [42, 43]. For instance, Fig. 10 shows a model aircraft mimicking

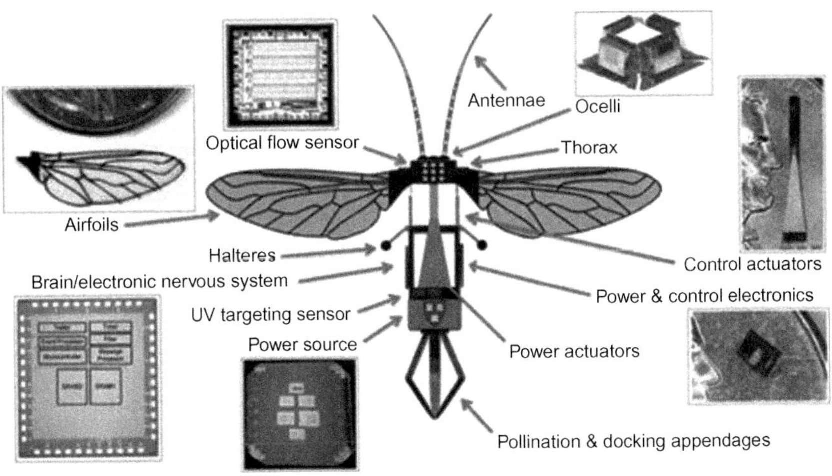

Fig. 8 Schematic of a robobee to show the main components of a micro air vehicle.

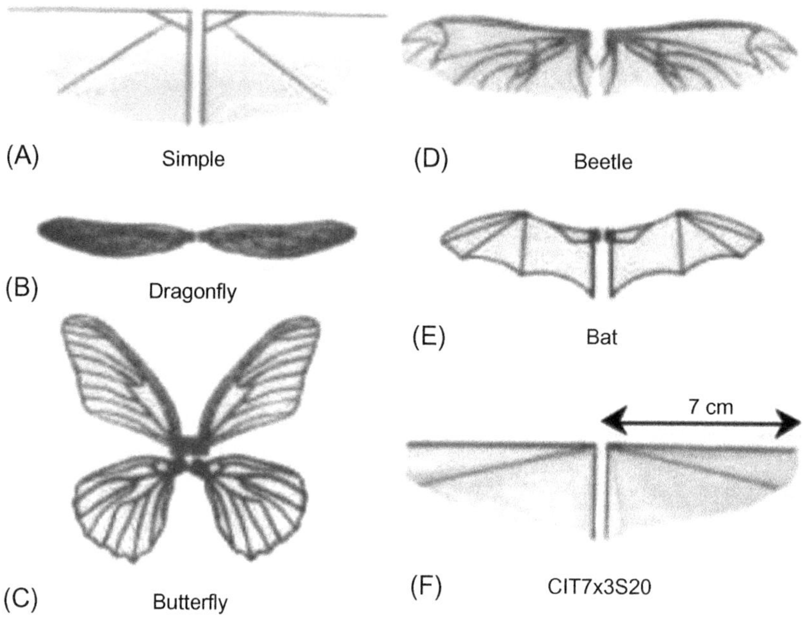

Fig. 9 Ti alloy MEMS wings for micro air vehicle.

Fig. 10 A photograph showing a flying model aircraft.

the F-3 "Demon" fighter plane developed by the McDonnell-Douglas Corporation in the 1950s and 1960s [44].

Model aircraft can be divided into two basic groups: flying and nonflying [42, 43]. Nonflying model aircrafts are mainly designed for display and product promotion purposes, whereas flying model aircrafts are mostly developed for racing, hobby, or research, ranging from simple gliders made of paper and/or polymer to powered ones made of a variety of materials such as wood and/or metals. Some of the flying model aircrafts can be large, especially when they are used to study the flight properties of a full-scale, real design to be developed.

Regarding the global commercial market for air modeling, there is no clear indication about how big it is. Rough estimation suggests it's an industry worth of ~$1 billion [45]. People involved in the business can be hobbyists and/or those from airplane companies who wish to use air modeling as gifts for promotion and marketing.

4.2 Ti and Ti alloys in model aircraft

4.2.1 Structure of a model aircraft

Because model aircraft, particularly those high-end ones, normally represent a full-scale aircraft, its structure can be identical to normal airplanes, consisting of these key components (Fig. 11) [46]: main body, wings, engine, landing system, and remote controlling system, if they are designed to fly.

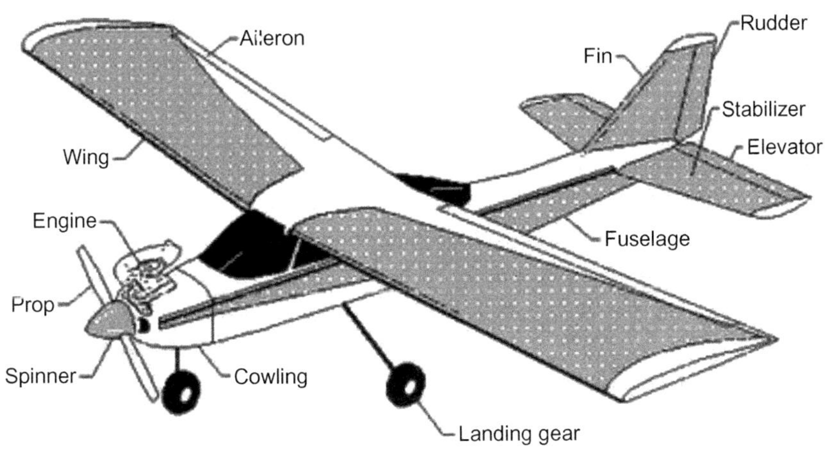

Basic trainer

Fig. 11 Schematic structure of a fixed-wing model aircraft.

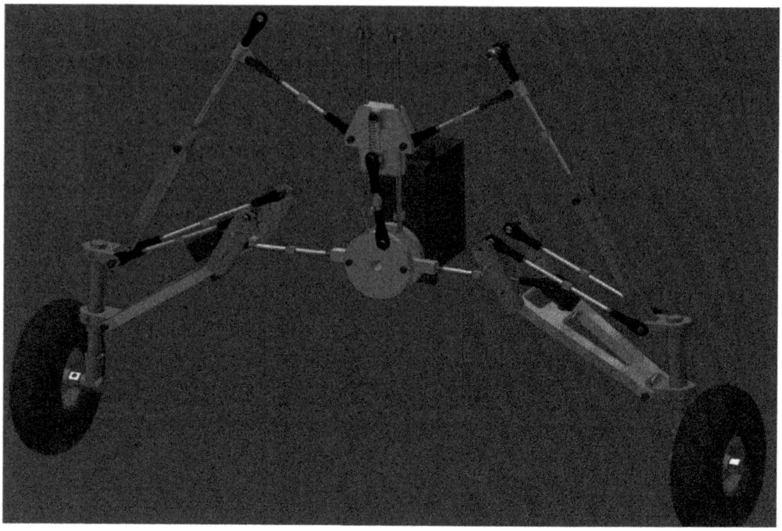

Fig. 12 Applications of Ti and Ti alloys parts in air modeling: as the turnbuckle for the landing gear.

Ti and Ti alloys are applied in model aircrafts in the following circumstances, see Fig. 12 and Table 1 for typical examples [30–32]:

- They are used as turnbuckle material for an Al 6061 landing gear kit, to increase the reliability of the whole structure, see Fig. 12 for a typical example [30].
- They are used as gear material to provide a torque of 35 kg high by merely using 63 g Ti alloy in the Savox SV-1270TG motor [31].
- They are used as screws made of pure Ti or column bolts made of Ti-6Al-4V alloy for assembling high-quality DIY model aircrafts [32].

5 Ti and Ti alloys in paramotor

5.1 Brief introduction to paramotor: Definition, classification, and market

Paramotor is also known as powered paraglider [47, 48]. Paramotor originated from an unpowered parachute, while later in the 1980s, internal combustion engine(s) was added to provide thrust, enabling higher and faster flight for passenger(s) on board. Modern paramotors are capable of flying at a speed of up to 75 km/h and stay in the air for several hours. The pilot controls the paramotor by throttle as well as changing the harness.

Generally speaking, paramotor can be divided into two types: foot launch and wheel launch. The former uses foot for landing and taking off, and the later uses wheels, see Fig. 13 for a typical foot launch paramotor [48].

It is estimated that there are about ~30,000 paramotor pilots worldwide, and the business is expanding stably as a hobby sport [49].

5.2 Products of Ti and Ti alloys in paramotor

5.2.1 Structure of a paramotor

As shown in Fig. 13, a foot-launch paramotor consists of the following key components [48]: frame structure, engine, parachute, and harness with integrated seat.

As examples, Ti and Ti alloys have found their applications in parmotors in the following two typical circumstances [33, 34]:

- They are used as the exhaust part by the French paramotor manufacturer, air conception, in their Ultimate 130cc engine, see Table 1 [33]. The main reasons to use Ti alloy are to reduce weight of the engine and, simultaneously, to realize better structural integrity by replacing steel with the stronger-but-lighter Ti alloy.

Fig. 13 A photograph of a paramotor to show some of its main components, namely the harness and the frame structure.

Fig. 14 A photograph of Ti alloy frame structure material for paramotor.

- They are used as the frame material for the Φ16-mm tubes in a paramotor brand called Dron developed by Airfer [34]. It is found that using titanium can reduce weight by about 1.5 kg per paramotor, see Fig. 14 and Table 1.

6 Ti and Ti alloys in air racing

Ti and Ti alloys have also found their applications in a type of manned air vehicle, air racing. Air racing is a highly specialized motorsport that involves airplanes or other types of aircraft to compete over a fixed course. The objective is to navigate the aircraft in the fastest possible time, meanwhile incurring as few penalties as possible [50, 51], see Fig. 15 for an illustration of air racing [52].

In air racing, a few hundredths of a second can separate winners from losers, so weight reduction is a serious concern, which is why low-density Ti and Ti alloys are being used. Their main applications are as screws and bolts to replace heavier steel ones (Table 1) [35].

Fig. 15 A photograph to show air racing.

7 Future applications and developments

From the overview presented in this chapter, one can find that the applications of Ti and Ti alloys in drones and the other flying objects show the following characteristics:

(a) Most applications of Ti and Ti alloys are as structural materials. These include key components like the mainframe structure for drones and paramotors (Table 1). The primary reasons for their applications are to reduce weight and to increase structural reliability, AS Ti alloys normally have better mechanical properties than many steels, meanwhile they stay much lighter in density. Ti-6Al-4V, whose melting temperature is about 1600°C and normal service temperature limit up to 230°C [5, 6] can also replace some lower temperature materials like Al alloys for elevated temperature applications such as in drones made for bush fire surveillance and/or rescue [27, 28].

(b) Ti and Ti alloys are also used as functional materials, for example, as the substrate material for a micro scanning mirror device [29]. This is due to the fact that Ti alloys have a suitable combination of Young's modulus and density ($\sim E/\rho$), therefore they are capable of sustaining high frequency rotation; meanwhile it is nonmagnetic, which helps avoid any interference with the permanent magnet of the mirror system [29].

(c) Ti-6Al-4V is the most widely applied Ti alloy. Occasionally other Ti materials such as Ti-3Al-5Mo-4.5V [26] and pure Ti [31] are also used.

(d) Conventional processing methods such as forging, casting, rolling, machining, etc. are dominant for making Ti and Ti alloys parts.

(e) Cost is still a main concern for the commercial scale-up of Ti and Ti alloys products in the field of GA. As discussed, Ti and Ti alloys show a great combination of low density, high strength, high ductility, and excellent corrosion resistance. The only obstacle and bottleneck issue to their wider applications lies in their relatively higher cost.

From a future perspective, advances in processing technologies and raw material development may further enable wider applications for Ti and Ti alloys. These include:

(f) Maturation of the 3D printing/additive manufacturing technology: AM emerges as highly automatic, highly individualized technique for processing Ti and Ti alloys [27, 28]. GA contains many personalization-prone business sectors. From this perspective, one can expect more 3D printed flying parts in the near future.

(g) Incoming of hybrid advanced processing technologies: For example, fused filament fabrication (FFF), another AM technique, can be used as a rapid prototyping tool to provide proof of concept via smaller amount of experimental runs [53]; mass production based on metal injection molding (MIM) [54, 55] can be subsequently followed to produce parts for, for example, model aircraft and/or paramoters. MIM has already demonstrated its capability to produce affordable parts with complex geometry and desirable properties, see Fig. 16 [56–58]. It is a mold-involved, mass-production processing technique, complementary to AM, which is best for fabricating small-number, customized products.

(h) Cost reduction in raw materials: This is pivotal for wider applications of Ti and Ti alloys. Efforts are being made toward this key target by waste recycling, powder modification, and other innovative pathways [59, 60].

(i) A wider selection and application of Ti materials: For example, TiAl intermetallics can be regarded as a special group of Ti alloys. They are unique for their high temperature performance, which can be up to ~800°C [5, 6, 14]. From this perspective, maybe 1 day one can produce TiAl micro turbine engines like the one shown in Fig. 17 [61, 62] by adopting 3D printing techniques such as EBM or SLM. Through providing these critical parts by advanced processing techniques such as 3D printing and additive manufacturing, products of Ti and Ti alloys may be able to embrace a larger consumer market in drones and the other four GA areas.

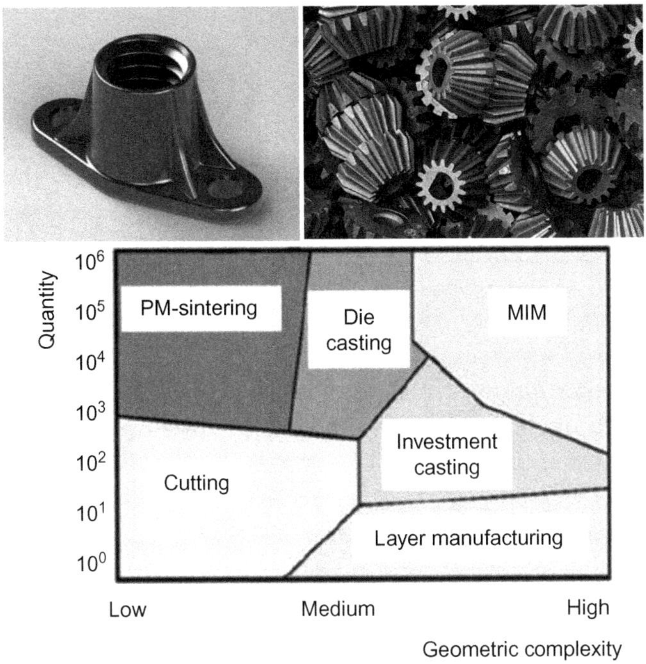

Fig. 16 (A) A photograph of an airplane fastener made by MIM, (B) nuts made by MIM, and (C) comparison between different processing techniques.

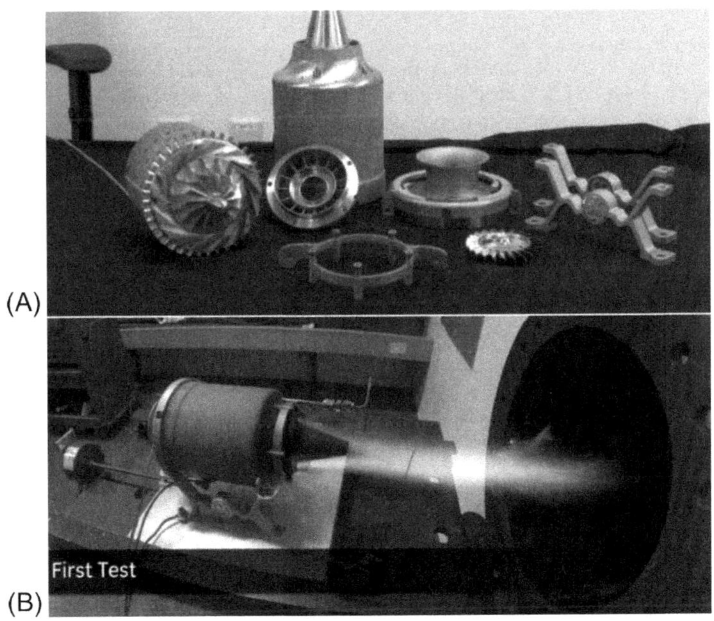

Fig. 17 (A) Parts of a 3D-printed micro turbine engine using EBM and (B) test-firing of the as-printed micro turbine engine. Materials used for the engine should be Ni-based superalloys.

8 Concluding remarks

This chapter has presented and analyzed the applications of Ti and Ti alloys in drones and other flying objects namely the micro air vehicle, model aircraft (aero modeling), paramotor, and air racing. By presenting typical, commercialized examples, we have shown that:

- Ti and Ti alloys have already been used in the aforementioned areas and demonstrated their benefits and uniqueness as structural and/or functional advanced materials.
- A greater future is well expected with the maturation of emerging processing technologies, including AM and MIM, as well as reduced cost on using Ti materials.

Acknowledgments

This work is supported by Shenzhen Science and Technology Innovation Commission [No. ZDSYS201703031748354], and National Science Foundation of Guangdong Province [grant number 2016A030313756]. M. Yan thanks the Humboldt Research Fellowship for Experienced Researchers for its support.

References

[1] D. Crane, Dictionary of Aeronautical Terms, third ed., Aviation Supplies and Academics, 1997, pp. 238–239.

[2] Strategic Review of General Aviation in the UK, 2006, pp. 52–53. https://publicapps.caa.co.uk/docs/33/StrategicReviewGA.pdf. (retrieved on the 29 June 2018).

[3] A. Balachandran, D. Karelia, J. Challa, Materials selection for unmanned aerial vehicle, Int. J. Mech. Eng. Tech. 5 (2014) 34–40.

[4] I. Inagaki, T. Takechi, Y. Shirai, N. Ariyasu, Application and Features of Titanium for the Aerospace Industry, Nippon Steel & Sumitomo Metal Technical Report 106, 2014, pp. 22–27.

[5] C. Leyens, M. Peters (Eds.), Titanium and Titanium Alloys: Fundamentals and Applications, Wiley-VCH, Weinheim, 2003.

[6] G. Lutjering, J.C. Willams, Titanium, second ed., Springer-Verlag, Berlin, 2007.

[7] M.F. Ashby, D. Cebon, Materials selection in mechanical design, J. Phys. IV 3 (1993) C7-1–C7-9.

[8] M. Yan, M.S. Dargusch, T. Ebel, M. Qian, A transmission electron microscopy and 3D atom probe study of oxygen-induced fine microstructural features in as-sintered Ti-6Al-4V and their impacts on ductility, Acta Mater. 68 (2014) 196–206.

[9] M. Yan, Y. Liu, G.B. Schaffer, M. Qian, In situ synchrotron radiation to understand the pathways for the scavenging of oxygen from commercially pure CP-Ti and Ti-6Al-4V by yttrium hydride, Scr. Mater. 68 (2013) 63–66.

[10] M. Yan, Y. Liu, Y.B. Liu, C. Kong, G.B. Schaffer, M. Qian, Simultaneous gettering of oxygen and chlorine and homogenization of the β phase by rare earth hydride additions to a powder metallurgy Ti-2.25Mo-1.5Fe alloy, Scr. Mater. 67 (2012) 491–494.

[11] M. Yan, M. Qian, M.S. Dargusch, Confinement of α phase and refinement of β phase by Fe and/or Pd in Grade 4 and Grade 7 Ti, MRS Commun. 4 (2014) 183–188.

[12] S.L. Lu, M. Qian, H.P. Tang, M. Yan, J. Wang, D.H. StJohn, Massive transformation in Ti-6Al-4V additively manufactured by selective electron beam melting, Acta Mater. 104 (2016) 303–311.
[13] M. Qian, W. Xu, M. Brandt, H.P. Tang, Additive manufacturing and postprocessing of Ti-6Al-4V for superior mechanical properties, MRS Bull. 41 (2016) 775–783.
[14] Y.H. Zhou, S.F. Lin, Y.H. Hou, D.W. Wang, P. Zhou, P.L. Han, Y.L. Li, M. Yan, Layered surface structure of gas-atomized high Nb-containing TiAl powder and its impact on laser energy absorption for selective laser melting, Appl. Surf. Sci. 441 (2018) 210–217.
[15] M. Hassanalian, A. Abdelkefi, Classifications, applications, and design challenges of drones: a review, Prog. Aerosp. Sci. 91 (2017) 99–131.
[16] I. Colomina, P. Molina, Unmanned aerial systems for photogrammetry and remote sensing: a review, J. Photogr. Remote Sens. 92 (2014) 79–97.
[17] J. Nicas, Drone Dogfight: Big Defense Firms vs, Techies, 2014, https://www.wsj.com/articles/drone-dogfight-big-defense-firms-versus-techies-1403891039 (retrieved on the 29 June 2018).
[18] D. Swinhoe, Commercial Drone Flight: Poor Rules, NASA, and the Future, https://www.idgconnect.com/abstract/10579/commercial-drone-flights-poor-rules-nasa-future, 2015 (retrieved on the 29 June 2018).
[19] A. Meola, Drone Market Shows Positive Outlook with Strong Industry Growth and Trends, https://www.businessinsider.de/drone-industry-analysis-market-trends-growth-forecasts-2017-7?r=USandIR=T, 2017 (retrieved on the 29 June 2018).
[20] B. Dipert, K. Chen, J. Dashwood, J. Clayton, S. François, A. Joshi, M. Sahi, M. Jacobs, Vision Processing Opportunities in Drones, https://www.embedded-vision.com/platinum-members/embedded-vision-alliance/embedded-vision-training/documents/pages/drones. (retrieved on the 29 June 2018).
[21] Global UAV, Global UAV Market Growth to $28.27 Billion by 2012, Driven by Civil, Commercial, Military and Security Sectors, http://idstch.com/home5/international-defence-security-and-technology/industry/high-growth-of-uav-industry-predicted-u-s-market-will-grow-to-15-billion-in-2020/, 2017. (retrieved on the 29 June 2018).
[22] Exocet, All Weather Low Cost VTOL UAV UAS Unmanned Drone with EOIR Gyro Stabilized Thermal FLIR HD Zoom Gimbal Turret Payload, https://www.x20.org/exocet-vtol-uav-uas-unmanned-drone-eoir-thermal-flir-hd-zoom-gimbal-payload/exocet-all-weather-low-cost-vtol-uav-uas-unmanned-drone-with-eoir-gyro-stabilized-thermal-flir-hd-zoom-gimbal-turret-payload-8/, 2016 (retrieved on the 29 June 2018).
[23] Niono, The World's First Titanium Alloy Frame has Been Listed, 4-Inch, 5-Inch and 7-Inch, http://www.niono.net/2018/01/01/the-worlds-first-titanium-alloy-frame-has-been-listed-4-inch-5-inch-and-7-inch/, 2018 (retrieved on the 29 June 2018).
[24] DJI Phantom 4, http://dynamicuav.com/products/dji-phantom-4/, (retrieved on the 29 June 2018).
[25] https://www.aliexpress.com/item/Titanium-Alloy-screw-M3X14-for-Mini-Drones-DIN-7991-Flat-Head-Hex-2-0-Driver-Polished/32848113231.html, (retrieved on the 29 June 2018).
[26] L.C. Ye, G.F. Zhang, Z. You, Large-aperture kHz operating frequency Ti-alloy based optical micro scanning mirror for LiDAR application, Micromachines 8 (2017) 120.
[27] A. Green, Titanium Propels 3D Printed Drone to Aid in Bushfires, https://blog.csiro.au/titanium-propels-3d-printed-drone-to-aid-in-bushfires/, 2015 (retrieved on the 29 June 2018).
[28] S.J. Grunewald, 3D Printed Titanium Helps Make This Drone Ready to Fight Fires, https://3dprint.com/110849/titanium-firefighting-drone/, 2015 (retrieved on the 29 June 2018).

[29] T.N. Pornsin-sirirak, Y.C. Tai, H. Nassef, C.M. Ho, Titanium-alloy MEMS wing technology for a micro aerial vehicle application, Sens. Actuators A 89 (2001) 95–103.
[30] CL-415 Retractable Main Landing Gear Kit, https://mraerodesign.securicom.ca/en/products/info.cfm?ProduitID=58, (retrieved on the 29 June 2018).
[31] Savox SV-1270TG 63g Digital Titanium Gear 35 kg, http://www.hobby-aeromodelling.com/index.php?main_page=product_infoandcPath=22_418andproducts_id=9310, 2018. (retrieved on the 29 June 2018).
[32] https://www.aliexpress.com/item/4Pcs-M4-x-10-15-20mm-Length-M4-Pure-Titanium-Alloy-Screw-Stigma-Model-Aircraft-DIY/32753711829.html (retrieved on the 29 June 2018).
[33] Air Conception Produce Titanium Exhaust for 130cc Engine Further Reducing Weight, https://paramotornews.com/2013/10/24/air-conception-produce-titanium-exhaust-for-130cc-engine-further-reducing-weight/, 2013 (retrieved on the 29 June 2018).
[34] Dron: New Paramotor Frame From Airfer, http://xcmag.com/tag/paramotor-frames/, 2017 (retrieved on the 29 June 2018).
[35] Red Bull Air Race 2017: Weight Saving and Flying Faster, https://www.fai.org/news/red-bull-air-race-2017-weight-saving-and-flying-faster, 2017 (retrieved on the 29 June 2018).
[36] H. Liu, S. Ravi, D. Kolomenskiy, H. Tanaka, Biomechanics and biomimetrics in insect-inspired flight systems, Philos. Trans. R. Soc. B 371 (2016) 20150390.
[37]. K.Y. Ma, P. Chirarattananon, S. B. Fuller, R. J. Wood, Controlled flight of a biologically inspired, insect-scale robot, Science 340 (6132) 603–607.
[38] J. Excell, The Rise of the Micro Air Vehicle, https://www.theengineer.co.uk/issues/aerospace-and-defence-2013/the-rise-of-the-micro-air-vehicle/, 2013 (retrieved on the 29 June 2018).
[39] D. Floreano, R.J. Wood, Science, technology and the future of small autonomous drones, Nature 521 (2015) 460–466.
[40] Visiongain, Small Unmanned Aerial Vehicle (UAV) Market Forecast 2015–2025, https://www.visiongain.com/report/small-unmanned-aerial-vehicle-uav-market-forecast-2015-2025/, 2015 (retrieved on the 29 June 2018).
[41] Nivedit, RoboBee is a Tiny Drone That Mimics the Intelligence of Bees, http://www.rcdronearena.com/2016/05/23/robobee-tiny-drone-intelligence-bees/, 2016 (retrieved on the 29 June 2018).
[42] M. Simons, Model Aircraft Aerodynamics, fourth ed., Nexus Special Interests, Swanley, 1999.
[43] S. Mayerowitz, Airline World's Tiny Secret: Infatuation With Model Planes, https://eu.usatoday.com/story/todayinthesky/2015/03/18/airline-worlds-tiny-secret-infatuation-with-model-planes/24947785/, 2015. (retrieved on the 29 June 2018).
[44] von Vuelos en Globo MX, Aeromodelling, https://www.pinterest.co.uk/vuelosenglobomx/aeromodelling/ (retrieved on the 29 June 2018).
[45] Q.W. Fu, A.T. Zhang, T.H. Wang, Aviation and Aeromodel, sixth ed., Aviationnow Press, Beijing, 2013.
[46] http://www.jetwings.info/aeromodelling.html (retrieved on the 29 June 2018).
[47] J. Goin, D. Pagen (Eds.), The Powered Paragliding Bible, Airhead Creations, 2006.
[48] J. Umenberger, A.H. Goktogan, System identification and control of a small-scale paramotor, in: IEEE International Conference on Robotics And Automation (ICRA), 2013, pp. 2970–2976.
[49] Paramotors, https://www.fai.org/page/paramotors (retrieved on the 29 June 2018).
[50] D. Berliner, The Big Race of 1910, Air & Space Magazine, 2010.
[51] History of the red bull air race, in: Red Bull Air Race World Championship, Press Kit Perth, 2008.

[52] National Championship Air Races, https://www.visitrenotahoe.com/event/national-championship-air-races-2018/, 2018 (retrieved on the 29 June 2018).
[53] S. Singh, S. Ramakrishna, R. Singh, Material issues in additive manufacturing: a review, J. Manuf. Process. 25 (2017) 185–200.
[54] T. Ebel, Metal injection molding (MIM) of titanium and titanium alloys, in: Handbook of Metal Injection Molding, 2012, pp. 415–445.
[55] T. Ebel, O.M. Ferri, W. Limberg, M. Oehring, F. Pyczak, F.P. Schimansky, Metal injection moulding of titanium and titanium-aluminides, Key Eng. Mater. 520 (2012) 153–160.
[56] M. Scharvogel, W. Winkelmueller, Metal injection molding of titanium for medical and aerospace applications, JOM (2011) 94–96.
[57] MIM Technology, http://www.zimmer-group.de/en/mainmenu/megamenu/col2/verfahrenstechnik/mim-bauteile (retrieved on the 29 June 2018).
[58] D.H. Ding, Z.X. Pan, D. Cuiuri, H.J. Li, Wire-feed additive manufacturing of metal components: technologies, developments and future interests, Int. J. Adv. Manuf. Technol. 81 (2015) 465–481.
[59] Y. Taninouchi, Y. Hamanaka, T.H. Okabe, Chlorination-volatilization behavior of titanium metal scraps during recycling using reaction-mediating molten salt, Mater. Trans. 57 (2016) 1309–1318.
[60] Y.H. Hou, B. Liu, Y. Liu, Y.H. Zhou, T.T. Song, Q. Zhou, G. Sha, M. Yan, Ultra-low cost Ti powder for selective laser melting additive manufacturing and superior mechanical properties associated, Opto-Electron. Adv. 2 (2019) 180028.
[61] A. Liszewski, GE 3D-Printed a Miniature Jet Engine That Runs at 33,000 RPM, https://gizmodo.com/ge-3d-printed-a-tiny-jet-engine-that-runs-at-33-000-rpm-1703922055, 2015 (retrieved on the 29 June 2018).
[62] P.L. Han, Private communications, 2019.

CHAPTER 6

Is it ethical to use advanced materials such as titanium in sports equipment?

F.H. (Sam) Froes
Titanium Industry, Tacoma, WA, United States

Contents

1 Introduction

Advanced materials exhibit mechanical and physical properties beyond the levels of widely used materials like steel, aluminum, and magnesium. Materials such as composites and monolithic metals such as titanium have had a major influence on increased performance of transportation systems such as in aerospace, automobiles, trains, and in oil and gas exploration/production industries and medical/dental devices. The significant features in material characteristics include strength and ductility, stiffness/modulus, increased temperature, forgiveness (a term that includes fracture toughness, fatigue-crack growth rate, etc.), and reduced density. Because of the higher cost of these advanced materials, the level of acceptance is industry dependent (Fig. 1).

In this chapter, the use of advanced materials such as titanium in sports equipment is addressed. At the highest level (professional,) millions of dollars

Titanium for Consumer Applications
https://doi.org/10.1016/B978-0-12-815820-3.00005-8

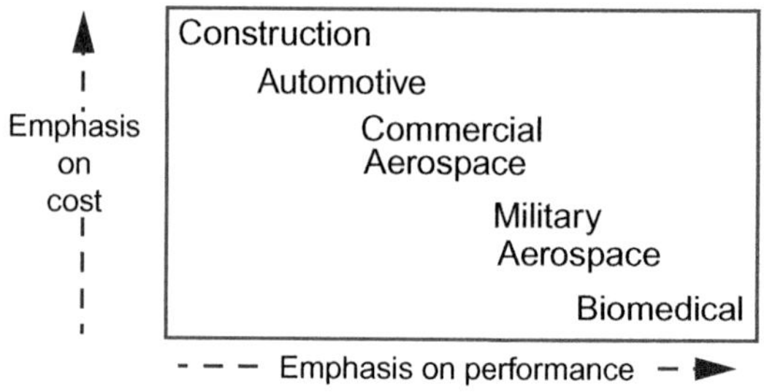

Fig. 1 The impact of cost in various industries.

are involved in sports with results depending on fractions of a second or tenths of an inch/cm. Even the dedicated amateur is willing to invest considerable amounts of money for performance enhancement. Thus the use of titanium or other advanced materials in athletics that can lead to improved performance is readily acceptable.

Over the past 100 years, the materials available for the fabrication of sports equipment have become much more numerous. No longer are there only naturally occurring materials including wood, twine, gut, and rubber; now there are high-tech metals, particularly titanium.

This chapter will first consider the question of sporting equipment manufactured specifically from advanced materials, with examples drawn from a variety of sports. Then we will look at how measurable (absolute) records have been affected by the introduction of these advanced materials.

2 Sports equipment design

For the best design of sports equipment, consideration must be given to various disciplines, including materials science, mechanical engineering, and physics. The issues to be addressed include improved performance and making user-friendly equipment to avoid the occurrence of injury. The optimum design also requires consideration of anatomy, physiology, and biomechanics. A simple definition of biomechanics is that it is the science of how internal and external forces affect the body. That is, it applies the laws of physics and mechanics to the body including joints, ligaments, and tissues when they experience loading (Fig. 2).

When sports equipment is designed, the behavior of materials such as titanium must be accounted for. These behavior characteristics include strength, ductility, density, fatigue resistance, toughness, modulus (damping), and never forget the cost.

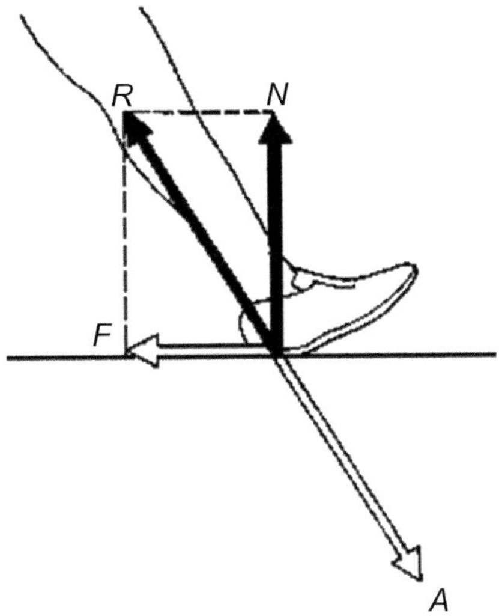

Fig. 2 The relationship between the action (*A*)/reaction (*R*) force, the normal (*N*) force, and the friction (*F*) force. *(Courtesy of K.E. Easterling.)*

Sports equipment fabrication, which often features creative designs, combines various types of materials together including metals, ceramics, polymers, and composites, with consideration given to biomechanical requirements. The attributes of different materials can be determined by evaluating specific properties, that is, by taking density into account. Thus, for a material with high stiffness for a low weight, the materials with the highest specific stiffness would be chosen. In this regard, cellular configurations look best compared to solid material materials because the density of the cellular materials is lower than the density of the solid. For the design of complex sports equipment, a well thought-out design criterion is needed, allowing the choice of the best materials to be made.

3 The impact of advanced materials

3.1 Running

In running events, shoes in particular have shown substantial improvements. However, it is the author's feeling that this improvement has been much more in the field of comfort/avoidance-of-injury in contrast to performance improvement. In the first Olympics in 1896 held in Greece, Spiridon Loues was the champion in the marathon (the distance of which was somewhat shorter than today's 26.2 miles, 42.2 km). Greece celebrated with Spirdon as he just made it in under 3 h with a time of 2 h, 59 min. In modern times, 100+ years later, the Olympic record for men is 2 h, 6 min, 32 s (Samuel Wanjinu of Kenya, Beijing 2008), more than a 30% improvement. The parallel Olympic record for women is held by Tiki Gelana of Ethiopia (set in London 2012 at 2 h, 23 min, 7 s). This woman's event has only been held for a little more than 20 years. The International Association of Athletics Federations (IAAF) now recognizes a world record for men of 2:02:57, by Dennis Kimetto of Kenya at Berlin on September 28, 2014. For women, the IAAF recognizes two world records: a record for "Mixed Gender" of 2:15:25, set by Paula Radcliffe (United Kingdom) on April 13, 2003, in London, and also by Paula Radcliffe on April 17, 2005, again in London, a "Women Only" record of 2:17:42 (Fig. 3). These improvements can be attributed to enhanced human performance.

Taking a look at the other marque running event, the competition for the title of fastest man on Earth (the 100-m dash), Usain Bolt (Fig. 4) of Jamaica reigns supreme with a world record time of 10.58 s in 2009 and a winning time of 9.81 s in the Rio Olympics (2016), the first man to win this event in three successive Olympics. This can be compared with a time of

Fig. 3 Paula Radcliffe, Women's Marathon World Record Holder. From https://en.wikipedia.org/wiki/File:Paula_Radciffe_NYC_Marathon_2008_cropped.jpg.

Fig. 4 Usain Bolt, the fastest man on Earth, in characteristic pose. He plans to use his speed to play soccer in Germany. From https://en.wikipedia.org/wiki/File:Bolt_2007.2.jpg.

10.8 s by Luther Cary, USA, in 1891. Again, this is due to an enhancement in performance due to human improvements.

There were no Olympic games in 1896 for those requiring prosthetic limbs. Recently paraplegic athletes can more than compete; they can outperform most of us who are not disabled.

First organized in 1948 by a small group of British World War II veterans, the parapalegic Olympic Games has matured today to become a very large international sporting event. The disabled athlete has competed extensively in marathons, with Kevin Counitan, who is blind, having achieved a total in excess of 150 worldwide events in a variety of marathon events.

Paraplegic Joe Gaetani uses the Springlite II (costing about $850) prostheses device, which was constructed from a thin carbon fiber/epoxy pylon allowing a good balance of stiffness and flex at lower density than conventional materials (1 kg compared to wood at 4.5 kg). He broke world records in both the 100-m sprint (12.23 s) and 200-m sprint (24.82 s) in 1992 at the Paralympics in Barcelona, Spain. The capability to walk or run increased distances with reduced fatigue and discomfort in comparison to more traditional equipment was provided by this new device. Energy-storage return aids performance and contributes to the capability to walk or run greater distances with much less fatigue and discomfort than with traditional devices.

Tony Volpentest was born without feet but won gold medals in the paraplegic Olympics in Barcelona (1992) in the 100-m sprint (11.63 s) and 200-m sprint (23.07 s). He runs on carbon-graphite feet built by Flex-foot, bolted to carbon-composite sockets that encase his legs. This configuration is essentially a springboard. The runner punches the track with each step, catapulting himself or herself forward more efficiently than if the person was running on two human feet. Recently, paraparlegic Oscar Pistoriushas ran with a similar device, which is stiff and springy. For disabled long-distance runners, in an event such as the marathon, the device is extremely shock-absorbing. It can only be concluded that, for the paraplegic runner, advanced design/materials significantly enhances performance.

In Rio (2016), J. Peacock of Great Britain won the T44 (there are numerous categories of disablements this one involved at least one leg fitted with a device similar to that of Tony Volpentest) 100 m in a time of 10.81 s. Again, a mechanical/advanced material device, such as Titanium, aids the runner's performance.

This is not a major contribution for the conventional runner but a vast influence on the paraplegic competitor.

3.2 Pole vault

In the 1896 Olympics pole vault event, the winning height of 10 ft, 6 in. was attained with a bamboo pole. The pole vault winning height then inched upward for 60 years, until records began to fall in the 1960s as competitors began to use aluminum poles. The world-class pole vaulter nowadays uses a highly sophisticated composite pole. This has given rise to winning Olympic heights in Rio (2016) of 19.78 ft (Men's Thigo Braz) and 16.56 ft (Women's Yelena Isinbayeva). Clearly in the pole vault, advanced materials have contributed significantly to performance.

The specific design property required for pole vaulting is given by

$$\frac{\sqrt{E}}{p}$$

Adding the requirement of durability, the carbon-fiber composite is the material of choice, with a bamboo as second choice. With the minimal-twisting requirement added, the carbon-fiber composite is even more attractive. Here, the concept is three layers of different fibers in combination with a high strength and high stiffness outer layer of carbon fiber and a webbing of intermediate fibers, and a built-in resistance to twisting provided by an inner layer of wound-glass fibers. The glass fiber is 80% longitudinal and 20% radial fibers.

There is a major contribution from advanced materials in pole vaulting.

3.3 Bicycling

A very efficient form of transportation is provided by bicycling, with the energy expended reduced compared to walking and greatly reduced to that of powered vehicles. This is recognized in China where in excess of 10 million bicycles are produced per year. The bicycle has been used for nearly 200 years, developing from the 1817 Draisienne (called a walking device) to the present-day Rover safety bicycle formulated in 1885 by J.K. Starley.

Greatly improved bicycles have resulted from the use of advanced materials, along with improved aerodynamics consideration. The winner of the 1989 Tour de France Greg LeMond (3420 km) won by only 8 s, an achievement attributed to a clip-on handlebar extension and an improved aerodynamic configuration.

The modern-day bicycle has been improved by the use of spoked wheels, the chain geometry, pneumatic tires, and the addition of a number of accessories (e.g., seats, brake levers, and pedals). The two most significant advances are in the frame and wheels.

The bicycle is basically a modified space frame similar in concept to that found in bridges, cranes, etc. The bicycle consists of a diamond frame. This diamond frame and the connected alternate-structure cross frame are fabricated using thin-walled components of tubular shape designed to resist combinations of tension, compression, bending, and torsional stresses when being ridden. The bicycle's requirements and possible construction materials are summarized in Table 1 where titanium is favored from strength considerations.

Table 1 The materials requirements for a bicycle

Requirement	Possible material
Light (low density)	Carbon-fiber composites
Stiffness	Aluminum
Strength	Titanium
Toughness	Magnesium
Fatigue	Resistance steel
Corrosion resistance	Metal-matrix composites

Material requirements for bicycles can be summarized as those that minimize bending of a cantilever beam, giving a mass defined by

$$M = \text{Constant} \times P / E$$

The best material choice, referring to Table 1, should exhibit a high specific strength, P/E. Carbon-fiber composites look attractive, as does titanium.

On June 7, Sir Bradley Wiggins broke the World Hour Record with the help of 3D-printed, super lightweight, and aerodynamic handlebars developed at The University of Sheffield's Mercury Centre, UK, in conjunction with Italian bicycle manufacturer Pinarello. Wiggins broke the world record for the longest distance that a cyclist has ridden within an hour on the track, after spanning 218 laps—54.53 km—of the Lee Valley VeloPark in London. The

Other concepts featuring titanium include hybrid frames of carbon-fiber-reinforced composites combined with titanium.

The improvement in the Olympic pursuit records brought about by the use of advanced materials can be judged from the fact that, in 1964, Daler was individual pursuit champion in the 4000 m recording a time of 5 min, 5 s, whereas in 1992, Boardman won the same event in 3 min, 22 s.

Advanced materials have made a major contribution to bicycles.

3.4 Tennis

Tennis is a sport in which absolute achievements are difficult to define. How do we compare Rod ("The Rocket") Laver (Fig. 5) with modern tennis players (such as Roger Federer or Andy Murray) with their high-tech rackets?

Up until about 1995, tennis rackets were constructed from wood (ash, maple, and okume). Metal frames, fabricated from steel, titanium, or aluminum, were then introduced. Currently, composite and titanium rackets are the weapons of choice due to their efficiency in propelling the tennis ball at great speed, combined with accuracy while damping undesirable vibrations that can lead to tennis elbow.

A modern tennis racket is designed to make the size of the sweet spot in the central part of the racket larger, which results in low or no shock

Fig. 5 Rod Laver, demonstrating his ability with a wooden tennis racket. How much better would this all-time great have been with a high-tech modern racket?

to the player's elbow upon impact with the ball. This is dependent upon the frame stiffness and the handle and head size. The International Tennis Federation has recently placed a restriction on racket size. Tennis rackets today are manufactured mainly from monolithic metals including titanium and metal-matrix composites. A titanium racket is shown in Fig. 6.

The conclusion must be that advanced materials have significantly affected the game of tennis.

3.5 Golf

It is impossible to gauge past achievements with those of today in absolute terms. Clubs have improved significantly, making it impossible to compare, say, Bobby Jones using hickory shafts with Dustin Johnson, Freddy Couples, or Tiger Woods who nowadays play golf with clubs consisting of a graphite-epoxy shafts and heads made from an oversized hollow piece of titanium.

Driving distance is an absolute criterion that can be related to the material of construction. For the first 26 rounds of the 2018 season,

Fig. 6 Head titanium Ti6 tennis racket head.

Tony Finau averaged 329 yards with his drives, considerably in excess of the distances recorded as recently as 10 years ago. A complicating factor here is the advances made in golf balls where those that go the longest distances are now banned because they would make current golf courses obsolete.

Fig. 7 summarizes how the driver has developed. Overall, the weight has been reduced whereas the length of the club is longer, from 109 to 114 cm. Grip and shaft weight have decreased, from 165 to 115 g. The head weight has not changed at about 200 g, but the hollow titanium construction allows the head to be much larger, with its mass concentrated at the outside of the hitting face (Fig. 8). The net result is a club that allows longer distance (the longer arc providing greater club head speed) and also a straighter (bigger sweet spot) shot.

So, for golf, we must conclude that advanced materials have made a very large impact, particularly in the distance achieved with the driver.

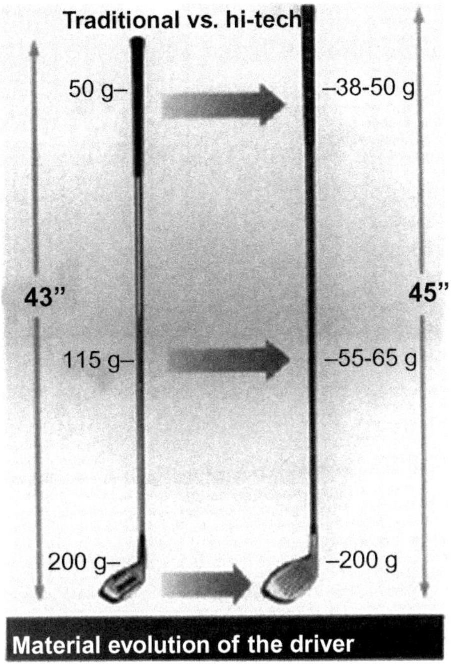

Fig. 7 The evolution of the golf driver.

Fig. 8 A golf driver with a 460-cc hollow head with weight distributed around the periphery of the head resulting in a large "sweet spot."

3.6 Baseball/softball

Major League Baseball banned aluminum bats because they would make baseball stadiums obsolete (an excessive number of home runs). However, in softball, aluminum bats are accepted, and titanium bats can have a major influence on softball. Titanium bats have a bigger sweet spot than wood or aluminum bats resulting in a higher speed off the bat. However, this increased speed results in more injuries to infielders (who cannot react quickly enough to the faster speed of the ball), and this has resulted in the Softball Association banning use of titanium bats. Again, an advanced material, particularly titanium, has made its presence felt.

3.7 Lacrosse and horseshoes

As with the sports discussed earlier, titanium has made its mark in lacrosse (Fig. 9) and horseshoes, both applications taking advantage of the high strength and low density of titanium. However, performance improvement with the use of titanium is not conclusive.

3.8 Darts and ice skates

Even in darts, the low density of titanium is an advantage (Fig. 10). There are also applications in ice skates. In both cases, the improvement in performance with the use of titanium is inconclusive.

4 Ethical considerations

Where the influence of advanced materials on sporting performance can be judged in absolute terms, significant improvements have been made by their use. However, use of advanced materials in sports equipment gives rise to

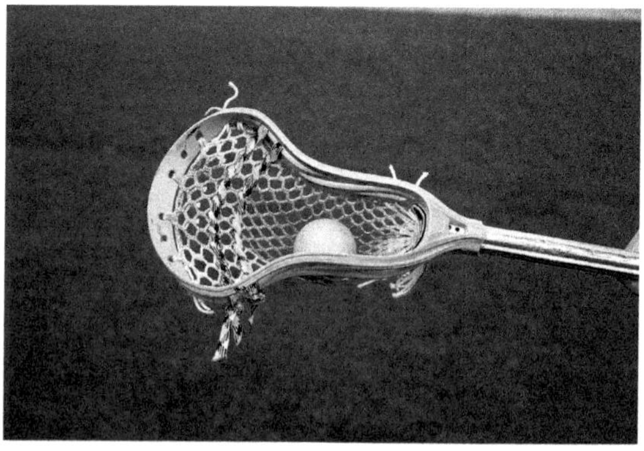

Fig. 9 The shaft or crosse of a lacrosse stick made from titanium.

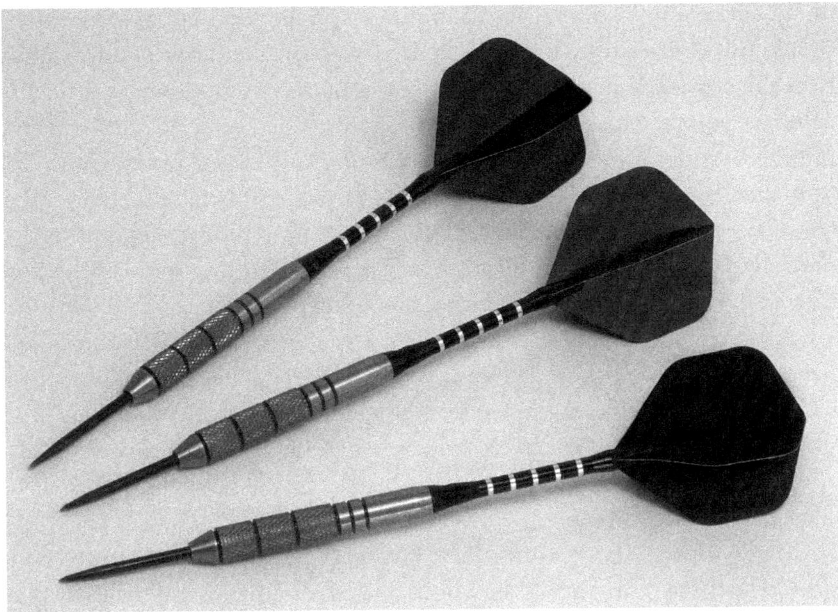

Fig. 10 Titanium darts.

ethical[a] questions. Improved behavior can be reached by allowing the use of advanced materials, but should a line be drawn, or should indiscriminate use of advanced materials be allowed? The answer, proposed by a young student in a class at the University of Idaho, is that if the patent is assigned to you, then only you can utilize this material or design.

A second question then arises: Should competition at the highest level only be affordable to the elite, resulting from the high cost of the necessary equipment? This in turn suggests the most controversial issue: If we approve a specific class of materials/designs in a particular sport, but we do not allow other classes of materials/designs, will we actually favor a particular class of people to excel in that sport?

Vaulting poles fabricated from carbon fibers, spiral-tailed javelins, special dimple patterns on golf balls, stiffer tennis rackets with carbon-fiber materials, titanium bicycles and the egg position, discuses constructed with their weight distributed close to the perimeter, and America's Cup yachts (this

[a]The dictionary defines *ethics* as "pertaining to acceptable principles of right and wrong governing conduct…in a group." Other, "looser" definitions include "no rule against it, so it must be OK" or "if you're not caught, there is nothing wrong." You choose what you like.

one is too complex, forget it) all promote "further and faster". But where should this end? Can we be sure that men/women are in competition rather than the advanced materials? We do not require pole vaulters to return to bamboo pylons, but what about darts guided electronically, heat-seeking missiles to shoot grouse (according to a young lady in one of my University of Idaho classes, this is OK as long as the grouse is still edible), bicycles that are solar-energy-enhanced, and golf balls that are terrain-following so that they always find the lowest local elevation on the putting surface (the bottom of the hole)? Clearly these ideas are not acceptable.

Prosthetic devices, presented earlier, used by Gaetani, Volpentest, and Pistorius, allow the athlete to increase his speed to greater than a competitor basically running on two human feet (at a cost of as much as $7000). With a similar device, 2016 100-m Rio Olympian Usain Bolt (in a time of 9.84 s) could probably break 9 s. Should nothing in excess of natural springiness be allowed in competition?

Many sports rules committees, facing the high-cost question, consider that their sport should be affordable to all athletes, rather than only those with deep pockets. Initially disc bicycle wheels were banned from Olympic competition due to their high cost, but with time, the wheels decreased in cost to a point at which they were affordable to all competitors, and they were then deemed acceptable. Should a single hull design be specified for both canoes and kayaks (see later)? The tiny glass beads dispersed in rigid foam in the Hoyt bow gave it increased stability compared to conventional wooden bows but at a significantly increased cost. Should this improved bow be allowed in competitions? Greatly reduced change in behavior with variation in temperature and greatly reduced moisture absorption with the Hoyt bow has resulted in winning scores of 1350 (out of 1440) contrasting with about 1100 30 years ago with a wooden bow—however at a price.

For the paraplegic, operating a wheelchair made from titanium means high technology and high cost. Here, we have bike-like wheels, and material use consisting of carbon fibers and titanium straight from the aerospace industry. Nowadays, there exists a chair for each sport including basketball, racing, and even tennis. In the latter sport, chairs are constructed with sharply angled back wheels permitting rapid side-to-side movement. For basketball, high seats are produced for forwards, while for guards have increased chair slant permitting quick turns. (Costs can be as much as $3300.)

Can legislation regarding advanced material use in sports equipment have a major effect on who wins a particular event? An elite rower averages around 2 m in height and weighs about 95 kg. In contrast, elite kayakers are about 15–30 cm shorter and weigh almost 14 kg less. These characteristics relate to the boat dimensions; a rowing boat can be as long as the rowers wants to make it, thus favoring strength with weight a secondary consideration (longer boats are able to distribute more weight over an increased area while not riding too low in the water, which leads to a significant increase in drag). In direct contrast, the kayak's length is restricted.

With a javelin redesigned promoting lightness and improved aerodynamics, this new design necessitated finesse to maximize flight distance. Uwetlohn of East Germany (1984) mastered this finesse and projected the javelin beyond 100 m, endangering spectators and other athletes at the far side of the stadium. In 1986, the new design was banned leading to a major reduction in the world record.

So do advanced materials including titanium have a place in sports equipment, and is it ethical to use these advanced materials?

Further reading

[1] D. Bjerklie, High-tech Olympians, Technol. Rev. 96 (1993) 22.
[2] K.E. Easterling, Advanced Materials for Sports Equipment, Chapman and Hall, London, 1993.
[3] G. Lea, On your bicycle, Economist 330 (1994) 94.
[4] N. Russell, High-tech sport, Economist 324 (1992) 17.
[5] R. Wirhead, Athletic Ability and the Anatomy of Motion, Wolfe Medical Publication, London, 1989.
[6] E. Moritz, S. Haake (Eds.), Engineering of Sport 6 Volume 1: Developments for Sports, Publisher Springer-Verlag, New York, 2006. http://www.springer.com/us/book/9780387317731.
[7] M. Estivalet, P. Brisson (Eds.), The Engineering of Sport 7 Proceedings Volume of the ISEA 2007, Publisher Springer-Verlag, Paris, 2008.
[8] Braghin, et al. (Eds.), The Engineering Approach to Winter Sports, Springer, 2016. 179.00, 978-1-4939-3019-7.
[9] Physics of Baseball & Softball, Cross and Rod, Springer, 2011. $199.00, 978-1441981127.
[10] Subic, Materials in Sports Equipment, vol. 2, Woodhead (Elsevier), 2007. 376 pp., 9781845691318.
[11] Y. Hong, Routledge Handbook of Ergonomics in Sport and Exercise, Taylor & Francis, 2014. 608 pp., HB, $245.00, 9780415518635.
[12] Fuss, Subic, Strangwoodd, and Mehta, Routledge Handbook of Sports Technology and Engineering, Taylor & Francis, 2013. 528 pp., HB, $245.00, 9780415580458.

CHAPTER 7

Sporting goods: Components for bicycles

M. Ueda, M. Ikeda
Faculty of Chemistry, Materials, and Bioengineering, Kansai University, Osaka, Japan

Contents

1 Introduction

Titanium and its alloys have high specific strength, good balance between strength and elongation, excellent corrosion resistance, and good biocompatibility. Therefore, they are used in a wide variety of applications, for example, blades in jet engines, landing gear in airplanes, artificial bones in humans, frames of glasses, watches, and so on. Also, these materials are widely used in sporting and outdoor goods, for example, golf clubs, climbing irons, ice ax, and bicycles. Bicycles are one of the attractive and suitable goods for titanium, for example, the handlebars and seat posts of mountain bikes are made from Ti-15-3 [1]. This chapter aims to discuss several applications of titanium and its alloys to bicycle components.

Titanium for Consumer Applications
https://doi.org/10.1016/B978-0-12-815820-3.00007-1

2 History of the bicycle [2]

The first bicycle prototype without "pedals" was developed by Karl Friedrich Christian Ludwig Freiherr Drais von Sauerbronn, a German inventor, in the early 19th century. That bicycle was called a "Draisine" in German (Fig. 1) [3]. Of course, the Draisine was made of wood, not metal. In middle of the 19th century, the Velocipede was developed by Pierre Lallement in France, with "pedals and crank to the front wheel." That is called a Michaux-type bicycle, although there are many divergent views to this story. That bicycle was made from wood, and only the wheels were made of steel. Also, that bicycle was called a "bone-shaker" because of the uncomfortable ride. In latter period of the 19th century, the "Penny Farthing," (Fig. 2) [3] also called the "Ordinary", was developed in England, which was made of iron and consisted of a big front wheel and small rear wheel.

Furthermore, the penny farthing had ball bearings, spokes and solid rubber tires. A few decades after that bicycle, the Rover Safety Bicycle (Fig. 3) [3], the prototype of modern bicycle, was born in the UK. That safety bicycle had front and rear wheels that were the same size, a pedal connected to the rear wheel with a chain, and pneumatic bicycle tires. In the 20th century, the bicycle became low in cost because of mass production, so anybody could buy one. Today, there are many types of bicycles: racing bicycles, BMX bikes, mountain bikes, and so on. Available materials for bicycle have also extended, for example, Al, Mg, and Ti alloys, and Composite.

Fig. 1 Example of Draisine [3].

Fig. 2 Example of penny farthing or ordinary [3].

Fig. 3 Example of Rover Safety Bicycle [3].

3 Classification of the bicycle [3]

Fig. 4 [3] shows the classifications of bicycles. A City Cycle is a bicycle of serious view to "Design"; basket, kickstand, and so on are standard options, and a derailleur is also available for the bicycle. Recently, a bicycle with an auxiliary electric motor is widespread among the general public. The design of a Treking bike is almost the same as the standard city bicycle. This bike is available for multipurposes. On a mountain bike (abbreviated MTB), suspensions, disc brakes, and fat tires are equipped. This bike is suitable to ride on bad condition road, for example, unpaved roads, steps, and so on. A Road Bike has racing handlebars, thin tires, and is light weight. This bike is suitable for long distance cycling. Fig. 5 [3] shows the names of bicycle parts (MTB/ROAD).

4 Titanium components of the bicycle

4.1 Flame [4]

Tube frames (Fig. 6) [4] are made from commercial purity titanium that adds higher oxygen because of higher tensile strength, about 750 MPa. Flat plate parts for connection of frame, for example, the Rear End, are produced from a Ti-6Al-4V alloy that possesses higher strength due to accidents, for example, falls. Those frames are produced by a cutting process, lost-wax casting, or an additive manufacturing process.

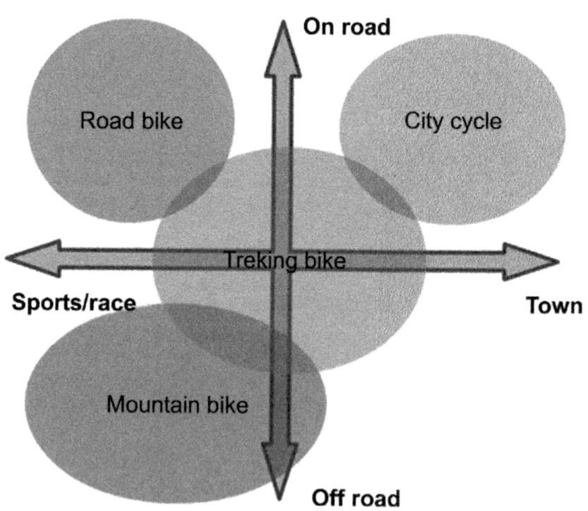

Fig. 4 Example of classification of bicycles [3].

Part	Name
①	Fram
②	Front fork
③	Saddle
④	Seatpost
⑤	Tire
⑥	Wheel
⑦	Stem
⑧	Handlebars
⑨	Gearshift
⑩	Brake
⑪	Front hub
⑫	Pedal
⑬	Crack
⑭	Chain wheel
⑮	Front derailleur
⑯	Chain
⑰	Cassette
⑱	Rear derailleur
⑲	Rear dropout

Fig. 5 Example of names of bicycle parts (MTB/ROAD) [3].

Fig. 6 One example of standard tube frames [4].

4.2 Screws (fasteners) [4]

Almost all screws are made with Ti-6Al-4V alloy. Those are mainly produced by forging and rolling, although are not done by cutting, nowadays. Surface modification is very important to improve wear resistance and seizure resistance. In generally, surface oxidation, diamond-like carbon coating, and wide peening and cleaning are treated on the surface. Of course, the screws also used in wide applications as fasteners other than just for bicycles.

4.3 Components (gear and hub) [3]

Components Gear (Fig. 7) [3] and Hub (Fig. 8) [3] need higher strength and lower weight, therefore titanium and its alloys that are higher in strength than aluminum and its alloys and lighter than steels are adopted. These components are mainly used in higher class racing bicycles. Beta-type titanium alloys that possess higher strength than other types of titanium alloys are adopted as materials for those parts. Those components are made by cutting and die forging.

5 Problems of cold working in titanium alloys for bicycle

5.1 Seizure and adhesion by deficiency of lubricant [3]

During cold forging, titanium parts are deficient in lubricant by local heating of those deformed parts that have low thermal conductivity. When heating

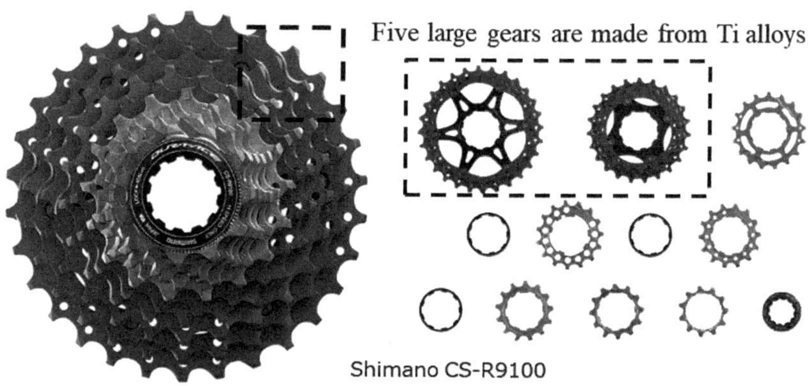

Fig. 7 Cassette sprocket gears [3].

Fig. 8 Rear free hub [3].

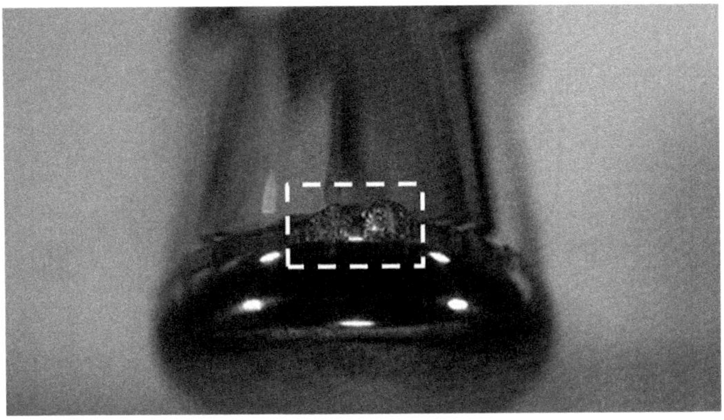

Fig. 9 Seizure of titanium alloy on die [3].

the parts, seizure of titanium occurs on dies and metal molds. Because there are not solid lubricants that chemically bind to titanium, lubrication cannot suitably conduct to titanium parts. In the present, multiple lubricants, zinc phosphate coating and oil lubrication responds to lubrication of titanium. To suppress heating to avoid this problem, the cutting speed is suitably controlled, that is, slowly controlled. Fig. 9 shows seizure of part of a die.

5.2 Cracks and breakage of titanium parts [3]

When the titanium part that contacts to die and then cold-forged was cold-forged again, cracks occur or the reforged part is partially broken. Cracks occur by cold forging at a lower deformation ratio than that of steel. Therefore, many annealing treatments are needed to forge at room temperature. Those processes need a longer time to cold forging and incur a higher cost.

5.3 Difficulty of cutting [4]

Cutting of titanium and its alloys is more difficult than steel. Thus, it is necessary that the number of sawteeth are higher and the thickness of the saw is larger. After cutting, there are many burrs on cut parts.

6 Conclusions

Although titanium and its alloys are classified as difficult metallic materials to cold forge, many researchers and engineers have actively developed suitable forging process for those alloys. In the future, if titanium alloys become low cost and if there are improvement in their cold-forging ability, titanium alloys will be more widely used for components of bicycles.

Acknowledgment

The authors specially thank the Shimano Inc. and the TiG Co., Ltd.

References

[1] R.R. Boyer, R&D and applications developments in the titanium industry in the USA, in: P.A. Blenkinsop, W.J. Evans, H.M. Flower (Eds.), Proceedings of the Eighth World Conference on Titanium, Titanium'95 Science and Technology, Birmingham, UK, October 22–26, 1995, 1996, pp. 41–50.
[2] T. Ambrose, The History of Cycling in Fifty Bikes, Rodale, New York, 2013.
[3] Private communication, Shimano Inc.
[4] Private communication, TiG Co., Ltd.

CHAPTER 8

The beauty of titanium: Enhancing residential design utilizing titanium elements

Ellen A. Semeniuta
Grandis Titanium, Rancho Santa Margarita, CA, United States

When people think of titanium, they historically associate it with the aerospace and chemical industries, and lately with golf, bicycles, race cars, and jewelry. There are a few magnificent feats of architecture finished with titanium including Frank's Gehry Guggenheim Museum in Bilbao, Spain; the National Grand Theater in Beijing, China; the Taipei Arena in Taipei, Taiwan; and the Oita Bank Dome in Japan. In this chapter, I would like to show you a single family residential building where titanium was utilized to enhance its design.

Welcome to California.

You are introduced to titanium at your very first sight. You enter the property through a titanium gate (Fig. 1). The light ornamental design of the scrolls, curves, and leaves gives a lighter impression because of the titanium. Round tubes are used for the scrolls, and square tubes are used for the gate frame.

Coming up the hill, you see the guest house. The garage door catches your eye (Fig. 2). Four titanium sheets stamped with little squares are applied to the door panels. The design is simple. Its beauty radiates from the titanium itself. Besides its resemblance to the panels of a spacecraft, the titanium lends other useful characteristics: strength, energy efficiency, and environmental safety. You never will see rust or corrosion on this door.

Further up the hill, you approach the ground level of the main house. Its garage door has a design similar to the guest house's (Fig. 3). The door is slightly bigger, but the same stamped little squares are used for the panels' decoration.

Across from the garage you can see the pool house. A bridge connects the main house and the pool house letting you on to the pool house roof. There are a few architectural elements on the roof of the pool house made

Titanium for Consumer Applications
https://doi.org/10.1016/B978-0-12-815820-3.00009-5

Fig. 1 Entrance gate.

Fig. 2 Guest house garage door.

Fig. 3 Main house garage door.

from titanium: the protective railing around the pyramid skylight, the drainage grid straps, and the vertical tubes between the balusters on the deck parapet railing.

A skylight (Fig. 4) is a very open architectural element that should stay open, but placed in the middle of the walking deck, it must be protected. Titanium is the best material to use for the railing around the glass pyramid structure. It is strong and doesn't need a lot of structural elements for support, yet keeping the glass pyramid open. The color of the titanium railing is also the best match for the glazing.

The drainage grids (Fig. 5) for the channel drain on the pool house roof are made from titanium because of its anticorrosive characteristics. They were custom-made per my design to accommodate the special setting of the marble tiles on the flat roof.

Another titanium element on the deck are the tubes between balusters (Fig. 4). Per the California Building Code, the space between any elements

Fig. 4 Titanium guardrail around skylight and titanium tubing between balusters.

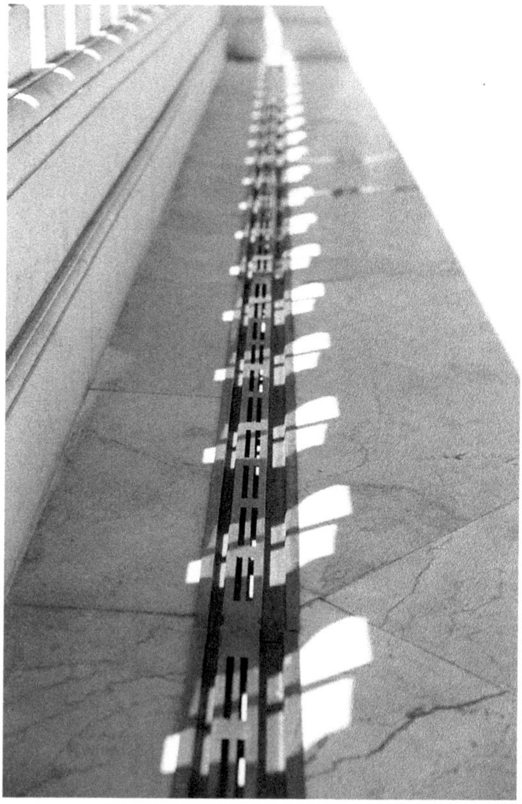

Fig. 5 Channel drain on pool house roof deck.

on the deck/balcony railing should be no more than 4 inches. The whole of Italian architecture with all its balconies and bridges would break this rule! There are a couple of options to comply with California Building Codes on the balcony design with the balusters. The first option is to build the balusters without any spacing between them. Another option is to put something between the balusters. I thought the best option was to put titanium tubes between the balusters. Its primary benefit is structural; the titanium posts weigh significantly less than the precast balusters. Additionally, they are aesthetically pleasing, allowing one to see through the home's many balcony railings, making a big difference when it comes to enjoying the views from inside the house (Fig. 6).

The main entrance to the house has a spectacular 16-by-22-foot glass portal, covered with an extremely delicate design of "leaves" and "birds" made from titanium rods (Figs. 7 and 8). It definitely doesn't carry the function of a protection grid but instead emphasizes the scale and beauty of the entrance. In spite of the large dimensions of the entrance door, the titanium material used for the ornamental design makes this architectural element extremely light and beautiful.

Upon entry, the magnificent two-story foyer welcomes you. Before you are two of the most attractive design elements of a foyer, both made with titanium - an oval medallion on the floor and the railing perched on the grand stairs.

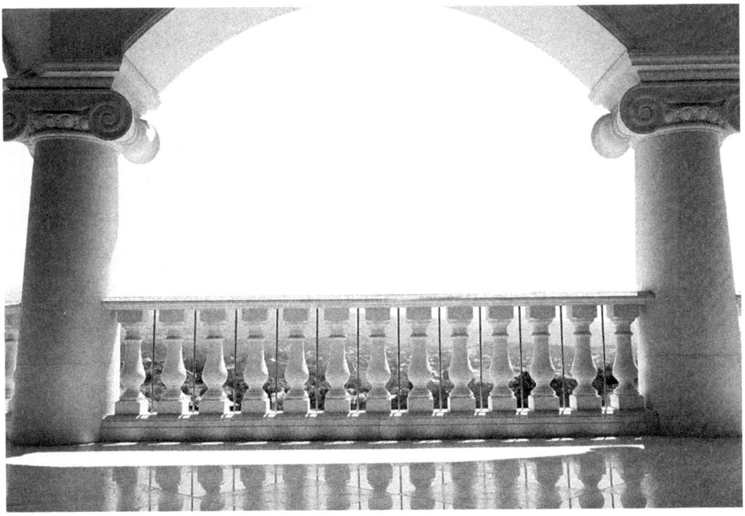

Fig. 6 Titanium tubing between balusters on balcony.

Fig. 7 Main entrance, exterior.

The grand stairs flow up on two sides making a balcony at the top (Fig. 9). On the left side of the staircase, the titanium railing continues to the gallery, jutting out slightly to form a Juliet balcony (Fig. 10). The metal ornaments of the railing have more traditional design in comparison with the aforementioned floral design of the main entrance due to the safety, "4 inches" restrictions. This railing has more density in scrolls, but the titanium material still creates a very light image of the metal work, both in terms of weight and aesthetic (Fig. 11).

The 12-by-8-foot oval medallion, incorporated in the marble floor of the foyer, is located between the two wings of the grand staircase (Fig. 12). The titanium elements in the medallion correspond to those of the titanium railing. The medallion itself has an asymmetrical scheme, and the blend of marble, onyx, granite, and architectural quality titanium (CP grade 2) creates a great artwork. All the elements in the medallion, stone slabs, and titanium sheets were cut in the stone shop using waterjet techniques.

Fig. 8 Main entrance, interior.

Fig. 9 Grand stairs.

Fig. 10 Juliet balcony (left); gallery and left side of grand stairs.

Fig. 11 Fragment of grand stairs.

Fig. 12 Medallion with titanium and stone combination.

Down the hall, you approach the center of the house, where you can enjoy a 360-degree view of the property from the hall and bar area. The titanium sheets were used as a decorative element infused into the millwork of the bar cabinets. The combination of mirrors and titanium surfaces creates a look of elegance, culminating in a classic design of dark millwork finished cabinets and blue-black granite countertops (Fig. 13).

Fig. 13 Titanium elements above mirrors at bar's millwork.

The aforementioned titanium sheets were used also on the face of the soffits in the game room on the ground floor, right below the bar. The flexibility of the sheets proved to be an asset as they were used on the complex vertical surfaces of the soffits. The unique look of titanium on the face of the soffits emphasizes the distinctiveness of the ceiling's design. Overall, the titanium works beautifully with the simplistic art nouveau design of the game room (Fig. 14).

The two zones—the bar on the main floor, and the game room on the ground floor—are connected by a curved staircase, where the railing has a simple, straight titanium tube design. The minimalism of the design in combination with titanium's beauty creates a general impression of refinement (Fig. 15).

These are not the only staircases located within the house, yet all secondary stairs have the same simple design of the bar—game room staircase. The straight titanium tubes are used on all the secondary staircases, galleries, and balconies. As mentioned before, this simplistic design and the beauty of titanium create a look of elegance while remaining very light (Fig. 16). Titanium brackets to support the wood railing were designed for some of the stairs that require a railing along the wall. Titanium was used for the brackets to increase the look of sophistication (Fig. 17).

In some areas of the house, the titanium applications played not just a decorative role but also served as a utilitarian purpose. This allows titanium's

Fig. 14 Titanium sheets on the face of soffits.

Fig. 15 Titanium tubing on the staircase from bar to game room.

beauty to have an additional benefit, such as for the custom brackets needed for the wood railing or to support a marble sink in a custom-designed steam room (Fig. 18). This specialized product was needed because of the custom sink and the unique outlines of the benches in the steam room. Titanium is also essential as it is especially resistant to the steam and hot temperatures metals would be exposed to in the room.

The outstanding environmental characteristics of titanium are very beneficial for outdoor application, and titanium was the first choice to use in the design of exterior elements. Titanium was utilized in several exterior balconies as well. The railing on the semicircle balcony of the master bedroom has a combination of a few different elements in its design (Fig. 19). The middle portion has a leafy look, and on each side straight tubes alternate with tubes and rings. The simple pattern does not distract from the view of the village below.

Fig. 16 Titanium tubing on staircase to attic.

Fig. 17 Titanium support brackets.

Fig. 18 Titanium support for marble sink in steam room.

Fig. 19 Titanium railing on master bedroom balcony.

The railing on the balcony outside the children's bedrooms is made from rectangular shape tubing and shares the same combination of straight tubes and rings (Fig. 20).

A titanium railing was used as well to guard the observation deck above the tennis court, letting you sit and watch a tennis game unencumbered by the railing (Fig. 21).

Fig. 20 Titanium elements in the railing on children's balcony.

Fig. 21 Tennis court observation deck railing.

The flat bars were used as handrails for outdoor stairs at tennis court and oval observation deck (Fig. 22).

As mentioned before, in the process of the home's construction, titanium was given a utilitarian purpose in addition to its aesthetic implementations. It was used for vent creation (Fig. 23) and in some parts of construction for the flashing and for specific areas of the roof (Fig. 24).

Titanium sheets were used in a few exterior elements as well: the roof of the dormer windows in the attic, shaped in a half-barrel design with the ribs following the circumference for the aesthetic, and for the joint connection between the sheets (Fig. 25). The titanium sheets also come into play on the chimney shrouds on the main building (Fig. 26) and the BBQ pavilion (Fig. 27). The shape of the shrouds is very simple, but the beauty of material makes them very attractive and underlines the uniqueness of the design.

Fig. 22 Rooftop sun deck titanium hand rail.

Fig. 23 Titanium vent and flashing.

Fig. 24 Titanium flashing at master balcony.

Fig. 25 Titanium sheets on dormer windows.

Fig. 26 Titanium chimney shroud at main house.

Fig. 27 Titanium chimney shroud at BBQ.

Other pieces of the roof made from titanium are the dormer vents. Titanium dormer vents are located on the roof of the main house (Fig. 28) and the guest house (Fig. 29). Part of the beauty of using titanium on the roof and related elements is titanium's weather-resistant characteristics. Copper rusts and corrodes over time to a dull green color and is difficult to maintain. Titanium continues to look brand new even as centuries pass. This look is exemplified in titanium's paradoxical ability to be both shiny and matte. It has luster but does not sparkle. This is key for the view from the main house down to the neighborhood as the guesthouse and barbecue structures, with many titanium elements on the roof, are located below. The use of titanium on the roofs of these two buildings doesn't cause people observing from above to be blinded by any potential glare. Finally, the titanium elements simply blend well with the pale green color of the slate roofing.

Fig. 28 Dormer vents on main house roof.

Fig. 29 Dormer vents on guest house roof.

One of the most impressive applications of titanium material in outdoor design is the waterfall (Fig. 30). Five titanium sheets, 4-by-12 foot, applied on the concrete retaining wall make an extremely beautiful composition with the artificial rock formation and surrounding landscape. The waterfall serves two purposes: the first is as landscape decoration, and the second one is as an outdoor screen. Titanium has a very good surface to reflect images. Even during bright sunny days with water running on the titanium wall, the quality of the projected images is outstanding.

Another fun titanium object on the property is a cylinder-shaped ingot that serves as a beer table by the outdoor sauna. This piece weights about 2 metric tons, was pulled out of scrap, and settled forever in the beautiful landscape (Fig. 31).

Working with titanium was a challenge because the contractors were all dealing with the material for the first time in their practice. The sheet metal company made all the roof elements and indoor soffits. The iron ornamental company made all the railings. The garage door company stamped sheets with designed squares and created panels for the doors. Despite their inexperience with the specific metal, they produced magnificent results.

All titanium materials for the construction was provided by Grandis Titanium company, as well as any technical assistance.

Fig. 30 Titanium waterfall/projection screen.

Fig. 31 Titanium beer table.

I hope for all homeowners and designers who desire to use titanium in their project success. Any little production troubles you may come across are worth it when you see the beauty of the finished product; the airy lightness and elegance of titanium will make your home very special.

CHAPTER 9

Application examples and their applied technology for architectural material

Kazuhiro Takahashi
Nippon Steel Corporation, Chiba, Japan

Contents

1 Introduction

Due to its excellent corrosion resistance, titanium has been used for an increasing number of applications at chemical, power, and desalination plants, and for heat exchangers, civil engineering projects, and building construction, among other uses. Titanium is very suitable as metal for roofs, walls, and other exterior uses of buildings because of its light weight, small thermal expansion coefficient, and extremely high corrosion resistance. Since the 1970s, commercially pure titanium sheets (hereinafter referred to as titanium sheets) have begun to be applied to the roofs and walls in Japan. Titanium sheets as architectural material were first used in buildings in highly corrosive coastal areas and before long expanded to public buildings

Titanium for Consumer Applications
https://doi.org/10.1016/B978-0-12-815820-3.00014-9

required to have higher durability "long life," such as museums, grand theaters, shrines, and temples. The reasons for adopting titanium to architecture were corrosion resistance first, then design quality, and finally the high-class image provided by titanium [1–7]. It can be understood that the architects recognize the unique color and texture of titanium, which is different from typical plating and coating. Therefore, the landscape of the building is very important [8–11]. The technology of titanium sheet processing for architectural material from the viewpoint of enhancing the landscape, various surface finishing, improvement of discoloration resistance and suppression of pocket waves, such as uneven undulations and distortions during roll forming, have been energetically developed [11–16]. Details of each technology will be introduced in this chapter.

2 Various surface finishes

Antiglare quality has been specifically required for architectural material used as roofs and walls from the standpoints of landscape and esthetic appearance [11–15]. Minute unevenness on the surface is effective in diffusing the reflection of sunlight to suppress glaring. The means of forming minute unevenness are via mechanical or chemical methods, or a combination of both. Roll-dull, pickling, combination of roll-dull and pickling, and alumina shot blasting, as well as anodic oxidation of these finishes, are widely used as typical surface finishes of titanium sheets. Generally, the lower the glossiness (Gs45 or Gs60) and the higher the whiteness (L^*), the more superior the antiglare property of the metal surface is. L^* is an index from darkness to whiteness in the $L^*a^*b^*$ color system of CIE1976. Glossiness and L^* are in good agreement with the impression that the human eye receives from the surfaces.

In the following, the manufacturing method and features of the surface finishes will be explained [10, 11, 13–15]. Fig. 1A, B, C, and D show optical micrographs of representative surface morphology of roll-dull, pickling, a combination of roll-dull and pickling, and alumina shot blasting, respectively [13–15]. Fig. 2A, B, C, and D show application examples of titanium sheets used for roofs and walls of buildings: (A) National Grand Theater in Beijing, China; (B) Shimane Art Museum in Shimane-pref., Japan; (C) Sensoji Temple in Tokyo, Japan; and (D) Kyushu National Museum in Fukuoka-pref., Japan, respectively [13, 15, 17].

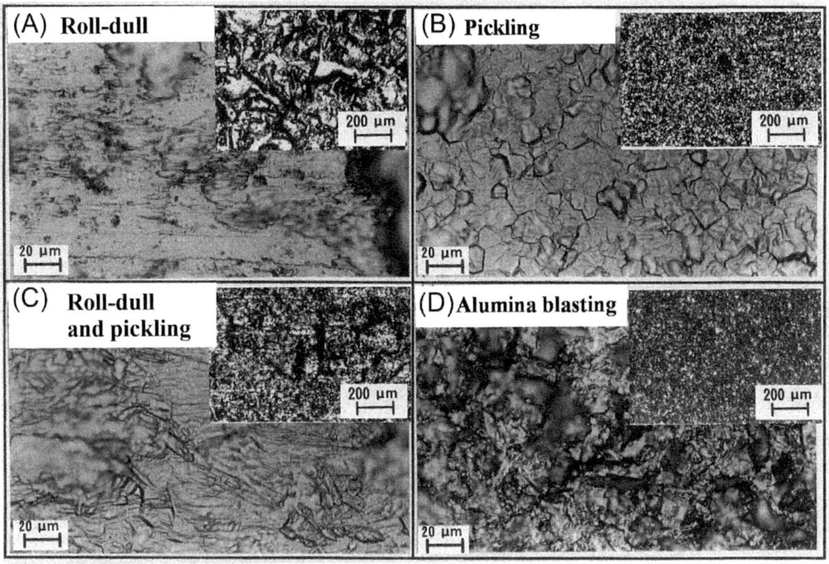

Fig. 1 Micro-morphologies of titanium surface finishes for architectural material.

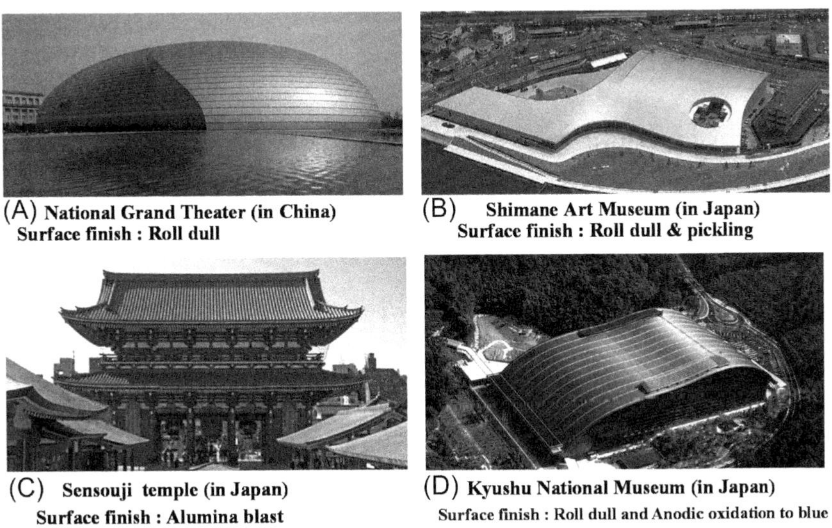

Fig. 2 Application examples of titanium sheets for architectural material.

2.1 Roll-dull finish

Roll-dull is a mechanical method of finishing. Unevenness on the surface of the work-rolls for cold rolling is transferred to the surface of the titanium sheet [10, 11, 13–15]. The roll-dull has a somewhat metallic luster and a unique texture provided by the unevenness on the surface, as shown in Fig. 1A. Buildings to which the roll-dull finish has been applied are Kyushu Bank Dome (Japan), Hangzhou Grand Theater (China), National Grand Theater "National Centre for Performing Arts" (China, as shown in Fig. 2A), Taipei Arena (Taiwan), etc.

2.2 Pickling finish

Pickling is a chemical method of finishing. Generally, a pickling solution for titanium is "nitric-hydrofluoric acid solution," which is an aqueous solution mixed with nitric acid and hydrofluoric acid. The unevenness on the surface formed by pickling is very finer than that of the dull-roll finish as shown in Fig. 1B. On the pickled titanium surface, very fine unevenness with the almost same size as that of titanium crystal grain is formed. The pickling finish gives us a relatively whitish and matte texture. Roofs and outer walls with the pickling finish are Suma Aqualife Park in Kobe, Tokyo International Exhibition Center "Tokyo Big Sight," Kumamoto General indoor pool "Aqua Dome Kumamoto," Minato-Machi River Place in Osaka (Japan), etc.

The surface morphology of a pickling finish largely depends on the composition of the acid and the crystal grain size of titanium [12–14, 16, 18–20].

Surface morphology affects the composition of the nitric-hydrofluoric acid solution and crystal grain size of titanium. The glossiness decreases, and the whiteness L* increases because the nitric acid concentration is lower in the case of constant hydrofluoric acid concentrations, as shown in Fig. 3. This effect is thought to be due to depressing the anisotropy of the dissolution rate depending on the HCP crystal orientation of titanium as the nitric acid concentration of pickling solution increases.

The surface morphologies of two kinds of titanium sheets having different crystal grain sizes are shown in Fig. 4 [13, 19]. Furthermore, by decreasing the crystal grain size, the unevenness on the pickled surface becomes finer. Consequently, the amount of unevenness per unit length is larger, and the glossiness is thus reduced [12–14, 16, 19, 20].

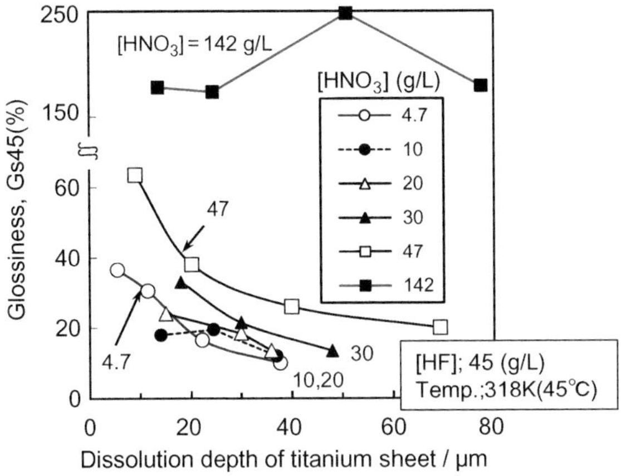

Fig. 3 Change of glossiness (Gs45) of titanium on dissolution depths when titanium was pickled in several nitric-hydrofluoric acid solutions.

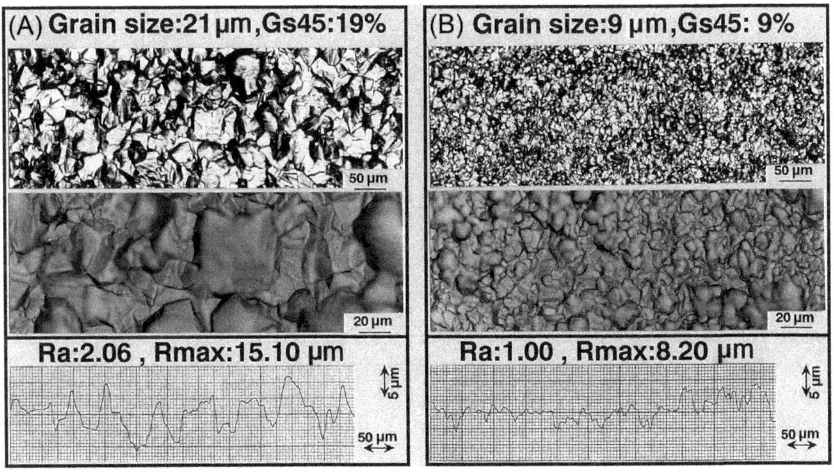

Fig. 4 Optical micrographs and roughness profiles of titanium surfaces pickled in nitric-hydrofluoric acid solution. Effect of grain size of titanium sheet.

2.3 Combination finish of roll-dull and pickling

Combining the roll-dull and the pickling methods makes it possible to produce a surface with features of both [13, 14, 19]. As shown in Fig. 1C, very fine unevenness of about 1 μm in size to enhance the whiteness is formed by an undulation of roll-dull finish. Applied buildings are Marinemesse Fukuoka, Mie Prefectural College of Nursing, Shimane Art Museum (Japan, shown in Fig. 2B), Malaga Convention Centre (Spain), etc.

2.4 Blasting finish

Alumina shot blasting is a mechanical method to blast the titanium surface. As shown in Fig. 1D, a relatively sharp and extremely fine unevenness is formed. This finish is characterized by a quiet color tone with very low glossiness, resembling that of Japanese clay roof tile. Roofs and outer walls with the blasting finish are Koetsji temple Main Hall (Kyoto), Tokyo National Museum "Heisei Hall" (Tokyo), Showa Hall (Tokyo), Sensoji temple (Tokyo, shown in Fig. 2C), etc. in Japan. At the Sensoji temple in Tokyo, the mass of the entire roof has been reduced to one-fifth by applying titanium roof tiles. As a result, safety measures such as earthquake resistance of temple buildings was improved [21]. The blasting finish partly supports Japanese traditional beauty.

2.5 Coloring "anodic oxidation" finish

In addition to the previously discussed titanium surface finish, many kinds of vivid color tones, such as blue and gold, can be obtained by anodic oxidation [8, 9]. The principle of this coloring is an interferential color depending on the thickness of the oxide film. In the anodic oxidation method, appropriate voltage is applied to the titanium sheet as an anode in an electrolytic solution, such as phosphoric acid aqueous solution. The thickness of the oxide film can be controlled from about 20 to 400 nm by adjusting the voltage. In this way, various colors are produced [22–24]. Buildings to which anodic oxidation colors have been applied are Hanayashiki Golf Club House (Japan, Color: Blue), Kyushu National Museum (Japan, Color; Blue, as shown in Fig. 2D), Hotel Marqués de Riscal (Spain, Color; Pink & Gold), etc.

To harmonize with the copper roof or steel roof of existing buildings, some surface finishes that are anodized after blast treatment have been developed. The copper roof would appear to have the green-blue of copper rust "verdigris," and the steel roof would appear brown, called corten, which is the color of stable iron rust. If the copper roof was damaged by acid rain, the blasted and anodized titanium sheets could be adopted in some temples and shrines in Japan [25–27]. These surface finishes were applied to Daitokuji Oubai-in Temple (Kyoto; Color: Verdigris), Kitano Tenmangu Treasury (Kyoto; Color: Verdigris), Nara National Museum (Nara, Color: Brown), etc., in Japan.

3 Discoloration resistance in atmospheric environment

3.1 Discoloration phenomenon of architectural titanium as exteriors

There have been no reports of pitting or crevice corrosion of titanium sheets used for building roofs, walls, etc. thus far; however, there have been some cases of discoloration on titanium sheet surfaces after long exposure to the atmosphere. The color change exceeded 10 in terms of the color difference ($\Delta E^{*}ab$). A titanium oxide film on the sheet surface was found to be thicker at the discolored surface by a small amount, by only tenths of a nanometer or more. The discoloration was caused by growth of titanium oxide films on the surface, which causes interferential colors depending on the thickness of the films [12, 20, 28–36].

The color difference, $\Delta E^{*}ab$, is calculated by the following equation: $\Delta E^{*}ab = \{(\Delta L^{*})^2 + (\Delta a^{*})^2 + (\Delta b^{*})^2\}^{1/2}$, according to the $L^{*}a^{*}b^{*}$ color system of CIE1976. Here, ΔL^{*}, Δa^{*} and Δb^{*} mean the differences in the values, respectively, of L^{*}, a^{*} and b^{*}, before and after the exposure test or the immersion test or aging.

3.2 Environmental, material factors and improvement on discoloration of titanium

As an environmental factor, low pH due to acid rain enhances discoloration of titanium. As Fig. 5 shows, the effects of solution pH on discoloration of titanium sheets and the discoloration $\Delta E^{*}ab$ remarkably

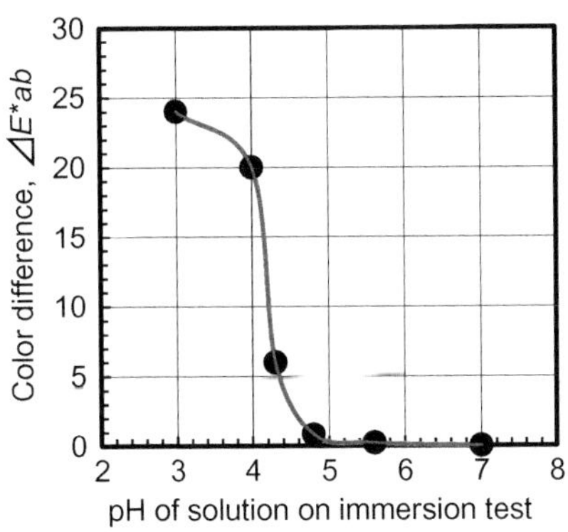

Fig. 5 Effect of pH in solutions on discoloration of titanium sheets in immersion test at 333 K (60°C) for 14 days.

occur below a solution pH of about 4.5 [17, 29, 32, 33]. From this result, it is clear that acid rain with low pH dramatically induces discoloration of titanium. It is considered that titanium ions dissolved into acid rain deposits as titanium oxides through hydrolysis reaction, resulting in the growth of the oxide film, which cause interferential color [17, 29, 32, 33]. As a matter of course, the higher the temperature of environment, the faster the reaction, thus the discoloration is enhanced. Also, Hisano et al. reported on the discoloration mechanism, that ultraviolet irradiation (UV) and hydrogen peroxide (H_2O_2) produced by photo-electrochemical reaction with UV induces the growth of the oxide film on a titanium surface [31].

Concerning material factors affecting discoloration, titanium carbide (TiC) or fluorine (F) on the surface of titanium sheets exposed to an acid solution accelerates the growth of oxide film [15, 17, 20, 28, 29, 32, 33, 35, 36]. TiC dissolves more than 10 times more easily in sulfuric acid solution than titanium does. Therefore, the dissolved TiC quickly forms titanium ions. Fluoride ions easily form soluble complexes with titanium, and furthermore F serves to accelerate the dissolution of titanium in an acidic solution, such as acid rain.

Based on the previously discussed material factors, vacuum-annealed titanium sheet product that has a decreased amount ofTiC on its surfaces and a pickled product in nitric-hydrofluoric mixed acid solution that has oxide films containing reduced amounts of F were developed [15, 17, 20, 28, 29, 32, 33, 35, 36]. Fig. 6 shows the results of accelerated discoloration tests on the two developed products [15, 17]. Photographs on the right-hand side of the graph show the appearances of some specimens after the tests. It is clear that the developed products in Fig. 6C and D demonstrate far more stable resistance to discoloration. In Okinawa Prefecture (Japan), where it is a subtropical climate, the color difference of the developed products after exposure to the atmosphere for 4 years were less than 1. Superior discoloration resistance has been confirmed as well as results of the accelerated discoloration tests [35, 36].

As mentioned previously, the anodic oxidation coloring of titanium obtained by an interference of oxide film are widely used for architectural materials. These colored sheets using the developed products exhibit the same stable discoloration resistance as those of the metallic color, both during the accelerated discoloration tests and during actual atmospheric exposure in Okinawa [35, 36].

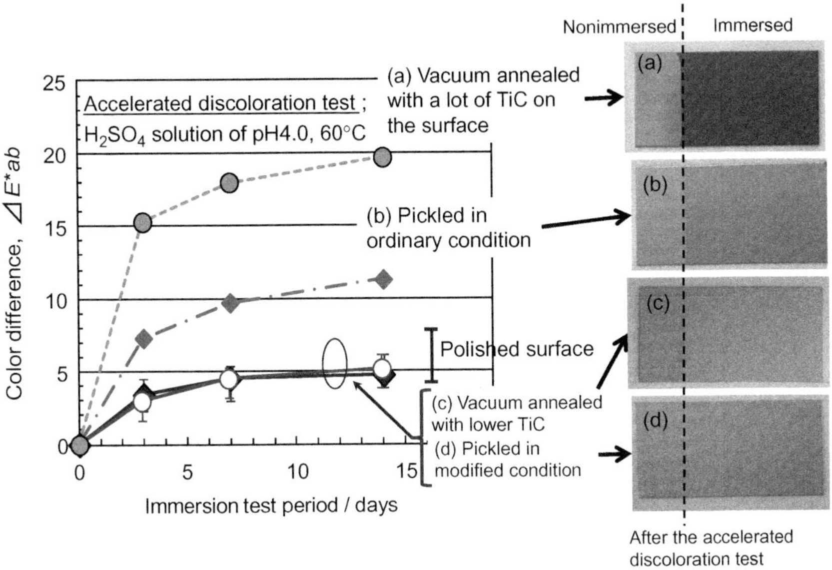

Fig. 6 Changes of color difference of immersion test on pickled or vacuum-annealed titanium sheets. Accelerated discoloration test: pH 4, 333 K (60°C).

3.3 Discoloration removal and cleaning techniques

Techniques removing discoloration and contamination for titanium exteriors, such as roof and wall materials, have also been developed [16, 34, 37]. Actually, there are cases where these techniques were applied at the Guggenheim Museum in Bilbao, Spain, etc.

4 Suppression of pocket waves during roll forming

Generally in making metal roof, a coiled strip of metal is roll formed into an approximately L-shaped cross-sectional shape to join both edges of the roll-formed metal roofs by welding or overlapping. It is known that uneven undulations and distortions, called pocket waves, are generated in the width-center unformed portion (web portion) of the roll formed coil strip, as shown in Fig. 7 [13, 14]. It is important to reduce the propagation of strain from the bent portion on edges of coil strip. In the steel roofs, increasing the proof stress or the yield elongation suppressed pocket waves [38, 39]. Crystal grain refining of titanium sheet and work hardening by skin pass rolling using dull roll are adopted to titanium roofs as the method of suppressing the same pocket waves as the steel roofs [11–14, 16, 40].

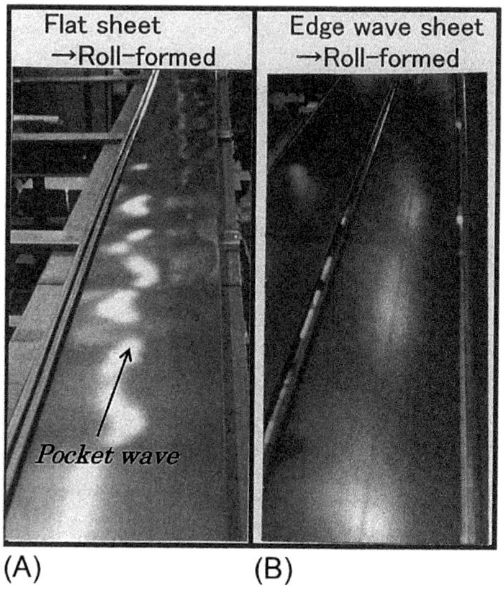

Fig. 7 Pocket wave during roll forming.

Furthermore, it is very effective when applying the waves in advance to the edge of the titanium sheet coil. The longitudinal shrinkage distortion occurring at the edges bended by the roll forming is absorbed by these pre-waves [13, 14].

5 Summary

Presently, titanium sheets with various surface finishes are widely used for buildings all over the world. In appreciation of their excellent corrosion resistance, attractive appearance, good resistance to discoloration, and suitable construction techniques, titanium sheets are expected to expand their application further as a durable and lightweight exterior finishing material for buildings.

References

[1] M. Fujii, Titanium Jpn. 42 (1994) 130–135.
[2] K. Kinoshita, Titanium Jpn. 43 (1995) 272–276.
[3] T. Kaneko, Titanium Jpn. 44 (1996) 254–257.
[4] K. Ashihara, Titanium Jpn. 45 (1997) 214–218.
[5] K. Kinoshita, Titanium Jpn. 46 (1998) 319–323.
[6] T. Kaneko, Titanium Jpn. 47 (1999) 323–326.
[7] K. Kinoshita, Titanium Jpn. 48 (2000) 336–338.
[8] H. Satoh, J. Surf. Finish. Soc. Jpn. 43 (1991) 1014–1019.

[9] T. Yashiki, Y. Yamamoto, KINZOKU, vol. 67, Metal Technology (Japan), Prs. by AGNE Gijutsu Center, Tokyo, Japan, 1997, pp. 151–156.
[10] M. Ishii, K. Kinoshita, K. Kimura, Titanium Jpn. 48 (2000) 106–112.
[11] K. Kimura, J. Surface Finish. Soc. Jpn. 51 (2000) 803–806.
[12] T. Yashiki, Y. Miyamoto, A. Okamoto, R&D Kobe Steel Eng. Rep. 49 (3) (1999) 35–38.
[13] K. Takahashi, Titanium Jpn. 50 (2002) 303–308.
[14] K. Takahashi, KEIKINZOKU, J. Jpn. Inst. Light Metals 55 (2005) 637–641.
[15] K. Takahashi, Titanium Jpn. 61 (2013) 120–125.
[16] T. Yashiki, Y. Miyamoto, Y. Yamamoto, A. Okamoto, E. Yoshikawa, K. Yanagisawa, in: G.L. By, J. Albrecht (Eds.), Ti-2003 Science and Technology, Wiley-VCH Verlag GmbH & Co., Hamburg, 2004, pp. 3103–3110.
[17] K. Takahashi, M. Kaneko, T. Hayashi, J. Tamenari, K. Tokuno, K. Kinura, I. Muto, Nippn Steel Sumitomo Metal Tech. Rep. 106 (2014) 28–33.
[18] K. Takahashi, T. Yamazaki, T. Nishijima, H. Shimizu, Nippon Steel Tech. Rep. 85 (2002) 64–70.
[19] K. Takahashi, T. Yamazaki, T. Nishijima, H. Shimizu, Nippon Steel Tech. Rep. 85 (2002) 71–76.
[20] K. Takahashi, M. Kaneko, T. Hayashi, J. Tamenari, H. Shimizu, in: G. Lütjering, J. Albrecht (Eds.), Ti-2003 Science and Technology, Wiley-VCH Verlag GmbH & Co., Hamburg, 2004, pp. 3117–3124.
[21] S. Ando, Titanium Jpn. 66 (2018) 79–85.
[22] T. Yamaguchi, K. Miki, ALUTOPIA, vol. 11, KALLOS Publishing Co. Ltd., Japan, 1987, pp. 41–48.
[23] T. Yashiki, H. Ohyama, AGNE Gijutsu Center in Japan, KINZOKU, vol. 66, Metal Technology, Japan, 1996, pp. 1061–1068.
[24] K. Yamagishi, Bull. Iron Steel Inst. Jpn. 2 (1997) 1061–1068.
[25] Y. Takahashi, Y. Tadokoro, I. Muto, N. Tagomori, S. Hihoshi, Nippon Steel Tech. Rep. 62 (1994) 9–15.
[26] N. Tagomori, H. Kihira, K. Kinoshita, T. Nakamura, S. Soeda, Proceedings of JSCE Corrosion '94, Japan Society of Corrosion Engineering, 25–27 May 1994, Tsukuba, Japan, 1994, pp. 201–204.
[27] H. Kihira, R. Matsuhashi, S. Soeda, N. Tagomori, K. Kinoshita, T. Nakamura, Proceeding of 42nd Japan Corrosion Conference, 25–27 September 1995, Sapporo, Japan, 1995, pp. 23–26.
[28] M. Kaneko, K. Takahashi, T. Hayashi, I. Muto, K. Tokuno, K. Kimura, 15th International Corrosion Congress, 22–27 September 2002, Granada, Spain, Paper No. 26, 2002.
[29] M. Kaneko, K. Takahashi, T. Hayashi, I. Muto, K. Tokuno, K. Kimura, Tetsu-to-Hagané, vol. 89, The Iron and Steel Institute of Japan, 2003, pp. 833–839.
[30] T. Yashiki, Corros. Eng. Sci. Technol. 38 (2003) 147–152.
[31] Y. Hisano, K. Takasawa, N. Hara, K. Sugimoto, Zairyo-to-Kankyo 52 (2003) 199–209.
[32] K. Takahashi, M. Kaneko, T. Hayashi, I. Muto, J. Tamenari, K. Tokuno, Tetsu-to-Hagané, 90, The Iron and Steel Institute of Japan, 2004, pp. 278–285.
[33] M. Kaneko, K. Takahashi, T. Hayashi, K. Tokuno, I. Muto, Materia Jpn., Bull. Jpn. Inst. Metals Mater. 43 (2004) 61–63.
[34] A. Pelayo, P. Cano, M. Vaquero, in: G.L. By, J. Albrecht (Eds.), Ti-2003 Science and Technology, Wiley-VCH Verlag GmbH & Co, Hamburg, 2004, pp. 3111–3116.
[35] M. Kaneko, K. Takahashi, T. Hayashi, K. Tokuno, J. Tamenari, Xi'an International Titanium Conference, 16–19 October 2005, Xi'an, China, 2005.
[36] M. Kaneko, K. Takahashi, T. Hayashi, K. Tokuno, J. Tamenari, Mater. Perform. 45 (2) (2006) 38–42.

[37] Features of Titanium Building Materials of Nippon Steel & Sumitomo Metal Corp, http://www.nssmc.com/en/product/titan/catalog/, 2012.
[38] T. Hira, H. Abe, K. Nakagawa, Y. Koizumi, S. Saeki, J. Jpn. Soc. Technol. Plast. 20 (1979) 933–939.
[39] M. Kiuchi, J. Jpn. Soc. Technol. Plast. 21 (1980) 301–308.
[40] Y. Miyamoto, S. Hawa, ISIJ Int. 31 (1991) 863–869.

CHAPTER 10

Additively manufactured titanium artworks

Tingting Song[a], Scott Mayson[b], Shudong Luo[a], Bing Liu[c], Milan Brandt[a], Ma Qian[a]

[a]Centre for Additive Manufacturing, School of Engineering, RMIT University, Melbourne, VIC, Australia
[b]School of Design, RMIT University, Melbourne, VIC, Australia
[c]Research Institute for New Materials Technology, Chongqing Powder Metallurgy Institute, Chongqing University of Arts and Sciences, Chongqing, China

Contents

1 Introduction

Titanium (Ti) and its alloys have been used to make artworks since the early 1950s when the titanium industry was established in the US (1948), the UK (1851), Japan (1952), and USSR (1954). Their excellent intrinsic resistance to corrosion imparts long durability to titanium artworks [1]. In addition, they offer high strength, sufficient ductility, and relatively low density (~4.5 g cm^{-3}). Finally, their excellent biocompatibility eliminates any serious concern with health and safety. Also, very few people seem to be allergic to titanium. This means that almost everyone can wear titanium jewelry without fear of allergy, especially those who suffer from allergic reactions from wearing gold or silver.

In general, titanium is difficult to machine. In addition, molten titanium is extremely reactive. It reacts with most gas atmospheres and most metallic and nonmetallic materials that would otherwise serve as crucible materials in melting. Besides, it is challenging to maintain the flow of molten titanium

Titanium for Consumer Applications
https://doi.org/10.1016/B978-0-12-815820-3.00017-4

over severe changes in dimension or direction within the mold cavity. As a result, casting of titanium, even to a simple shape, is still a complicated endeavor. Additive manufacturing based on the use of either titanium powder or wire is therefore particularly suited to the manufacturing of titanium products including titanium artworks. Historically, titanium artworks can be manufactured by (i) investment casting, (ii) conventional powder metallurgy, (iii) metal injection molding, (iv) machining, and (v) welding.

There are a variety of postmanufacture finishing processes that can be employed to improve the aesthetic appeal of titanium artworks. For example, titanium can be polished into a variety of finishes. The most popular finish is a more traditional titanium polish, but satin or brushed finishes are also options. In addition, titanium can undergo anodization [2]. Through heating or applying electric currents, depending on the temperature and length of heating time, the colors of anodized titanium can vary from yellow to pink or red to purple or black [2]. Titanium alloys that contain different alloying elements can display even more colors. For example, heating the Ti-5Al-5Mo-5V-1Cr-1Fe (wt%) alloy from room temperature to the range of 100–1000°C with an interval of 50°C can produce colors from silver white to light yellow to golden yellow to blue and then to light green and brownish black [3]. The color change on the alloy surface depends on the titanium oxides in the oxide layer. In this case, researchers have identified that "the silver white, yellow, and blue on the alloy with the oxidation temperature below 600°C were the results of TiO_2 white tint, TiO golden tint, and Ti_2O_3 blue color, respectively. The light green was the mixed color of TiO golden tint and Ti_2O_3 blue color in the oxidation temperature range of 600–700°C" [3]. This unique characteristic of titanium and titanium alloys can be utilized to add significant appeal to titanium artworks.

2 Examples of additively manufactured titanium artworks

2.1 Additively manufactured titanium artworks from RMIT University

Fig. 1 is an additively manufactured Ti-6Al-4V Larry Perkins Trophy. The details are specified as follows:

Name: Larry Perkins Trophy.

Client: Formula 1 Australian Grand Prix Corporation.

Event: Perpetual trophy for Melbourne 400 Supercars Race.

Design Director: Dr. Scott Mayson, RMIT School of Design.

Fig. 1 Larry Perkins Trophy made from carbon fiber and additively manufactured Ti-6Al-4V spokes.

Manufacturer: Advanced Manufacturing Precinct, RMIT University.

Specification: Width 500 mm × depth 500 mm × height 90 mm.

Production/Realization Technology: The Perkins trophy was designed in virtual reality consisting of 2×2 Carbon Fiber twill shell with five additively manufactured generatively designed titanium spokes that consist of an agent-based algorithm forming structures. Specifically, the five spokes, 20 nameplates, and the center-naming disk were manufactured in Ti-6Al-4V by selective laser melting (SLM, SLM250 machine) process. Those 20 titanium nameplates that visually mimic race car rear defusers are interchangeable over a perpetual lifetime where the drivers' names that win the Formula 1 Australian Grand Prix Melbourne 400 can be engraved over time. In addition, Fused Deposition Modeling (FDM) 3D printing was used to manufacture the backplate to hide internals and assist with the holding of the trophy.

In July 2018, the Larry Perkins Trophy picked up an international design award: A'Design Award [4].

The RMIT Hall of Fame Award in Fig. 2 is another additively manufactured example at Advanced Manufacturing Precinct, RMIT University. The detailed information is as follows:

Name: Hall of Fame Award.

Client: RMIT University International.

Event: Various International Awardings.

Design Director: Dr. Scott Mayson, RMIT School of Design.

Manufacturer: Advanced Manufacturing Precinct, RMIT University.

Specification: A varying agent structure that changes in size and density.

Function and Production Technology: The RMIT Hall of Fame Award is computationally designed as a varying agent structure that changes in size and density according to a set of particular geometric conditions. The resultant geometry was additively manufactured in Ti-6Al-4V alloy by the SLM process at RMIT's Advanced Manufacturing Precinct and hand-finished by Dr. Scott Mayson. The Award is awarded to the six

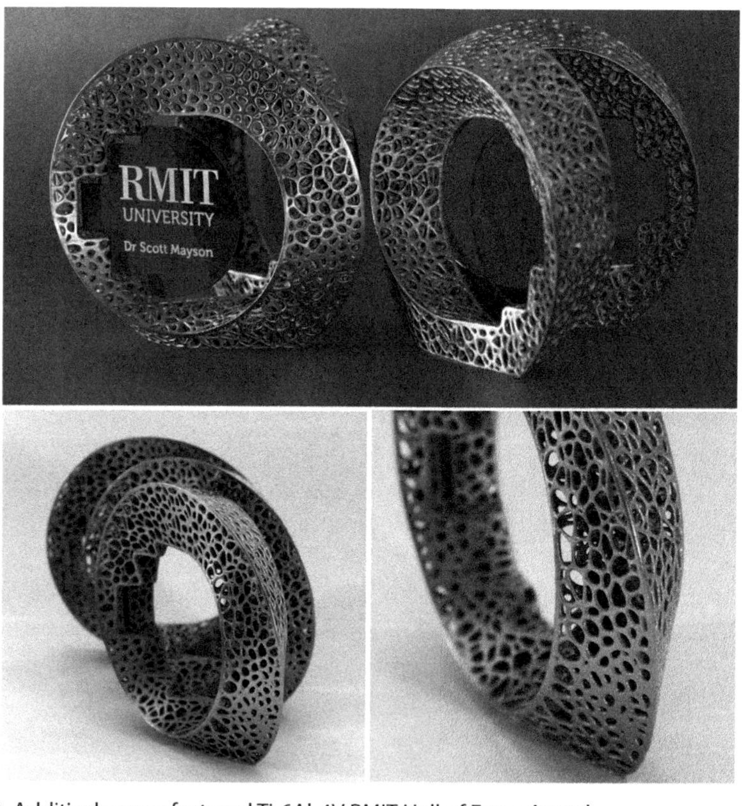

Fig. 2 Additively manufactured Ti-6Al-4V RMIT Hall of Fame Award.

highest-achieving Chinese and Global Agencies. RMIT University, as a Global University in Design, Technology, and Enterprise, commissioned Dr. Scott Mayson to design such a unique lightweight artwork that embraces the university's mission statement.

The RMIT University Mace (Fig. 3) was also 3D printed in Ti-6Al-4V by the SLM process in Advanced Manufacturing Precinct, RMIT University, and the detailed information is given as follows:

Name: RMIT University Mace.

Client: RMIT University.

Commissioned: RMIT Chancellor Dr. Ziggy Switkowski.

Event: Melbourne and Vietnam Ceremonial Occasions.

Design Directors: Dr. Scott Mayson, RMIT School of Design and Associate Professor Roland Snooks, RMIT Architecture and Urban Design Team.

Manufacturer: Advanced Manufacturing Precinct, RMIT University.

Specification: 1200 mm in length and weighing 1.6 kg.

Production/Realization Technology: The mace is digitally constructed via a multi-agent swarm algorithm that is digitally optimized for SLM in Ti-6Al-4V alloy without requiring support material. Additional algorithms were written to remove non-buildable geometry and replace it with a geometry the SLM could manufacture. The mace totals 1200 mm in length and weighs 1.6 kg. Due to manufacturing constraints, the mace was built in four sections of variable length with unique saddle joints

Fig. 3 RMIT University Mace additively manufactured in Ti-6Al-4V alloy by the SLM process.

as shown in Fig. 3. Subsequently, the sections were laser welded at each perimeter by Dr. Mark Edgoose, RMIT Gold & Silver-Smithing. Additional selective edge and primary surface polishing was completed by hand at RMIT University by Dr. Scott Mayson.

The next titanium artwork additively manufactured in Advanced Manufacturing Precinct, RMIT University is the Coat of Arms shown in Fig. 4.

Name: Coat of Arms.

Client: RMIT University.

Commissioned: RMIT Chancellor Dr. Ziggy Switkowski.

Event: Melbourne and Vietnam Ceremonial Occasions.

Design Directors: Dr. Scott Mayson, RMIT School of Design and Associate Professor Roland Snooks, RMIT Architecture and Urban Design Team.

Manufacturer: Advanced Manufacturing Precinct, RMIT University.

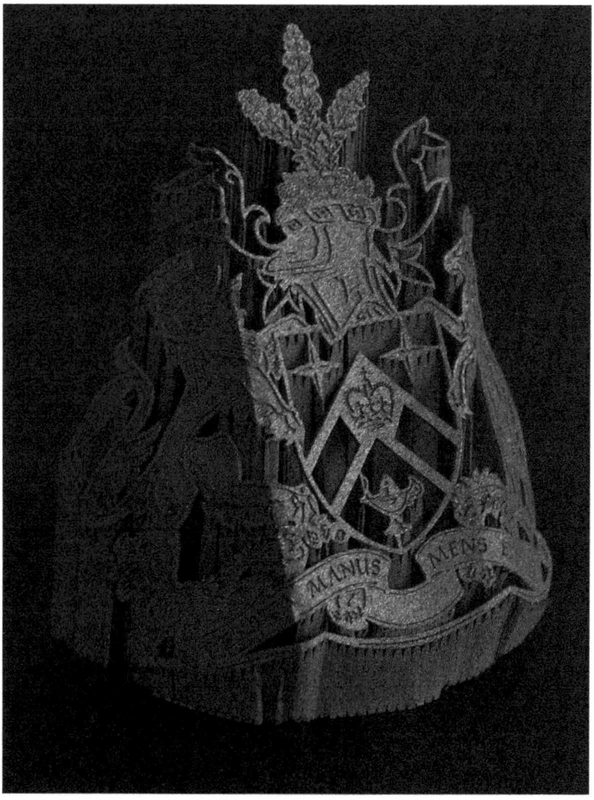

Fig. 4 Additively manufactured Ti-6Al-4V Coat of Arms by the selective laser melting (SLM) process.

Production/Realization Technology: The coat of arms is a digitally sculpted form derived from the original coat of arms oil painting at RMIT University. This 3D sculpted geometry was formed over the original digital skins of Melbourne & Vietnam Mace as shown in Fig. 3. It adorns the motto "Perita manus, mens exculta"—a skilled hand, a cultivated mind—which continues to be used by the university today. The coat of arms was similarly additively manufactured in Ti-6Al-4V at Advanced Manufacturing Precinct at RMIT University. The resulting form was hand-polished at RMIT, and Titanium Nitride coated at Sutton Tools—Surface Technology Coating Division.

2.2 Additively manufactured titanium artworks and jewelry from PROGOL3D®

PROGOL3D® is a leader in 3D printing of precious metals and innovative titanium jewelry. Fig. 5 displays a range of additively manufactured (by SLM) Ti-6Al-4V objects with complex design, wearability, hollow features, minimum thicknesses, and ultralightweight.

Fig. 5 Additively manufactured (by SLM) Ti-6Al-4V objects with complex design, wearability, hollow features, minimum thicknesses, and ultralightweight.

2.3 Additively manufactured wearable hollow/thin Ti-6Al-4V rings from PROGOL3D®

Fig. 6 shows two hollow Ti-6Al-4V rings printed by SLM. Specifically, the hug ring in Fig. 6B is hollow inside with a wall thickness of 0.3 mm. The two little holes allow Ti-6Al-4V powder extrusion for the hollow structure. The lower limit of the metal walls is given by the thickness of the single trace of laser melting, according to the minimum required robustness of the jewel. The minimum thickness of the melting trace depends on the type of laser printer, the construction parameters, and the type of metal powder [5]. For titanium alloys, the practical minimum thickness is in the range of 0.1–0.2 mm [5]. In addition, thanks to the local additive nature, the overall geometry of a design has little influence on the thickness and minimum sections of the jewel. This feature has allowed the manufacturer to successfully print rings with small wall thicknesses, down to 0.2 mm for extended areas of the raw jewel, as in the case of the printed

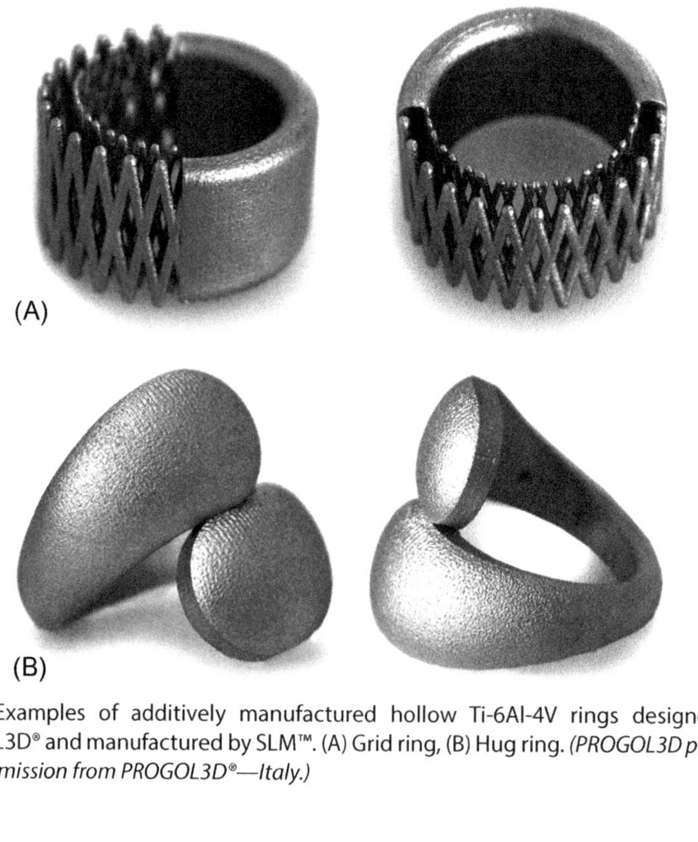

Fig. 6 Examples of additively manufactured hollow Ti-6Al-4V rings designed by PROGOL3D® and manufactured by SLM™. (A) Grid ring, (B) Hug ring. *(PROGOL3D projects with permission from PROGOL3D®—Italy.)*

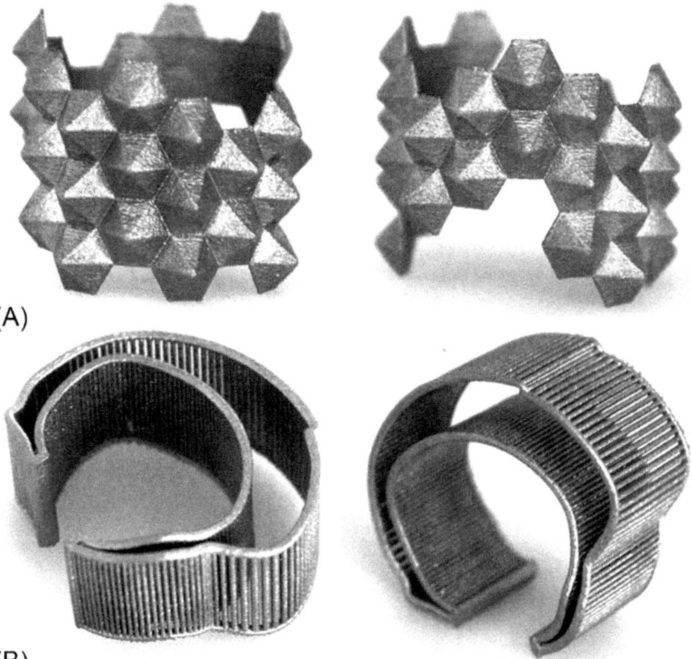
(A)
(B)

Fig. 7 Examples of additively manufactured Ti-6Al-4V rings with minimum thicknesses designed by PROGOL3D® and manufactured by SLM™. (A) Beehive ring, (B) Mat ring. *(PROGOL3D projects with permission from PROGOL3D®—Italy.)*

Ti-6Al-4V Beehive ring in Fig. 7A. It has also allowed items with grid or mat structures, such as the grid rings in Figs. 6A and 7B, obtainable only by additive manufacturing.

2.4 More additively manufactured Ti-6Al-4V artworks from PROGOL3D®

Fig. 8 gives more examples designed by PROGOL3D® and manufactured by SLM™ showing the perfect integration of creative design and advanced manufacturing technologies. Specifically, the bean bracelet in Fig. 8E arises from a concept of contrasts, conjugated in the form of surfaces. It is in spired by the concept of rafting in which the Dunes feature one completely smooth side, and the other one is texturized by small waves due to the sliding of the sand grains. The designer has translated this project into a jewel, creating a ring that on one side is completely polished and on the other will not undergo any further finishing after SLM printing.

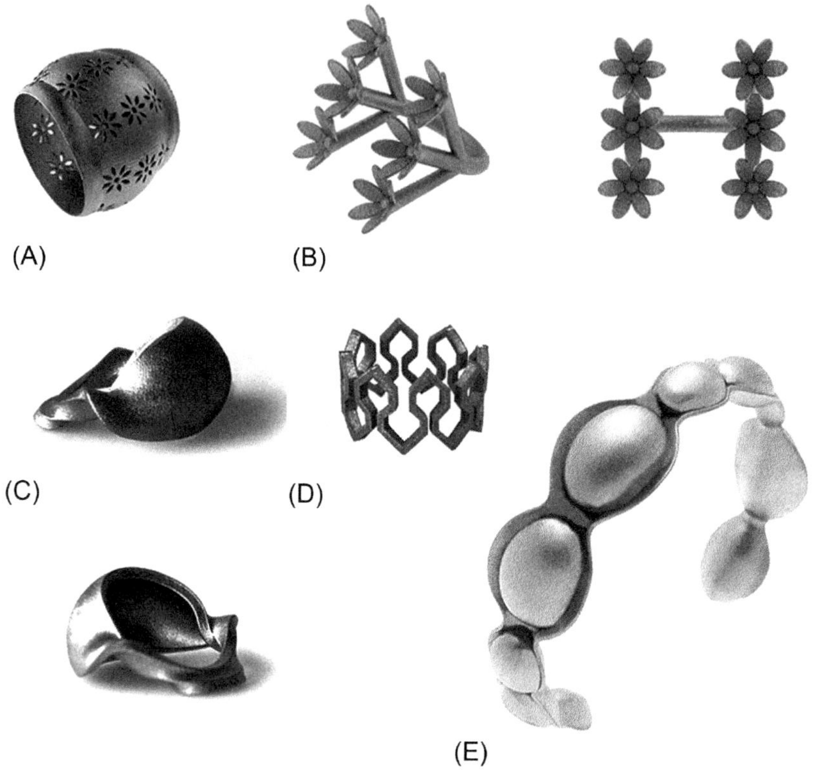

Fig. 8 Examples of additively manufactured ultralight Ti-6Al-4V artworks with complex designs designed by PROGOL3D® and manufactured by SLM™. (A) Pirovre ring, (B) Geranium ring, (C) Tissue ring, (D) Octopus ring, (E) 45 Ti ring. *(PROGOL3D projects with permission from PROGOL3D®—Italy.)*

2.5 Other selected additively manufactured Ti-6Al-4V artworks

Additively manufactured titanium artworks have also become a fancy choice of souvenir and personalized gifts. Fig. 9 shows a few such examples. Fig. 9A is an additively manufactured terracotta warrior, 110 mm tall, manufactured in Ti-6Al-4V using the selective electron beam melting (SEBM) process by Xi'an Sailong Metal Materials Co., Ltd., China, where Xi'an City is the hometown of the Terracotta Army of Emperor Qin Shihuang. The twisted Ti-6Al-4V lattice ring (Fig. 9B), 60 mm high, was also manufactured by Xi'an Sailong Metal Materials Co., Ltd. using the SEBM process. It demonstrates the manufacturability of SEBM toward lattice structures [6].

The Ti-6Al-4V signature pen in Fig. 9C with a length of 126 mm and a diameter of 11 mm, was 3D printed in two pieces using the SLM process

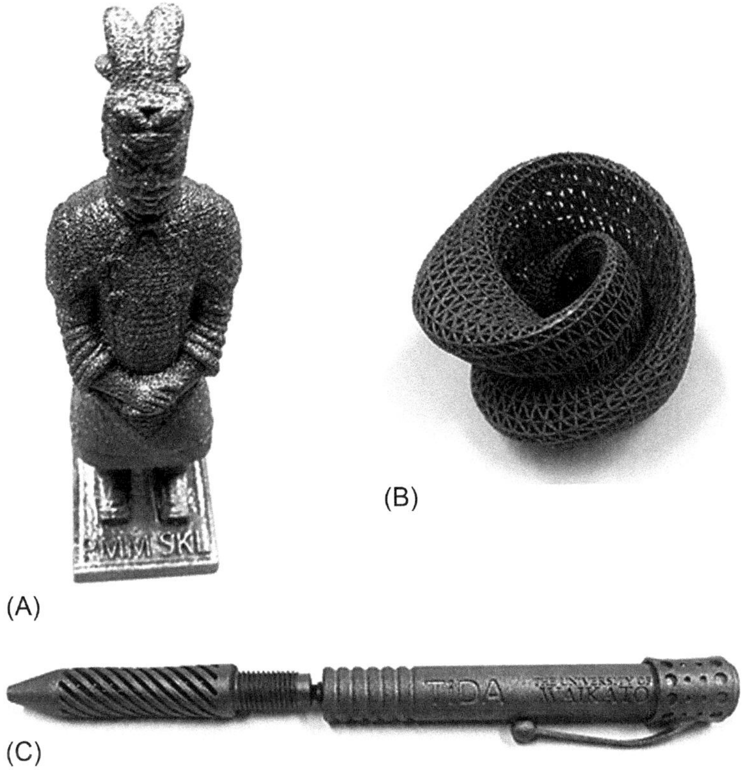

Fig. 9 (A) Additively manufactured Ti-6Al-4V terracotta warrior and (B) twisted Ti-6Al-4V lattice ring by selective electron beam melting (SEBM). (C) Ti-6Al-4V signature pen additively manufactured by SLM process. *((A, B) With permission from Xi'an Sailong Metal Materials Co., Ltd., China; (C) With permission from TiDA, New Zealand.)*

by The Titanium Industry Development Association (TiDA), New Zealand, collaborating with the University of Waikato. It is light, shiny, and personalized illuminating the potential of 3D printing in the growing landscape of customized design and manufacture.

3 Concluding remarks

Titanium and titanium alloys offer medium-to-high strength, sufficient ductility, light weight, durability, biocompatibility, and modifiable color that designers and artists appreciate. On the other hand, the rapid development of metal additive manufacturing technologies enables designers to take full advantage of their imagination in the creative design of artworks. The selected examples discussed in this chapter demonstrate the impressive creativity of the designers and the exceptional manufacturability of metal additive manufacturing, either by selective laser melting or selective electron beam melting. It looks promising that the needs and market for 3D printed titanium artworks will continue to grow in the future.

Acknowledgment

This project is supported by the Australian Research Council (ARC) through LP140100608.

References

[1] A. Balyanov, J. Kutnyakova, N.A. Amirkhanova, V.V. Stolyarov, R.Z. Valiev, X.Z. Liao, Y.H. Zhao, Y.B. Jiang, H.F. Xu, T.C. Lowe, Y.T. Zhu, Corrosion resistance of ultra fine-grained Ti, Scr. Mater. 51 (3) (2004) 225–229.
[2] C. Wadhwani, M. Brindis, M.T. Kattadiyil, R. O'Brien, K.-H. Chung, Colorizing titanium-6aluminum-4vanadium alloy using electrochemical anodization: developing a color chart, J. Prosthet. Dent. 119 (1) (2018) 26–28.
[3] W. Peng, W. Zeng, Y. Zhang, C. Shi, B. Quan, J. Wu, The effect of colored titanium oxides on the color change on the surface of Ti-5Al-5Mo-5V-1Cr-1Fe alloy, J. Mater. Eng. Perform. 22 (9) (2013) 2588–2593.
[4] https://competition.adesignaward.com/design.php?ID=67337, (Accessed January 7, 2019).
[5] D. Zito, V. Allodi, P. Sbornicchia, S. Rappo, Why should we direct 3D print jewelry? A comparison between two thoughts: today and tomorrow, 2017. Available from: https://www.progol3d.com/Documents/SFS2017_GB.pdf, (Accessed May 25, 2019).
[6] X.Z. Zhang, H.P. Tang, M. Leary, T. Song, L. Jia, M. Qian, Toward manufacturing quality Ti-6Al-4V lattice struts by selective electron beam melting (SEBM) for lattice design, JOM 70 (9) (2018) 1870–1876.

CHAPTER 11

Development of titanium consumer products: From titanium art to tumblers

Takuji Horie
HORIE Corporation, Niigata, Japan

Contents

1 Technology development method

When I was 38 in 1984, I quit the company I had been working for to turn over a new leaf, and I then established a new company with an ambition. I founded the new company with no one but my wife, which was my 10th place of employment.

I think my approach to technology development is different from other people. First, I do not investigate data on the metal in advance. Instead, I begin by processing the metal on a trial basis and learn about it through trial and error. I face various metals in a really simple way and finally manipulate them at will. That is how I achieved development of titanium etching and anodic oxidation technologies, in-house construction of a processing plant, development of diffusion bonding technology, in-house production of mass production equipment, and development of various titanium consumer products and a sales network in Japan. However, needless to say, I could not reach my goal without proceeding with in-house development of processing equipment. I do not understand how I could succeed in such development, but I think it all began with the fact that I was full of curiosity and had been lucky enough to find a job at a metal processing company after graduating from junior high school.

Titanium for Consumer Applications
https://doi.org/10.1016/B978-0-12-815820-3.00004-6

My development method is simple and clear. If there are any problems, I will carefully observe the work method of others, devise how to improve them, and give them a try without hesitation. You may think that you can solve problems by referring to reference materials, but such an easy way of troubleshooting is not worthy of being called development. It is far from technology development, unless the work efficiency is improved by at least 20% to 100%, for example, by drastically changing the work method.

Of course, everyone fails at first, but I kept taking steps forward believing that failure is a stepping stone to success. Therefore, I would surely succeed after enough failures, as it is inevitably necessary to overcome barriers to achieve development, and necessary to repeatedly overcome those same barriers to achieve even truer development.

Repeated failures will help in accumulating experience, increasing the probability of success, and accelerating the speed of day-by-day development.

For example, my metal diffusion bonding method was accidentally discovered during heat treatment of titanium using an electric furnace. At that time, I only thought, "What a problem! How can I prevent it?" However, after about half a year of study, it became fully controllable, such as joining and separating the target parts easily in the same furnace.

By stacking and diffusion-bonding 16 etched titanium foils with this technique, we have commercialized a necklace top with a fine mesh visible through in a 2 mm-thick layer (Photo 1).

As an aside, during the presentation of this product, a professor emeritus of Tokyo University told me that my technique was diffusion-bonding, and

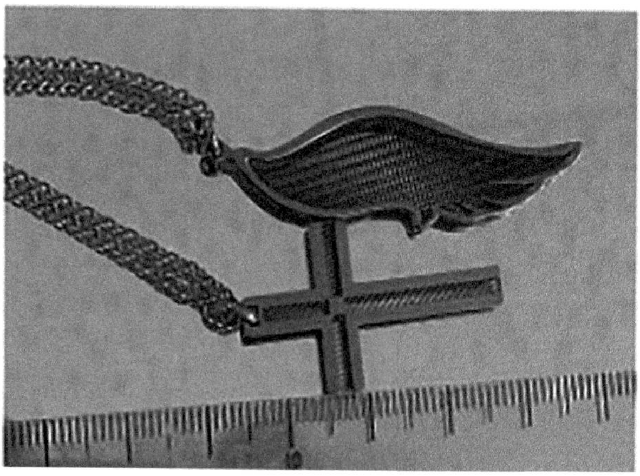

Photo 1 Necklace top made by stacking and bonding 16 titanium foils (with a mesh seen through it).

Photo 2 In-house manufactured electric furnace (Inner size: 700 mm × 800 mm × 3000 mm).

he asked me how I had succeeded in this processing. I remember when I replied that I had begun by testing a small stainless steel pan and commercialized it by producing production equipment in-house; he was speechless with my answer (Photo 2).

We have also developed anodic oxidation coloring technology for titanium using the same approach.

The "Horie's Basic Titanium Color Chart (124 Colors)" is prepared for controlling coloration of titanium in-house, and each of the 124 colors in this chart is divided into 40 counts and can be digitally reproduced by 100% (Photo 3).

2 Road to start-up

While I was enjoying the environment where I could learn various metal processing methods and improve my skills while repeatedly changing jobs, the forged products manufacturer (commonly called a blacksmith) I was working for decided to establish a new electroplating business. So I was fortunate enough to participate in building up a new electroplating factory with all employees starting from improving the equipment at hand. In those days, the electroplating industry had a very poor yield because of workers'

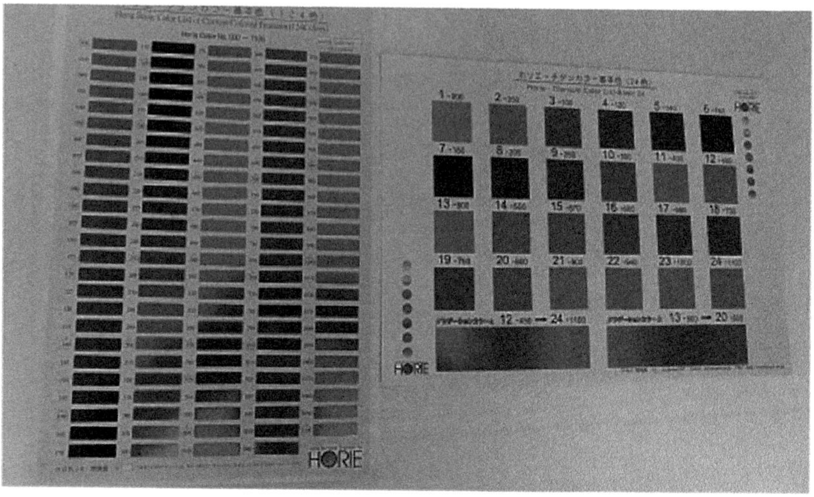

Photo 3 In-house color chart. Prepared in two versions: 24 colors and 124 colors (theoretically, 124×40=4960 colors).

lack of knowledge, resulting in a number of defective products. Our company was also facing a crisis of survival, mainly because of a lack of chemical knowledge.

A plating engineer from Tokyo told me that plating is chemistry, and if I had the right knowledge, I could eliminate defects. He gave me a Pocket Plating Dictionary, saying, "Memorize this." I have taught myself since then and even came to be able to translate chemical formulas.

About a year later, plating defects almost disappeared, our company management stabilized, and I came to be more and more fascinated with the depth of chemistry. I was immersed in chemistry and also developing original alloy plating and others, but I never imagined at that time that my involvement in electroplating would have a great impact on the subsequent development of titanium.

3 Start of titanium consumer product development

In early days, we were giving advice to local metal products manufacturers, helping them solve their problems, and contracting processing work. Therefore we received various consultations. Consequently, our business changed every year, and around 1988, our titanium product development suddenly started.

One day when the snow began to melt in this snowy country, two men visited our company. They were development engineers from a major metal

Photo 4 The first cylindrical part I formed. 0.5 mm plate × ϕ175 × 170 (H) (around 1988).

manufacturer, and they had thin plates of titanium with them. They asked me to process the plates into a cylindrical shape.

They asked, "Do you know the material called titanium?" I promptly answered "Yes, I do," and completed the cylindrical parts they requested in 2 hours after they left. That was the start of my titanium product development (Photo 4).

I had used titanium anode baskets for plating baths while I was engaged in electroplating, as previously mentioned, so I had a knowledge of titanium, such as it had strange properties, was very expensive, and extremely hard.

Incidentally, I heard that they had not been able to make this cylindrical part even after trying for 2 years. They probably had studied the properties of titanium in too much detail and been too cautious to proceed to the next step, I guess.

I sent them the sample products the next day after completion, and before long, their chief titanium engineer came to me and asked me for development. Their request was to develop consumer products using titanium.

Since then, I have been devising and producing products directly linked to people's lives and also developing processing technology. This year, 2019, is my 31st year, and I hope to continue development for the rest of my life.

Our first trial was development of cups. Naturally, we faced a series of failures in processing them using drawing dies for stainless steel. We could only form a few, even when using 1000 kg of titanium.

However, I did not want to give up at such an early stage. So I reversed my way of thinking and examined why a few had been formed successfully.

Photo 5 Outdoor mugs. 0.5 mm plate × ϕ70 × 170 (H) (around 1990).

Then we examined the type of lubricating oil, processing speed, annealing method, and others factors, then managed to learn forming technology after a year of struggle (Photo 5).

Next, I started to develop coloration development technology applying the anodizing method in which electroplating techniques were greatly helpful. In a sense, preparation of chemicals is my forte, so the development was fun for me.

4 Development of titanium art

We had developed titanium mugs, climbing stoves, tableware, earrings, and other products one after another. One day, I happened to see a picture in a painting shop and thought that it might be possible to form a color gradation by means of titanium coloration, just like airbrushing. So I decided to develop paintings making use of titanium coloration.

To tell the truth, my experience in plating was fully utilized in this development. While I was engaged in plating for further differentiation of our plated products, I tried to make beautiful decorative panels by drawing patterns on panels by applying several types of plating.

We purchased 42 types of resist ink available on the market at that time and performed comparative tests examining their resistance to gold, silver, copper, nickel, chromium, and other metals. None of them could withstand a strong acid, such as chromium plating solution, under high temperature and high current density conditions.

It is absolutely necessary that resist ink is resistant to chemicals and able to peel off easily. However, we were lucky enough to find so-called tricks to overcome such problems, and it led to subsequent development.

To differentiate titanium art from ordinary paintings, we devised a technique to make a fish-scale pattern glitter when seen from various angles by melting the pattern with an acid wash, then produced about 12 types of art (Photo 6).

Applying this technique, we also succeeded in cutting thin plates into shape by applying resist masking to both sides of a plate then melting it with an acid wash, but even at that point, I did not know it was the technique generally called etching. Being able to etch titanium was a milestone for our company. By applying etching techniques, we can easily cut parts into various shapes and in large quantities. Being able to process parts without using molds or dies enabled us to accelerate subsequent development of consumer products as well as the evolution of decoration technology for titanium products.

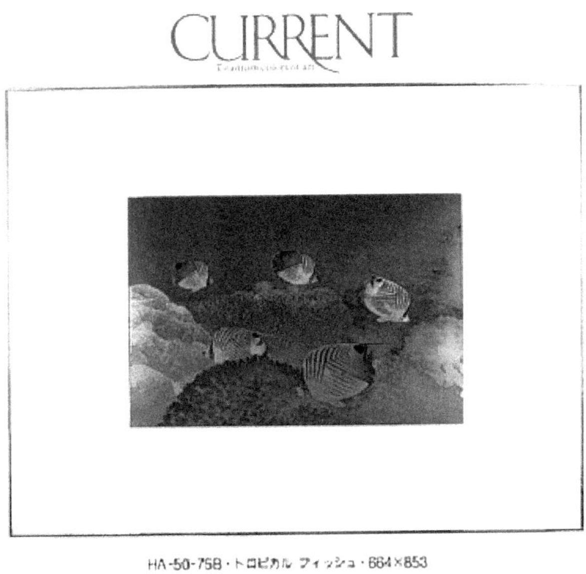

Photo 6 Titanium art. Lovely Fish (664 mm × 853 mm) (around 1992).

5 Development of double-wall titanium tumblers

The development of tumblers can be said to have derived from etching techniques. Etching is a technique to cut out small parts. We produced a skeleton model of the Zeppelin airship by cutting a 1 mm-thick titanium plate into shape by etching.

I planned to sell it by mail order in Japan, but I noticed that we had produced it without obtaining permission from the manufacturer of Zeppelin. So I visited the Zeppelin Museum in Germany, carrying the model with me, and explained the background of production. I finally obtained sales permission (Photo 7).

I was surprised at the fine taste of German beer during this business trip, and I seriously wondered how I could manage to enjoy it even in Japan, then I came to the conclusion that titanium would not change the taste because it is used as an implant material.

Of course the most important condition for enjoying the taste of beer is the taste itself, but the temperature of the beer and material of the container are important as well. Titanium is not soluble in food ingredients, and therefore it does not change the taste of food or drink. So what about other metals? After tasting with cups of various materials, I found that metals other than titanium cause changes in taste specific to each.

With regard to the relation between temperature and taste, beer manufacturers say that the optimum temperature to enjoy beer is around 5°C.

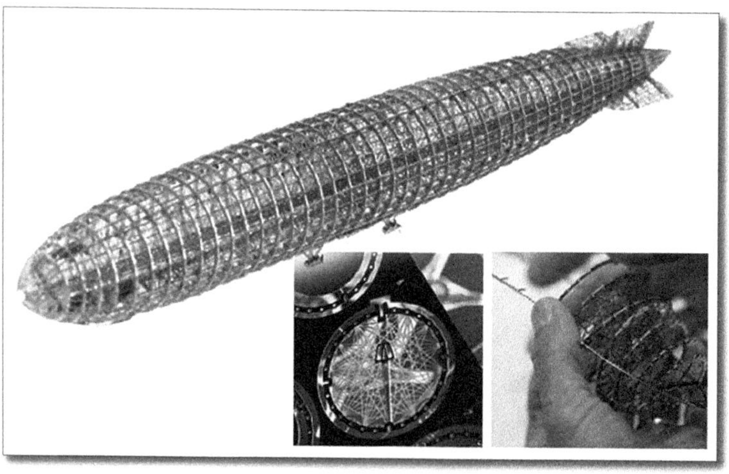

Photo 7 Zeppelin 1/500 model. $\varphi 60 \times 465$ (L) (around 2005).

Then I thought that it would be nice to make a tumbler that makes beer at the optimum temperature feel even colder, and I was convinced that a double-wall structure could be technically possible.

We repeated prototyping, believing that a titanium tumbler having a double-wall structure would make beer at the optimum temperature feel colder to the drinker due to the difference between the temperature the drinker feels by hand and that by tongue. However, our initial prototypes had a TIG-welded rim, which was inevitably thick, and therefore they could not provide good taste as we had expected.

Then we changed the welding method of the rim to resistance welding and succeeded in reducing the rim thickness to 0.7 mm. This thickness is less than 0.8 mm, which is the thickness of usuhari glass, and I am convinced that this is the minimum thickness that will not cut the user's hands.

By repeating tasting with various beers and finally adding an anodizing treatment, we succeeded in providing our intended taste of beer. We repeated tasting, also asking our fellow beer lovers to try our tumblers; we promoted catalog sales, but it did not sell well at a price of 10,000 yen. Because this product had originally arisen from my personal idea or taste, I thought, "That's fair enough." However, in the Spring of 2011, I became acquainted with a department store wholesaler, and he decided to sell our tumblers at large department stores in Tokyo and Kyoto. After that, our sales channels have spread to department stores nationwide. This is how our double-wall titanium tumblers arisen from my personal idea or taste were fortunately recognized in the Japanese market (Photo 8).

Photo 8 Double-wall titanium tumblers (Tenmoku). 270 and 350 mL types.

6 Fight for sales

One of the biggest barriers to developing titanium consumer products was how to make them sell well. For 5 to 6 years after we started development, even the existence of titanium products was not known, so we could only feel our way forward.

After establishing anodizing technology, we developed and commercialized titanium art, which sold to some extent as metal manufacturer souvenirs. When we commercialized our titanium art, a metal manufacturer taught us how to exhibit and publicize our products at exhibitions in Tokyo. Since then, we have been doing so for more than 25 years (Photo 9).

What is most important for making our products remembered by customers at exhibitions is to keep exhibiting them at the same place at the same exhibition. By doing so every year and proposing new products every time, it inevitably attracts customers' interest. Then we can start conversations and get information on what products are in demand.

Around our third year of exhibition, customers started to talked to us, and since around our sixth year, our products have been adopted little by little, which keeps driving us to the next proposal of new products. We must put the saying "Little and often fills the purse" into practice.

Photo 9 Our exhibition in early days. Tokyo Gift Show 2000 Fall (at Tokyo Big Sight).

Photo 10 Exhibition in Spring of 2019 (at Tokyo Big Sight).

Our development method of titanium products was to commercialize whatever products, exhibit them at exhibitions, listen to customers' opinions, and reflect on them toward our next products. I think there has been a lot of wasteful or inefficient effort, but we had no choice in doing so to make our comfortable-to-use products recognized and finally bought by many people (Photo 10).

7 Conclusion

We do not mind and welcome others imitating our titanium products. The market for titanium consumer products is still extremely small, so we hope more and more companies will give it a try and participate in the development of titanium products.

Titanium consumer products have been finally accepted by the Japanese market through the sales of outdoor goods and tumblers. The current market size of titanium consumer products in Japan is assumed to be around 4 billion.

Nevertheless, it is certain that titanium consumer products are fewer in types and number compared to other daily metal products, so we hope many companies will agree with and participate in the development of titanium consumer products, where there is still room for further development.

CHAPTER 12

Bone regeneration on implants of titanium alloys produced by laser powder bed fusion: A review

I. Yadroitsava[a], A. du Plessis[b], I. Yadroitsev[a]

[a]Department of Mechanical and Mechatronics Engineering, Central University of Technology, Bloemfontein, South Africa

[b]Physics Department, University of Stellenbosch, Stellenbosch, South Africa

Contents

1 Introduction

Bone regeneration on biomaterials is a topic of increased interest due to the possibility of creating implants that existing bone can attach to. Titanium alloys are very suitable for this purpose and have been proven to be biocompatible, while also having suitable mechanical properties [1]. Moreover, titanium alloys are one of the most widely investigated materials for metal additive manufacturing (AM), which means that custom-designed implants can be manufactured reliably by AM. The AM technology best suited to this

Titanium for Consumer Applications
https://doi.org/10.1016/B978-0-12-815820-3.00016-2

is powder bed fusion; laser powder bed fusion (LPBF) in particular holds some advantages. In this chapter, we investigate the current state of the art for the production of pure Ti and Ti6Al4V implants by LPBF, focusing on the requirements and capabilities for osseointegration. The first section discusses bone architecture and requirements for bone implants in general, identifying different requirements for different types of implants, and describing the various bone growth processes in more detail. This is followed by a section on surface structuring; the surface morphology and chemistry strongly influence the initial stages of bone growth, making it critical to the success of such implants. The next section describes porous structures, also known as lattice structures, which allow bone in-growth and attachment. This section describes various requirements for such lattices (pore size, lattice design, etc.) in terms of bone growth and summarizes what has been achieved thus far. The following section describes the mechanical properties of titanium lattice structures of various types that have been produced by LPBF. The next two sections describe in detail successful in vitro and in vivo experiments, respectively, for bone growth on LPBF titanium alloys. Finally, a discussion section summarizes the current state of the art and highlights requirements for future research efforts.

2 Bone structure and growth

Bone has a complex hierarchical structure. Each level performs diverse mechanical, biological, and chemical functions. The hierarchical levels of bone include macroscale, microscale, submicroscale, nanoscale, and subnanoscale features (Fig. 1) [2]. On the macroscale level, cortical (compact) bone has

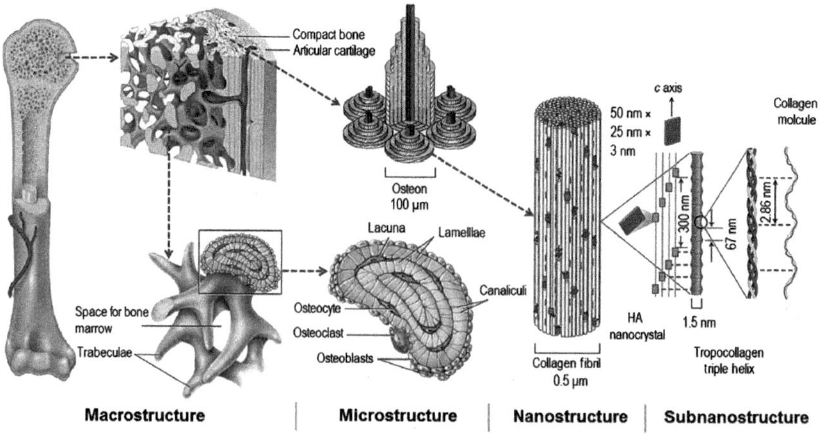

Fig. 1 Hierarchical structural organization of bone [2].

a high density and acts as a high-strength support structure, while internal trabecular bone is more elastic yet still providing strength and protection, with interconnected pore spaces for vascularization and flow of nutrients for continued bone remodeling. At the microscale, although cortical bone consists of lamellar structured osteons and haversian canals, spongy cancellous bone, which is also lamellar bone, is composed of an interconnected porous network of trabeculae. Nanoscaled bone structures include collagen fibers, whose main components are well-organized collagen fibrils [2].

The cortical bone thickness and the density and pore structure of the trabecular bone may vary significantly by location in the body and even within one bone, depending on local requirements. Near the joints, the trabecular bone is denser due to varying mechanical load requirements (strength and angles of loading), whereas in the central part of the bone, for example, the trabecular bone is less dense and less isotropic (more directionally aligned) [3]. This is illustrated in Fig. 2 using 3D microCT images of different areas of trabecular bone in the human femur.

Bone regeneration can be fulfilled by three processes: osteogenesis, osteoinduction, and osteoconduction. Osteogenesis is the process of bone formation. Osteoinduction is the process of transforming undifferentiated mesenchymal cells into osteoblasts and the formation of ectopic bone in vivo. Osteoconduction is the process of bone growth on bioinert or

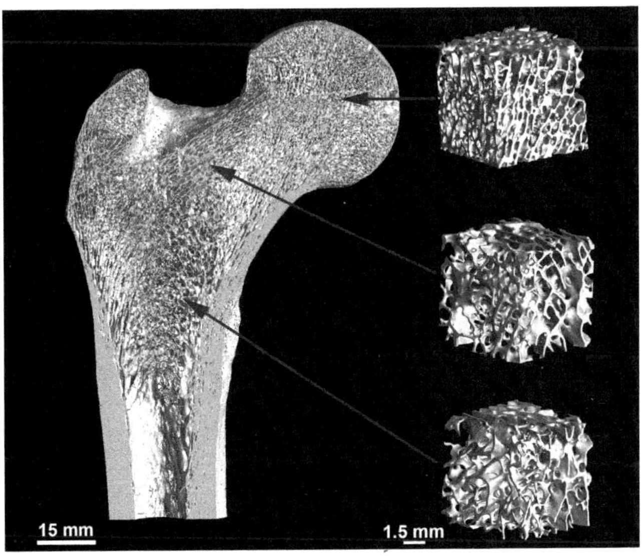

Fig. 2 Variation of trabecular bone structure by location, in this example, from human femur 26-year-old male.

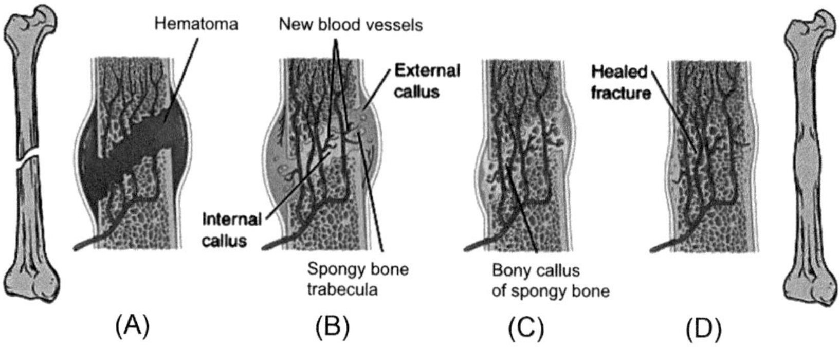

Fig. 3 Bone regeneration processes occurring due to bone fracture in absence of an implant: (A) break event followed by blood clot formation; (B) new blood vessels form, followed by osteoinduction—differentiation of cells into osteoblasts and formation of ectopic bone; (C) osteogenesis or formation of new bone through osteoinduction, creating new trabecular and cortical bone; (D) completed process and healed fracture [4].

physiological matrices, that is, the capability to allow new cell colonization, bone in-growth, and blood vessel formation (vascularization).

The bone healing process in absence of an implant involves multiple phases of healing: (a) in the inflammatory phase, a blood clot (hematoma) forms around the fracture; (b) in the reparative phase, granulation tissue and fibrin matrix form, new blood vessels are created, and differentiation of cells into osteoblasts and formation of ectopic bone occur; (c) osteogenesis or new bone formation occurs through osteoinduction, creating new trabecular and cortical bone; the remodeling phase can be very long—from months to years [4]. The process is better understood by an example of bone fracture and healing processes shown schematically in Fig. 3.

In the event of larger bone defects, the gold standard is still autografting (removing bone from another site on the same patient to aid in repair of the bone defect site) [5, 111]. However, this is usually limited due to the availability of only small sizes of graft material, and the possibility for complications and pain for the patient. The potential to use manufactured implants has prompted wide investigation of various biomaterials for this purpose. The additional advantage of produced and engineered implants is the possibility to tailor the exterior shape of the implant to perfectly match the bone defect site and even tailor the properties of the material itself using porous metal structures, which can further aid bone in-growth and attachment [6].

A recent review paper [7] on the use of various biomaterials in bone defect healing describes the bone healing process in more detail and describes the requirements for "ideal" biomaterials for this purpose. It mentioned that 3D printing has reached the clinical level but needs to consider all requirements for an ideal biomaterial. Martinez-Marquez et al. [8] described requirements or "critical quality attributes" for ideal custom 3D-printed bone prostheses and scaffolds as the summation of dimensional, mechanical, biological, and physicochemical functional characteristics. According to the previously described processes, these requirements might vary depending on the stage of bone growth, the location of the bone defect, or other factors; it is therefore a complex set of requirements.

The general consensus currently is that, for Ti alloys, the creation of a porous structure allows reduction of the elastic modulus of the implant material to closer match that of the bone alongside it. If this is not the case, the stress-shielding effect causes bone loss, loosening of the implant, and failure of the process. Besides lowering the elastic modulus, the porous nature allows for bone in-growth into the structure, further strengthening the bond between implant and existing bone. This process is thought to depend on pore size for bone growth and osseointegration. The interconnected nature of engineered porous lattice structures makes it an ideal material allowing for this process, with sufficient space for vascularization to occur. The details of different lattice types, their mechanical properties, and their surface properties are discussed further in subsequent sections.

3 Implant surfaces and osseointegration

3.1 Surface topography for bone implants

Bone may form by two different processes: contact and distance osteogenesis. For successful bone replacement, an ideal material has to facilitate high vascularization and direct osteogenesis on one hand, and promote osteochondral ossification on the other hand [9]. Biofunctionalization of the implants can be provided by different physical and chemical ways.

Nanosize, submicro, and microsurface patterning, that is, changing the topology and chemistry of the implant surface, influences the osteoblast adhesion, differentiation, orientation, and final osseointegration [10–17]. Webster and Ejiofor [18] studied 90%–95% dense compactions of Ti and Ti6Al4V from nanopowders with sizes of 0.2–2.4 μm and a micropowder with powder particulates >7.5 μm. It was found that human osteoblasts preferred surfaces with nanometer topology features (obtained from compactions of nanopowders).

According to Gui et al. [19], surface patterns for different bone implants can be classified as partially ordered, strongly ordered, and surfaces that have hierarchical patterns similar to bone features. Because ordered patterns lead to enhanced metabolic activities and osteogenic activity, and with regard to the high repeatability of the results, research interest to in vitro and in vivo studies of cell response for these structures is growing. To date, there are no proven and clear conclusions about the optimum geometry and scale of the required surface topography because micro-, submicro- and nanoscale topography all influence the cell response [19].

3.2 LPBF objects surface properties

The LPBF process has an inherent surface topography (or roughness), which may vary depending on the powder size used, the laser processing conditions, and the build angle (for example, for the struts in the lattice cellular implant). Due to the wide variety of systems used and lack of characterization methods for surface roughness of complex shaped implants, the effect of typical AM surfaces on bone regeneration has not been investigated in detail before, but it is expected that a rough surface is conducive to bone growth. An example is shown in Fig. 4 for a 10-mm solid cube from recent work advancing the standardization of testing procedures using X-ray tomography [20] and specifically for measuring the surface roughness using

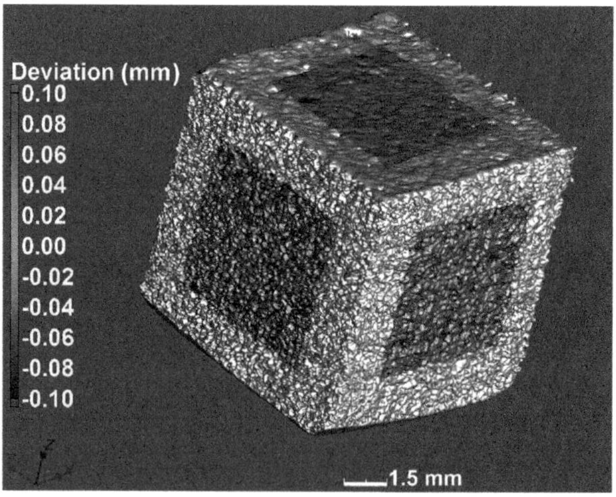

Fig. 4 Inherent surface roughness of the LPBF process. In this case, surface topography is shown on an as-built 10-mm cube of LPBF Ti6Al4V on two vertical surfaces and top surface. *Z* is the building direction. Top surfaces are usually much smoother than side surfaces, and this is clearly shown here. MicroCT data of this example was obtained from a recent round robin study [20, 21].

tomography [21]. For solid Ti6Al4V ELI samples that have been built at 30-μm layer thickness with an EOSINT M280 system, the average roughness was $R_z = 120–135$ μm for the side surface, and $R_z = 30–60$ μm for the top surface [22]. Clearly the surface properties of as-built surfaces might be conducive to bone growth in the as-built condition, but this requires further investigation. Also, the surface properties might vary depending on process parameters, powder size, scan speed, and hatch spacing, among various other parameters. As well, high roughness is a stress concentration factor accelerating crack nucleation, which influences the fatigue properties; this is important for load-bearing implants working in corrosive media, such as in the human body.

The inherent rough surface observed in Fig. 4 also transfers to lattice structures, which can be crucial for their application for bone in-growth, as surface modification techniques such as sandblasting and similar techniques will not penetrate the lattice structure, and the inherent process roughness may be one of the few ways to create a suitable rough surface on the inside of a lattice for implants. An inside view of a lattice sample is shown in Fig. 5, where the rough struts can be seen in this cube lattice with strut widths of 0.8 mm and spacing of 2 mm; the microCT data was obtained as part of a study that will be described in the next section [23].

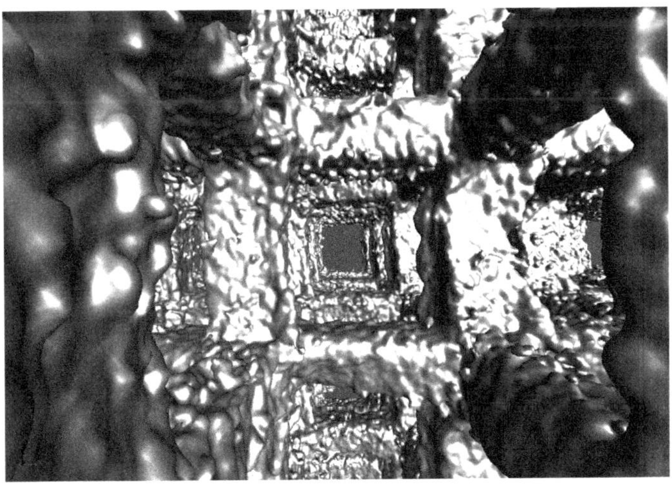

Fig. 5 Rough surfaces on the inside of a lattice produced by LPBF. Inside view by microCT. *(Data from A. du Plessis, I. Yadroitsava, D. Kouprianoff, I. Yadroitsev, Numerical and experimental study of the effect of artificial porosity in a lattice structure, in: SSF 2018—The 29th Annual International Solid Freeform Fabrication Symposium—An Additive Manufacturing Conference, Austin, Texas, August 13–15, 2018.)*

3.3 Functionalization of implant surfaces

Aside from the inherent surface roughness, various postprocessing methods can be potentially applied for cleaning and surface finishing of implants, which can change their properties. Xu et al. [24] evaluated in vitro and in vivo sandblasting, orderly arranged nanotubes, disorderly arranged nanotopology, acid etching surface modification, and as-built LPBF Ti alloy implants. It was found that micro-/nanotopography enhanced bone response in terms of cell morphology, proliferation, differentiation, bone implant contact, and bone bonding interaction [24]. Similar to Tsukanaka et al. [25], it was concluded that, for early mechanical stability, a rough surface is beneficial, but for osteoblast differentiation and bone formation, the surface has to be undergo a bioactive treatment. Park et al. [26] studied the effects of the Ti implant surface microtopography obtained by hydroxyapatite grit-blasting on the adhesion, proliferation, and differentiation of osteoblast-like cells. It was suggested that early osteoblastic cell differentiation in contact with modified surfaces could enhance bone healing at the early stage.

On the whole, approaches for producing multifunctional Ti-based surfaces for implantation are (1) inorganic coatings, (2) chemical surface treatments, and (3) functionalization strategies coupled with organic coating [16]. It was shown that mechanical integration (i.e., osseointegration) mutually with chemical bonding between bone and implant (osseocoalescence) can exhibit resistance to both shear and tensile loads [27]. Immobilization of bioactive molecules on the surface of the implant can be done by physical adsorption, chemically covalent bonding, or biomimetic incorporation of a bone-like mineral coating by surface activation after soaking the implants in special solutions [28, 29]. For example, it was shown that a fluoride-modified titanium surface is effective for osseointegration because the fluoride-modified titanium surface increased osteoblast proliferation, induced high-level expression of osteogenic markers, promoted bone mineralization, and enhanced bone formation [30]. Mg incorporated into fluoridated hydroxyapatite coating of dental implants showed a significant stimulating effect on osteoblastic cell responses [31]. Embedding silver nanoparticles into an oxide surface layer with plasma electrolytic oxidation in Ca/P-based electrolytes stimulates bone tissue regeneration and osseointegration of Ti6Al4V porous implants, preventing infection [31a]. The functionalization of LPBF implant surfaces requires additional accurate studies and certification for the production of multifunctional Ti-based implants that are reliable and promote rapid healing and long-term successful functioning.

4 Porous structures and osseointegration

Karageorgiou and Kaplan [32] analyzed biomaterial scaffolds for bone tissue engineering in vitro and in vivo. Different porous materials such as ceramics, metals, polymers, and composites show different influences on bone growth. Thus, similar approaches to all materials and porosity cannot be suggested as a general guide. Also loaded/unloaded implants can require different voids in terms of volume and shape. It was reported that the mean and 95% confidence interval for tensile strength of cortical bone in the longitudinal direction is about 135 [104;166] MPa; 205 [170;239] MPa in compression; and elastic modulus is 17.9 [10.1–25.7] GPa [17]. In the transverse direction, compact bones show 30%–70% lower strength and elastic modulus. Mechanical properties of cancellous bone vary both longitudinally and from one bone to another, and average values for compression strength varies from 2 to 5 MPa [33]. For comparison, nonporous 99.9% LPBF Ti6Al4V ELI samples in annealed condition show 900–1000 MPa ultimate strength in tension and about 1.9 GPa in compaction, with modulus of elasticity of 110–120 GPa [22, 34]. EOS GmbH (2009) (https://www.eos.info) reported LPBF Ti CP ultimate tensile strength of 455 $\pm$ 10 MPa and elastic modulus of 95 $\pm$ 10 GPa. Mismatch between the mechanical properties of the bone and a solid implant material can lead to stress-shielding phenomena [35], thus an interlocking interface is important as well as mechanical properties of a porous biomaterial.

Ti CP and Ti6Al4V cellular-lattice structures having an effective modulus of elasticity close to the bones are, in any case, a porous scaffold consisting of very hard material in comparison with real bone tissue. An analysis of the biomechanics of the loaded implant-bone system showed that, for the functioning of a bone that has grown into a porous metal scaffold, bone damage and the formation of scars on the metal-bone interface can occur under different loadings and load directions [36].

Wang et al. [2] reviewed the current state of topological and lattice design of porous metallic implants and the fabrication of such implants using AM. The review in detail examined approaches for scaffold design (representative unit cells and blocks), and demonstrated that topology optimization is a powerful digital tool that can be used to obtain optimal internal architectures for porous implants that not only satisfy multifunctional requirements but also mimic human bones. The following requirements are important for orthopedic implants: biocompatibility; suitable surface topology for cell attachment, proliferation, and differentiation; high porosity providing cell in-growth and transport of nutrients and metabolic products; reliable mechanical properties relevant to the anatomical loading; and eliminated stress-shielding effect [2]. All these requirements are also suitable for other types of internal nondegradable implants.

Recently, Mounir et al. [37] reported that, 1 year after implantation, porous Ti6Al4V ELI implants produced by electron beam melting did not show changes in the supporting bone. The hypothesis of the elimination of stress shielding with the application of porous Ti implants is thereby confirmed with high probability, but it requires careful study and longer period of observations.

Customized additive-manufactured Ti lightweight lattice implants with mechanical properties close to real bone and osseointegration capacity open new great perspectives in bone replacements. Learning from nature, because bone has a hierarchical structure, LPBF porous structures with micro-, submicro-, and nanofeatures have to be produced and analyzed.

5 LPBF lattice structures

5.1 Features of lattices manufactured by LPBF

With different AM methods—that is, laser or electron beam powder bed fusion, laser engineered net shaping, direct metal deposition, etc., different powder sizes and layer thicknesses, and different fabrication strategies that are used—different roughness, porosity, and even chemical composition will be present in manufactured implants from different processes. Even when manufacturing is done, for example, by LPBF, different orientations of the samples on the base plate, different building strategies, contouring, overhanging, scanning—all will lead to differences in the roughness in the scaffolds and small deviations from designed sizes. For example, in du Plessis et al. [23], the cubic lattice was designed with a total 15-mm width, 0.75-mm strut thickness, and 8 struts across one direction in total, resulting in 1.28-mm distance between struts and total 65% porosity. This design is shown in Fig. 6A. One set of samples was built in a vertical direction (Fig. 6B), other ones were at a 45-degree angle with supports (Fig. 6C). The differences in strut thickness, roughness, and microstructure is clearly visible in the cross-sections; higher roughness is present at the bottom areas of the horizontally built struts because overhanging occurred; different thicknesses of the struts in dependence of horizontal or vertical direction and also for Ti6Al4V different columnar prior beta-grains orientations are clearly present in as-built samples.

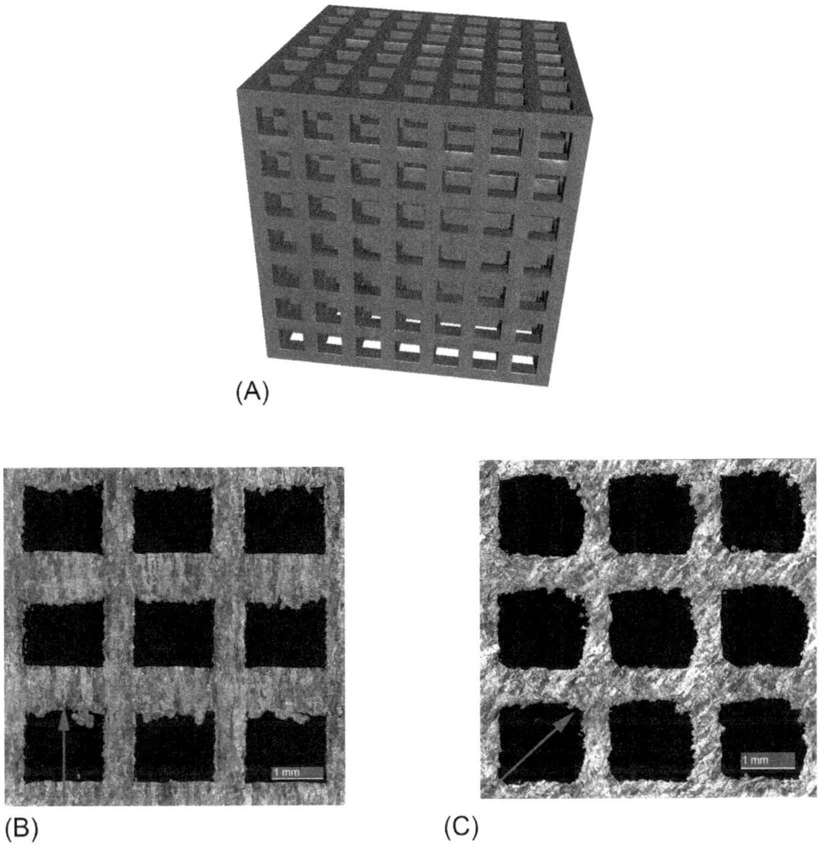

Fig. 6 Design of the cubic lattice structure (A) and cross-sections of LPBF Ti6Al4V ELI lattices manufactured in vertical (B) and 45-degree directions (C). Red arrows (light gray arrows in print version) indicate the building direction. Process parameters were similar in both cases.

The production of lattice structures by LPBF brings with it some challenges. In this section, we will discuss (a) manufacturing limits and features, (b) unwanted defects, and (c) minimum feature sizes. First, it is important to understand that there are inherent manufacturing limits to LPBF, despite the high complexity and freedom of design that is possible. Fundamentally, the method starts with a scanned laser track causing a molten track of material, followed by more alongside it. This means that a stable condition for the single track is critical in ensuring high integrity parts [38]. A single track is shown in Fig. 7 which clearly shows that the track height varies shown by color coding of microCT data. It is also seen in this image that partially fused particles attach to the sides of the track, which is also something that needs to be minimized and considered in final parts.

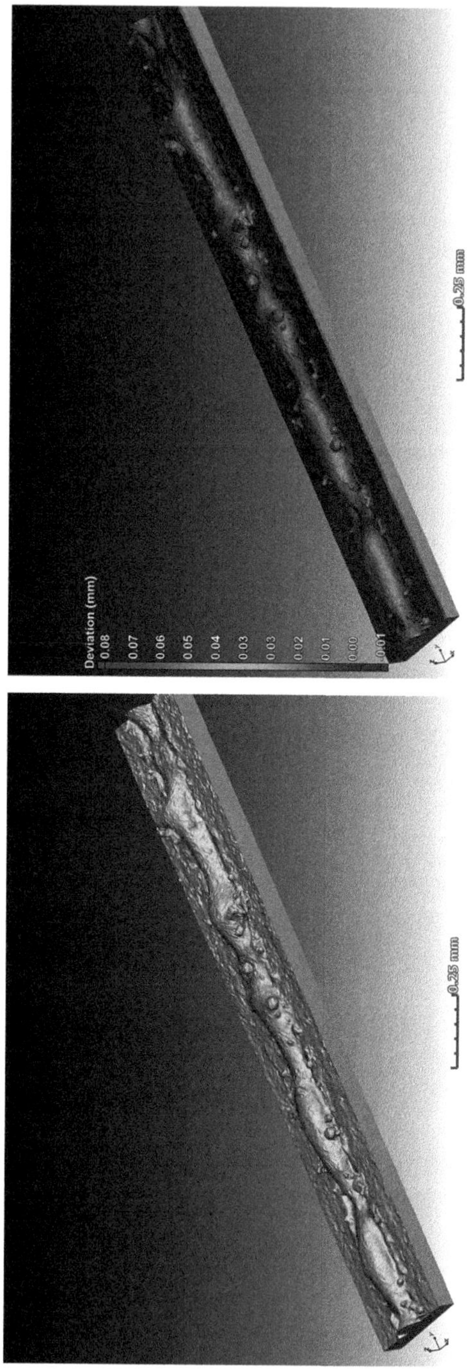

Fig. 7 Single laser track and surface profilometry with color coding of microCT data.

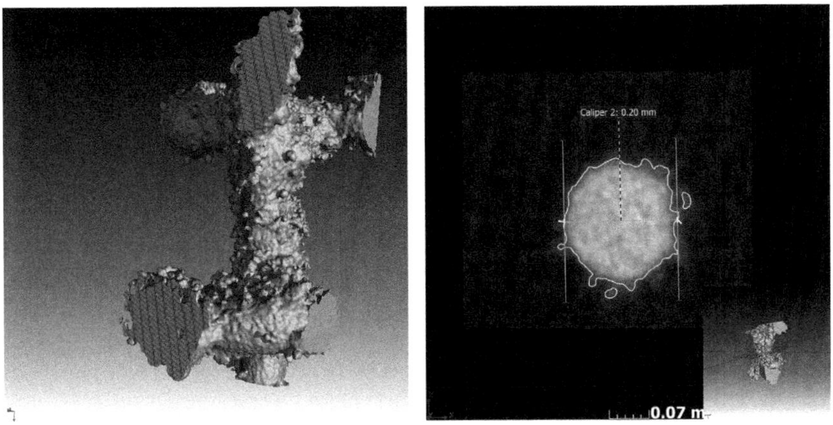

Fig. 8 Single strut of a microlattice structure built using the basic process parameters and scan strategy showing partially fused particles, rough surface, and approximately 0.2-mm strut thickness and cross-section of the strut. *(Data from recent publication A. du Plessis, D.-P. Kouprianoff, I. Yadroitsava, I. Yadroitsev, Mechanical properties and in situ deformation imaging of microlattices manufactured by laser based powder bed fusion, Materials 11 (2018) 1663.)*

The single track geometry effectively determines or limits the minimum feature sizes that can be produced. This is shown for microlattices from a recent study of the properties of microlattices [39]. Fig. 8 shows one internal strut and its cross-sectional view in microCT data. Besides these inherent limits, optimizing the scan strategy may offer a route to improve the minimum achievable feature sizes with a given system but requires further optimization. Such optimization was recently shown in works of Korn et al. [40, 41] and Ghouse et al. [42].

The second important concept is the formation of process porosity or other unexpected flaws in the part. This is always present to some extent, as shown in a recent round robin study of different LPBF systems [20]. In this round robin study, all parts were optimized and had very low total porosity levels (all $<0.5\%$), but in some cases the porosity was clustered, which can be detrimental for mechanical properties, especially in complex parts such as lattices. An example of a lattice structure with internal porosity in the struts is shown in Fig. 9, which is taken from a recent review of the use of microCT in AM [43]. In this example, there are some locations with up to 1% local porosity hotspots. This is due to the small feature size of the struts, where the inherent microporosity becomes more important than in bulk parts.

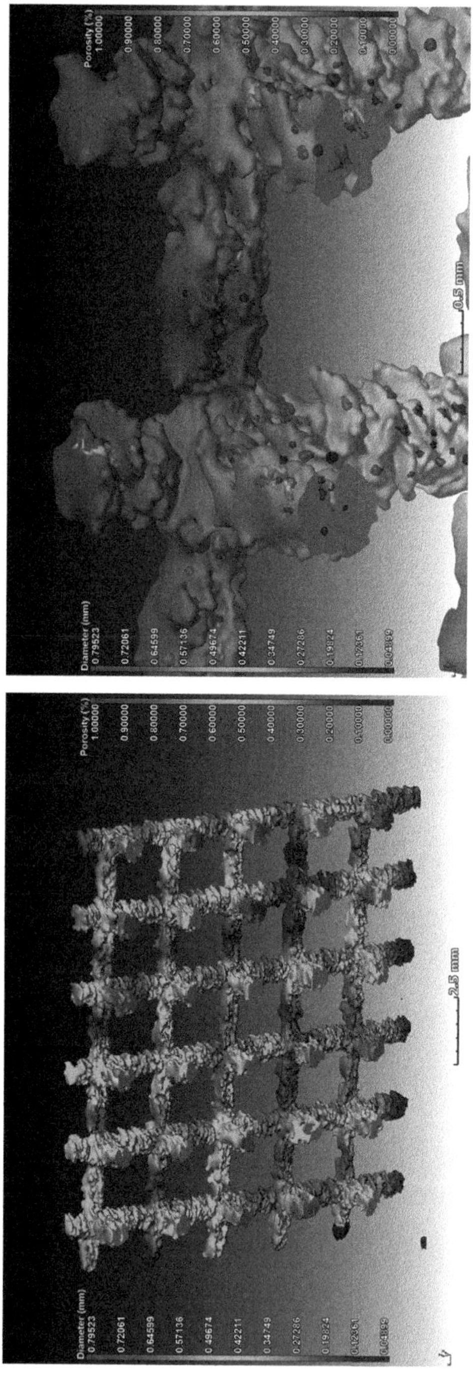

Fig. 9 Porosity inside the struts of a lattice (A). Local pore hotspot analysis shows areas up to 1% in local porosity (B). *(From A. du Plessis, I. Yadroitsev, I. Yadroitsava, S.G. le Roux, X-ray micro computed tomography in additive manufacturing: a review of the current technology and applications. 3D Print. Addit. Manuf. (2018), https://doi.org/10.1089/3dp.2018.0060.)*

Despite the presence of inherent porosity, recent work has shown that, for static loading, pores smaller than about 0.2 mm do not play a critical role on the static compressive strength of lattices of the LPBF Ti6Al4V ELI [23]. In this work, simulations and experiments were used to analyze the effect of defects, and it was shown that lattices with artificially induced pores with sizes up to 0.5 mm in 0.8-mm struts do not considerably affect the stress distributions and do not affect the mechanical compressive strength compared with reference lattices produced in the same process. This gives some confidence in the mechanical properties of AM lattices, despite their inherent porosity. Nevertheless, these pores may affect fatigue life and must be minimized; further work needs to be done on this topic.

Besides internal pores in struts, manufacturing imperfections such as rough surfaces, variations in strut thickness, staircase effect, and unwanted material left in the lattice are all problems that need to be given attention. Fig. 10 shows some unwanted additional inside of cube-shaped lattice structures, shown using microCT data and physical micrograph from electron microscopy.

Imperfect manufacturing occurs with differences in lattice strut thicknesses compared to the design, as shown in Fig. 11, where the view of the inner four units in a larger structure is compared with its design file; the red areas (light gray areas in print version) are parts where additional material

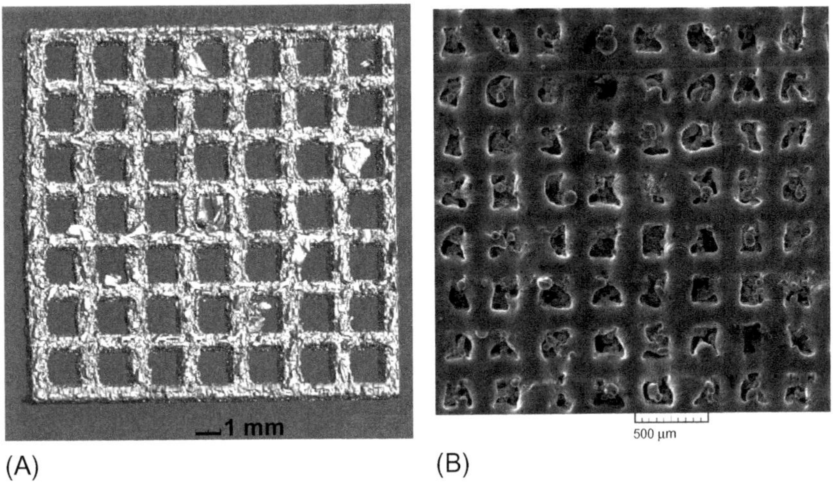

Fig. 10 Cube-shaped lattices. CT scan image of Ti6Al4V lattice with 1028-µm designed spaces between struts, head-on view (A), and SEM image of top view of CP Ti lattice with designed 150-µm pores (B) with unwanted material in lattices, which are impossible to remove without damaging the lattice itself.

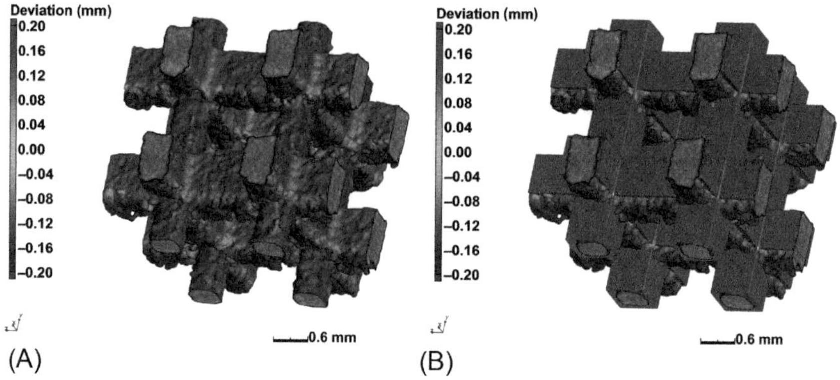

Fig. 11 Strut deviations shown on central section of lattice in cropped view, without (A) and with (B) overlap of CAD design file in gray.

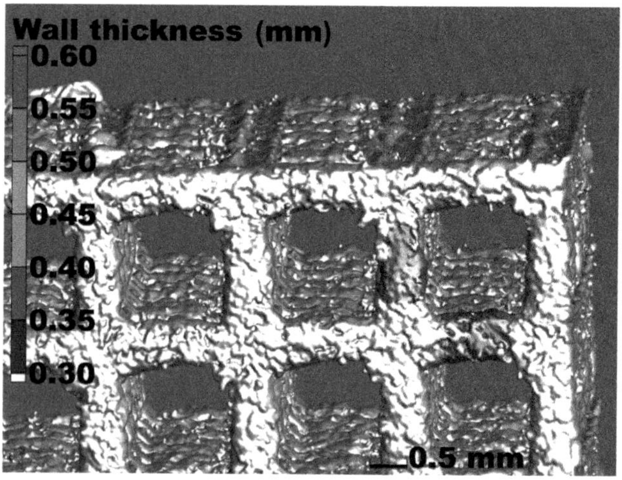

Fig. 12 Variations in strut thickness. In this case, the horizontal and vertical struts vary in thickness.

is present, and blue areas (dark gray areas in print version) are smaller than designed. It can be seen that vertically produced struts are narrower in both the *X* and *Y* directions.

The difference in LPBF lattice struts is shown further in Fig. 12 where thickness analysis color coding shows the vertical strut is thinner than the horizontal one. Calibration of the manufacturing system, layer thickness, process parameters in conjunction with the denudation effect, bonding, and melting particles from surrounding loose powder all play roles in the accuracy and roughness of lattice structures. These are some examples of manufacturing imperfections that occur, which become increasingly important as the lattice feature sizes decrease. As will be discussed in the next section in more detail, the ideal pore size for a bone replacement implant is a highly debated issue with some claiming > 400 μm is best, whereas others claim smaller is better. With pores in the region of < 1 mm, and considering the required elastic modulus, the resulting struts should be < 0.5 mm in most cases, which make all the previous considerations highly important.

All the previously discussed factors point to important considerations in improving the quality of AM lattices, also pointing to the need for process and quality inspection of these materials. The ideal method for this is X-ray microCT [43] as demonstrated by the earlier examples. It is also important to realize that the reported pore sizes may differ between different studies due to the complexity of defining a 3D pore size. One of the best methods is the maximal-sphere pore size, which measures the size of the largest sphere fitting into the given pore space [44, 45]. However, the cross-sectional diameter of an arbitrary pore area is often reported, which might make a direct comparison between different studies problematic, especially as an ideal pore size remains to be determined.

5.2 Mechanical properties of LPBF Ti CP and Ti6Al4V lattice structures

Structure-function relations can classify cellular material designs. Cellular materials in nature always attribute a structural benefit such as high specific stiffness under self-weight (honeycombs), resilience to compression forces or crack growth arresting (structures similar to Venus Flower Basket), high flexural (bending) stiffness (Toucan Beak's structures), etc. [46]. The mechanical properties of cellular structures are a function of the relative density, the cell's anisotropy, and the unit cell's architecture [47–50].

As indicated in the previous section, the accuracy and geometrical features of LPBF structures depends on laser power, layer thickness, scanning speed, scanning strategy, orientation, etc. Mechanical properties of AM lattice structures are all dependent on process parameters and building conditions, as well as material properties and postheat treatment [51–53]. To eliminate stress shielding, the mechanical properties ofTi alloy lattice structures are mainly driven by the need to match the elastic modulus of the adjacent bone to the implant. The mechanical properties of lattice structures can be predicted with reasonable accuracy by the models of Ashby-Gibson for open-cell foams; the elastic modulus is mainly a function of the density [54–57]. Based on this finding, a density of 0.35 for Ti6Al4V produces lattice structures with a theoretical elastic modulus of 14 GPa. This is in the range of cortical bone, whereas most AM studies to date where actual mechanical properties were measured reported approximately 20% lower values than the theoretical predictions. This can be attributed mainly to partially melted material on the surface, which do not contribute to the mechanical properties, or due to other build imperfections. What is important to realize is that, with a design density of 35%, it is possible to create various different types of lattice structure designs, and it is possible to vary the pore size of each design while keeping the density constant, which is achieved by increasing the unit cell size.

Fig. 13 shows a range of unit cell designs that can be produced by LPBF, which cover the typical "scaffold-based" designs and the newer "minimal surface" designs. Minimal surface designs were first introduced for additive-manufactured implant applications by Kapfer et al. [58], with various recent studies showing that they have numerous advantages over traditional strut-based designs [59–61].

The previously discussed designs were compared numerically (via simulations) in a study by du Plessis et al. [62] for the purpose of bone implant application. In this work, the designs were compared numerically based on design files only, which holds some advantages in making a selection for design, that is, these comparisons eliminate the added level of complexity of manufacturing imperfections, allowing a clearer comparison of the

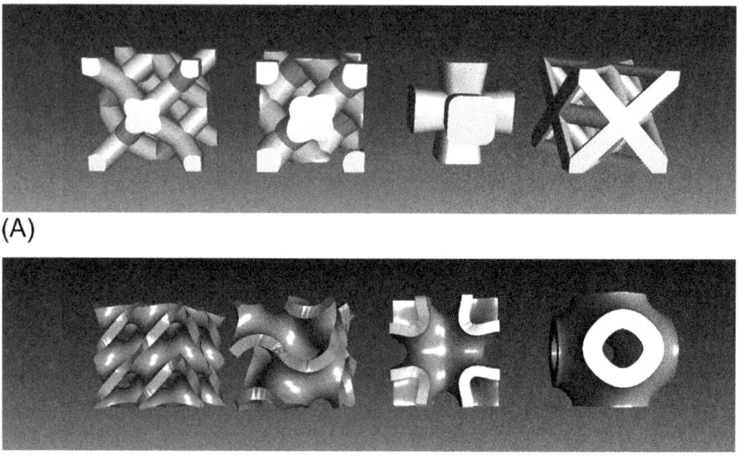

(A)

(B)

Fig. 13 Different designs of lattice unit cells covering (A) scaffold-based and (B) minimal surface designs.

advantages of each design type. It was shown that minimal surface designs at the density required for bone implants require very thin walls. This may be challenging for some LPBF systems, but because they are well connected and self-supporting, this may not be a limitation. It was also shown that the minimal surface designs have a superior angular performance on load simulation, making them more suitable for bone implants based on lower stress concentrations. Panesar et al. [63] also indicated that lattices with surface-based unit cells are more efficient at transmitting loads when compared to a strut-based design, essentially because of their higher degree of connectivity.

It was also shown that pore sizes vary considerably depending on the unit cell design selected among all designs; most interesting is that some designs have dual or triple pore size distributions, which might be an advantage for bone growth in different stages of growth. Permeability simulations showed little difference between models but showed that some designs that have high permeability have low tortuosity, which is presumably not ideal for bone implants. It is presumed that complex flow paths are required, which allow not only fast nutrient transport but also a complex flow to allow the nutrients to replenish all over the lattice structure and not just in high-flow paths. This concept is useful for the selection of a suitable implant lattice design.

According Maxwell's criteria, cellular structures can be divided into bend-dominated or stretch-dominated as shown in Fig. 14.

Additively manufactured lattice structures are typically of two types: bend-dominated or stretch-dominated, with shear failure or layer-by-layer

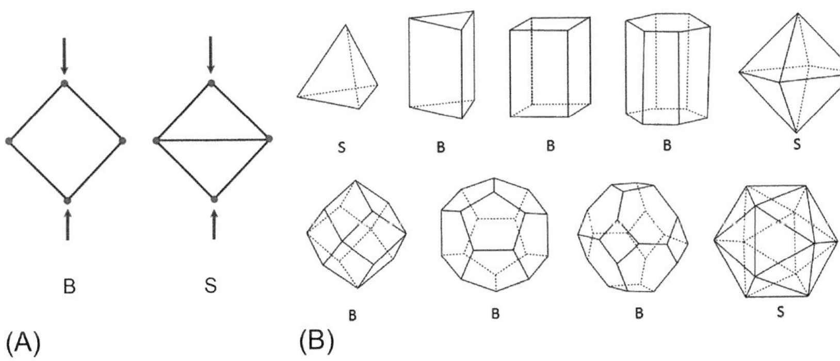

Fig. 14 Pin-jointed frames: (A) bend-dominated (B) and stretch-dominated (S), and (B) 3D polyhedral cells. Bend-dominated structures are indicated as "B" and stretching as "S" [64].

progressive failure, respectively [65]. In recent work on relatively thick-strut lattice structures, it was shown how the initial failures occur in that the failure initiated in the middle of the lattice in a stretch-dominated lattice or on the edge in the corner of the bend-dominated lattice [66]. The previously mentioned relatively thick struts are aimed at structural applications and the internal part of an implant (replacing solid material), whereas ideal surfaces and near-surface areas of bone growth implants require smaller pore spaces and correspondingly thinner struts.

Choy et al. [67] studied the mechanical properties of as-built strut-based cubic and honeycomb lattice structures produced by LPBF in Ti6Al4V. It was found that, when diagonal struts were in the plane view parallel to the compression direction, both structures deformed with abrupt shear failure across opposite corners of the lattice sample.

Annealed LPBF Ti6Al4V diamond-lattice structures for femoral-head repair implants were mechanically tested and proven for bone regeneration by Zhang et al. [68]. The elastic moduli and compressive strength increased with the size of the struts.

Liu et al. [69] indicated that optimized surface for AM lattices can play a role in regulating the relationship between density and mechanical properties. Changing in curvature decreased stress concentration resulting in significant enhancement of the structural strength of LPBF Ti6Al4V diamond lattices. Failure mechanism of porous titanium alloy samples with diamond-unit cells changed from bottom-up collapse by layer (or cell row) to shear band by topology optimization. The relationship between compressive strength and elastic modulus can be regulated by using different curvature in each unit cell [69].

Xiao and Song [70] studied graded lattice structures with rhombic dodecahedron-unit cells and found that specific strength and specific energy absorption of graded lattice structures are higher than those of uniform structures. It was suggested that different gradient modes and loading directions did not influence the energy absorption. Functionally graded lattice structures produced by AM show mechanical performance due to its high resilience to loading variabilities. Density gradients are potentially very important; various recent studies investigated the properties of graded lattice structures, including recent works by Liu et al. [71], Panesar et al. [63], and Han et al. [72].

Biomimetic design and high accuracy by LPBF provide a versatile way for reconstruction of individualized load-bearing metal implants [68]. But, as Feng et al. [73] indicated, the assessment of the mechanical properties of

LPBF lattice structures remain challenging due to their complicated geometries, which is time-consuming due to the requirement of numerous part fabrication and physical testing.

Other alloys of Ti are also very relevant to AM implant materials, as some improved alloys may be beneficial to the bone growth process, for example, Ta- and Nb-incorporating alloys [49, 52, 74–76].

A recent review paper by Ren et al. [77] focused on the fatigue properties of AM lattice structures with various design factors discussed in the context of fatigue properties of the produced lattice structures. Wang et al. [50] also indicated that fatigue property of lattice AM structures must be studied carefully because common causes of prosthesis cracks are fatigue breaks of the material due to long-term and repeated stresses in a corrosive environment.

As indicated in Tan et al. [85], to reach optimal cellular design is difficult since for such an optimization so many properties at once is required.

6 In vitro studies

Analysis of cell adhesion and mechanical interaction of the cell-substrate is a crucial consideration for bone replacements [78]. Cell adhesion phenomena can be divided into three stages: attachment to the substrate, flattening, and spreading of the actin skeleton with the formation of focal adhesion between the cell and its substrate. The affinity of cells to an implant controls functions and cell behavior, and as a result, the implant's life expectancy and effectiveness. Cell attachment can be affected not only by properties of the cells itself, but the topography, roughness, wetting behaviour, charge at physiologic pH of implant surface, chemical composition, etc. [11, 17, 30, 79].

Eriksson et al. [80] found that, at an early stage (1 week), wetting behavior of the substrate and surface energy play important roles in the healing process of Ti implants. After 3 weeks, the hydrophilic and hydrophobic Ti discs had very similar responses, and new bone formed directly at the surface of the titanium implants.

Fousová et al. [34] studied LPBF-annealed Ti6Al4V fully dense samples and porous samples for cytocompability with human osteosarcoma cells. It was found that, after 24 h' incubation, cells grew preferentially on flat surfaces on LPBF samples and avoided big adherent powder particles. Zhao et al. [81] studied cell affinity of porous titanium scaffolds fabricated by LPBF and stated that 1000-μm pore size is more suitable for cells to adhere to scaffolds because cell seeding is obstructed in smaller voids. It was found that, in 1000-μm scaffolds, cells were more flattened, better spread, and displayed more cells than 500-μm scaffolds. From a mechanical point of view, mechanical fatigue properties in 500-μm structures were better than scaffolds with higher porosity, and octahedron scaffolds were more preferable for fatigue life and static compressive properties.

Van Bael et al. [82] manufactured Ti6Al4V bone scaffolds with triangular, hexagonal, and rectangular pore shapes with designed sizes of 500 and 1000 μm, and tested the biological behavior of seeded human periosteum-derived cells for 14 days. At the early stage, smaller pore size and corners in lattices were more beneficial for cell growth, that is, living cells were densely distributed in the corners where cells could bridge (so-called "curvature-driven mechanism" of 3D cell growth [83]). After 2 weeks, DNA and metabolic activity were higher at 1000-μm pore samples, so it was suggested that graded structures combining small pores for initial cell attachment and larger noncircular pores to avoid pore occlusion are required; further studies are needed.

Wang et al. [84] also noted that, in 3D scaffolds, cells were observed suspended across struts and crevasses, and it was suggested that this can be attributed to curvature-driven growth to minimize surfaces of the living cells. No significant difference in terms of hBMMSCs proliferation was found among the diamond lattices with regular, irregular, gradient structure and truss frame with tetrahedral structure units; because the average pore size was the same (about 500 μm), the permeability and surface area were similar. Analysis of permeability in LPBF porous structures for bone tissue regeneration is very important when designing scaffolds because mass transport in vivo carries out through the diffusion process dependent on the morphology of the porous biomaterial and its permeability. Values of permeability of AM porous structures compatible with real replaced bone can increase the probability of successful healing after bone replacement [59, 84, 85].

Wally et al. [86] studied in vitro bone cell growth in Ti6Al4V regular lattices with a "spider web" design, diamond units, and graded porosity in dental implants for 1–28 days. It was noted that all implants supported bone cell growth. The difference was only in the mechanical properties and maximum calcium deposition found on scaffolds with 400-μm pores. Identical to Frosch et al. [87], there was no effect of pore size and shape on matrix production by MLO-A5 cells. This contradicts Markhoff et al. [88] in their study with open porous Ti scaffold designs on human osteoblast behavior; in static and dynamic cell investigations, it was indicated that cubic and pyramidal structures showed better collagen formation compared to diagonal lattices.

Wysocki et al. [88a] showed in vitro that, at an early stage, mechanical entrapment of the cells and their anchoring were more prominent in relatively small pores but hindered diffusion of nutrients in small spaces at later stages. When a 3D cell set-up—multicellular spheroid technique imitating an in vivo case—was used, it was found that surface areas colonized by MG63 cells migrating out of spheroids were two times higher on polished scaffolds. Contrary to this, Xue et al. [89] indicated that the cells preferred more irregular surfaces than polished ones; this conclusion was done when in vitro experiments with porous as-built structures and polished solid Ti samples were evaluated. Studies on co-culture osteoblast-like cells (MG63 cells) and ECs (HUVECs) of Ti showed that both hydrophilicity and surface roughness (S_a = 1–2 μm) enhanced osteogenic properties of Ti implants [90].

In work by Kokubo and Yamaguchi [91], it was mentioned that the initial apatite layer formation on the surface of the implant material might be critical to further bone growth, and that this layer can be used as an indicator for in vitro bone growth test success. The apatite layer formation was dependent on the heat and chemical treatments, which could be explained by the positive effect of a present surface oxide layer. In relation to in vitro test work on a Ti alloy containing Nb, Todea et al. [92] varied the chemical and heat treatments, and measured the formation of an apatite layer as an indicator of bone growth success.

A recent review of the topic of bone modeling in vitro [93] discusses the requirement to improve in vitro tests by simulating the sequence of events taking place in the bone regeneration process. This is demonstrated by the authors using combinations of different test conditions at different stages. Bone cell growth depends on chemical composition of material, wetting behavior, surface roughness, curvature of scaffolds, its inner structure, etc. It is a time-dependent process including cell-substrate and cell-cell interactions. Results received in vitro can be verified in vivo.

Table 1 summarizes various osseointegration studies to date on CP Ti and Ti6Al4V lattice structures produced by LPBF, indicating the lattice design type, pore size, and details of the type of experiments performed.

Table 1 CP Ti and Ti6Al4V alloy lattice structures produced by LPBF and osseointegration studies

Samples (material)	Treatment	Biological tests	Porosity (%)	Unit cell shapes	Size of pores (μm)	Reference
CP Ti						
Grade 1 Ti lattices	As-built	3 days of culture Cell affinity	63–84	Tetrahedron octahedron	500 1000	[81]
Pure Ti cylinders	heat treated	16, 26, 52 weeks, implanted in the dorsal muscles of dogs		Solid cylinders with channels	500 600 900 1200	[94]
Pure Ti lattice on solid plate or cylinders covered with a thin lateral wall	Heat-treated at 1300°C for 1 h	2, 4, 8 weeks, implanted into the metaphysis of the tibia of rabbits	61–66	Diamond structures	300 600 900	[95]
Bimodal CP Ti scaffolds	chemically polished in HF/HNO3	MG63 osteoblast-like cell	70	Core 200 μm, shell 500 μm Core 100 μm, shell 180 μm	100–500	[88a]
Ti6Al4V/Ti6Al4V ELI						
Scaffolds	As-built, acid etched	2 weeks, in vitro human periosteum-derived cells		Pore shapes: triangular hexagonal rectangular	500 1000	[82]
Cylindrical implants	As-built	4–8 weeks, implants into the lateral cortices of canine	50, 60, 70, 75	Tetrahedron Octet truss	500 770	[96]
Dental implants, conical and cylindrical		2 months, implanted in tibias of a sheep		Solid samples with hexagonal/diagonal internal structure	600 900 1200	[97]

Fully-dense, with dense core and porous surface into the depth of 1 or 2 mm porous sample with dense surface layer (1-mm thick) fully porous sample	Annealed at 820°C	In vitro, 1–5 days, cytocompatibility	0–79	Rhombic dodecahedron		[34]
Disks with bioinspired porosity	Surface modification by acid etching	In vitro, osteoblast-like cell and human osteoblasts	0;15–70	Bone-inspired porosity and micro/nanoscale surface roughness	177–653	[10, 98]
Structures with bioinspired porosity and surface modification by acid etching		In vitro, cranial bone onlay model with rats, 10 weeks				[13]
Dental implants with bioinspired porosity		10 weeks, implanted in the rabbit femur				[99]
Regular and graded porosity dental implants		In vitro, MLO-A5 cells 1, 4 days 1–4 weeks	51–83	"Spider web" design and diamond	300–650	[86]
Porous scaffolds		In vitro, culture hBMMSC, and in vivo 1–8 weeks Cylindrical implants with a cup implanted into distal femur of rabbits	70	Diamond crystal lattice: regular, irregular, gradient truss frame: with tetrahedral structure unit	500	[17]

7 In vivo studies

7.1 Regular structures

Size and shape of the pores, roughness, loading, and stresses, chemical composition of implant material, all influence on osseointegration of AM samples as it was shown with different models. Taniguchi et al. [95] studied osseointegration for LPBF heat-reated Ti diamond structures (300–900 μm in size) with rabbits. It was found that, after 2 weeks of implantation into the metaphysis of the tibia, plates with lattice structures with 600-μm pores showed better bone implant fixation ability, but after 4–8 weeks, all types of porous Ti revealed remarkably high fixation ability. Irregular shape of the pores, rough inside surface, as well as collapsed pores made it challenging to discern the true effect of pore size from these studies. The 300 μm pores were too small for efficient vascularization as was agreed with reported elsewhere that a minimum pore size for vascularization is 400 μm [101]. The 600-μm diamond structures showed the best mechanical properties and bone in-growth.

Accurate evaluation of the effect of pore size in LPBF samples can be difficult due to high roughness, attached particles, and variability in strut sizes. Arabnejad et al. [96] built tetrahedron and octet truss structures. After CT scans, it was found that real pore sizes were smaller than the designed structures because strut thickness and strut cross-section deviated in building direction, especially for overhang areas. Implanted samples into the lateral cortices of canines show that the amount of bone in-growth at 4 and 8 weeks was somewhat higher for the octet truss structure, but after 8 weeks, new bone had grown within all implants and completely filled the porous structure adjacent to the cortices. Bone tissue also was found in the implanted structures within the cancellous medullary canal. Liu et al. [103] studied in vivo solid as-built LPBF implants from Ti alloy and found that osseointegration took place differently in different places of the implant and growth rates varied with time.

Fukuda et al. [94] investigated LPBF Ti cylinders with channels implanted in the dorsal muscles of dogs for periods of 16, 26, or 52 weeks. It was noted that, even after 52 weeks, no bone formation has been recognized on the outer surface of the treated channel implants. Only at distances of 5 and 7 mm from the ends of 500-μm and 1200-μm channels, respectively, the highest osteoinduction was observed at an early stage. In whole, better osteoinduction was shown in smallest channels (500 μm) covered by hydroxyapatite. It was indicated that further investigations are necessary for pore sizes < 500 μm. Otsuki et al. [100] showed for powder sintered samples that interconnected pore throat sizes play a critical role in bone in-growth; narrow pore throats inhibited tissue differentiation in porous Ti implant. Taniguchi et al. [95] recommended diamond structures as a basic structure for orthopedic load-bearing applications. Obaton et al. [97] tested Ti6Al4V ELI diamond and diagonal lattice structures with different unit cell sizes (600, 900, and 1200 μm). Dental implants having DNA shape diamond internal structure have been screwed into the tibia cortical bone of sheep, and cylindrical specimens with diagonal structure were placed in a drilled bone. It was found that early-stage 2-month ossification was more prominent in 900 μm cells that in 1200 μm cells. In samples with designed 600 μm units, cells of the porous implants were closed by attached powder particles and ossification was difficult.

Tan et al. [85] analyzed cellular scaffolds for prosthetic orthopedic implants produced from different materials by laser and electron beam powder bed fusion, and suggested that an optimal pore size should be ~300–600 μm to provide mechanical strength, bone in-growth, high permeability, and vascularization. Chang et al. [102] investigated titanium meshes and bone growth in vitro and in vivo, and found that smallest pore sizes (~180 μm) show better cell differentiation and early bone growth, but larger pore sizes (300 and 400 μm) show better in-growth; this is explained as being due to better vascularization possible because of increased permeability. They concluded that combinations of pore sizes might be beneficial.

High roughness and open pores with sharp edges in LPBF implants can stimulate and promote contact osteogenesis. Wang et al. [84] studied Ti6Al4V regular, irregular, and gradient diagonal and tetragonal truss frame structures implanted into the limbs of rabbits. After 8 weeks of in vivo experiments, diamond regular and irregular structures showed better performance. It was found that, at an early stage, smaller pores with high curvature were more preferable for initial bone growth, but at later stages of bone regeneration, bigger pores had advantages, therefore there must be a balance between curvature growth effect and further vascularization.

7.2 Gradient and bone-inspired porosity

Cheng et al. [10, 13] tested Ti6Al4V implants with bone-inspired porosity by CT scans of a human femoral head retrieved from a hip replacement. It was found that an LPBF-scalable manufacturing method can be used for manufacturing increased osseointegration patient-specific orthopedic and dental implants and scaffolds. It was suggested that trabeculae-inspired porosity implants can be successfully placed even in areas with insufficient bone to induce vertical bone regeneration. Later, Cohen et al. [99] analyzed CT scans and histological cross-sections of solid and porous biomimetic implants in rabbit femurs after 10 weeks. It was found that LPBF porous implants enhanced cell response and mineralization. Vertical bone formation in the areas without a physical implant present was found. On the whole, trabecular bone-inspired porosity promoted bone growth and is a superior alternative to solid implants for bone-interfacing implants.

Figliuzzi et al. [104] described root analogue dental implants using CT scans with implants produced by LPBF from Ti6Al4V powder. After >2 years of clinical observations, it was found that a customized implant has perfect functional and aesthetic integration.

For different periods after implantation, bone recovery showed differences and varied response, so stabilized effects for different bones and types of the porous implants will be different. Also, healing in cortical and trabecular bone are different; cortical bone healing depends on osteonal remodeling, whereas trabecular healing relies on osteoconduction and new bone formation [105, 106].

Stresses and strains of different bones are different and also varies from individual to individual due to personal and lifestyle differences, anatomy, general health, age, and sex, which requires customization of the bone replacements and is therefore quite challenging [85, 104, 107, 108].

Murr [9] analyzed strategies for creating additively manufactured open-cellular metal and alloy structures for bone replacements. It was indicated that cellular structures serve as support or frames for living cells, thus its properties have to promote osseointegration, osteoinduction, and vascularization of living cells, as well as having high strength and fatigue characteristics. AM allows design and production of complex shapes and open porous structures with fatigue strengths compatible with the range of requisite bone fatigue strengths. It was shown that, for optimal mechanical properties of AM implants, cell migration, and interconnectivity for both bone cells and vascular cells, an optimal pore size is between 200 and 900 μm. However larger vascular structures require larger pore sizes, and in vitro optimization has not actually been experimentally determined. de Grado et al. [109] indicated that, for cell migration, 100- to 200-μm pores are enough, but 300- to 500-μm pores are recommended for the formation of capillaries. Tan et al. [85] also suggested optimal pores in the range of 300–600 μm in terms of bone in-growth, vascularization, mechanical strength, and permeability. Gradient structures with low density in the core and higher outer density and vice versa were proposed for different bone replacements [9, 110]. As indicated in Murr [9], the main question is how to transform a "porous environment" into a "living environment" where bone cells, tissues, and vascular structures receive enough oxygen and nutrients to function effectively. The osseointegration of an "ideal implant" is the first stage in long-term bone replacement; angiogenesis and vasculogenesis require additional efforts in living implant design and production through AM. In whole, all these works indicated that customized AM implants can be more accessible and functional and have a great future for advanced bone replacements.

8 Conclusions

The goal of this chapter is to provide improved understanding of the specific requirements for bone implants based on experimental and published literature to date, and hopefully contribute to the future design and production of optimized implants by LPBF with approved Ti alloys for biomedical applications, which can minimize healing time and produce highest quality long-term stability in patients. In this context, the complex mixture of bone regeneration process requirements seems to be summarized as follows:

- The initial stage of cell seeding and viability requires that there are sufficiently large available surface areas for cells to attach to and that these locations are conducive to initial stages of growth, that is, nano/microstructural roughness or structure and/or chemical content of the surface or the local environment. This initial stage leads to the formation of an apatite layer, followed by bone desorption, prior to further bone formation.
- Further bone growth requires that nutrients are delivered to the site of bone growth, which requires vascularization to take place. This vascularization requires space (reported as > 400-μm-wide pore spaces) and interconnected pore spaces (high permeability). However, if these spaces are too large, presumably bone growth is limited and may not entirely close the space, or may take a long time to do this. If the pore spaces are too small, the local replenishment of nutrients may be too slow and/or the pH may be incorrect, yielding inefficient bone growth rates.

Regarding the production of AM lattice structures for implants:

- LPBF scaffolds have to be used not in as-built conditions, but after stress-relieving/annealing heat treatment, as recommended by

international standards ISO and ASTM for biomedical applications of Ti alloys. Careful and detailed description of samples manufacturing and their evaluation have yet to be done.
- The microstructure and mechanical properties of titanium alloys, even after heat treatment, differ from the microstructure of cast and wrought alloys. Therefore, the study of the influence of microstructure in various types of lattices and their scanning/manufacturing strategy require special attention.
- Further analysis of fatigue properties and fracture analysis of LPBF lattice samples has to be done for optimal design of unit cells for load-bearing bone replacements preferably in a corrosive environment for understanding how roughness and microstructure influence on mechanical properties.
- There is no consensus on the size of the pores and the design of single cells for optimal osseintegration. Irregular structures, by alternating pores of 200–1000 μm in size, should be evaluated from the point of view of stable mechanical properties on one hand, and osseointegrative properties on the other hand.
- For in vitro studies, 3D cell setup looks more preferable in terms of evaluation of pore size and shape for bone growth.
- For in vivo studies, long-term observation needs analysis not only of bone in-growth into the implant, but also analysis of surrounding bone and tissues should be done.

Acknowledgments

This work is based on the research supported by the South African Research Chairs Initiative of the Department of Science and Technology and National Research Foundation of South Africa (Grant No. 97994). Samples were built in CRPM at Central University of Technology, Free State and the authors would like to thank Mr. Johan Els and Mr. D. Kouprianoff. We would like to thank the Division of Clinical Anatomy, Stellenbosch University for loaning the human bone specimen.

References

[1] D.F. Williams, On the mechanisms of biocompatibility, Biomaterials 29 (2008) 2941–2953.

[2] X. Wang, S. Xu, S. Zhou, W. Xu, M. Leary, P. Choong, M. Qian, M. Brandt, Y.M. Xie, Topological design and additive manufacturing of porous metals for bone scaffolds and orthopaedic implants: a review, Biomaterials 83 (2016) 127–141.

[3] K. Endo, S. Yamada, M. Todoh, M. Takahata, N. Iwasaki, S. Tadano, Structural strength of cancellous specimens from bovine femur under cyclic compression, Peer J. 4 (2016) e1562, https://doi.org/10.7717/peerj.1562.

[4] A. Hasan, B. Byambaa, M. Morshed, M.I. Cheikh, R.A. Shakoor, T. Mustafy, H. Marei, Advances in osteobiologic materials for bone substitutes, J. Tissue Eng. Regen. Med. 12 (6) (2018) 1448–1468.
[5] R. Dimitriou, E. Jones, D. McGonagle, P.V. Giannoudis, Bone regeneration: current concepts and future directions, BMC Med. 9 (1) (2011) 66.
[6] S.J. Hollister, Porous scaffold design for tissue engineering, Nat. Mater. 4 (7) (2005) 518–524.
[7] T. Winkler, F.A. Sass, G.N. Duda, K. Schmidt-Bleek, A review of biomaterials in bone defect healing, remaining shortcomings and future opportunities for bone tissue engineering: the unsolved challenge, Bone Joint Res. 7 (3) (2018) 232–243.
[8] D. Martinez-Marquez, A. Mirnajafizadeh, C.P. Carty, R.A. Stewart, Application of quality by design for 3D printed bone prostheses and scaffolds, PLoS One 13 (4) (2018) e0195291.
[9] L.E. Murr, Strategies for creating living, additively manufactured, open-cellular metal and alloy implants by promoting osseointegration, osteoinduction and vascularization: an overview, J. Mater. Sci. Technol. 35 (2) (2019) 231–241, https://doi.org/10.1016/j.jmst.2018.09.003.
[10] A. Cheng, A. Humayun, D.J. Cohen, B.D. Boyan, Z. Schwartz, Additively manufactured 3D porous Ti-6Al4V constructs mimic trabecular bone structure and regulate osteoblast proliferation, differentiation and local factor production in a porosity and surface roughness dependent manner, Biofabrication 6 (2014) 045007–045019.
[11] T.K. Monsees, K. Barth, S. Tippelt, K. Heidel, A. Gorbunov, W. Pomp, R.H.W. Funk, Effects of different titanium alloys and nanosize surface patterning on adhesion, differentiation, and orientation of osteoblast-like cells, Cells Tissues Organs 180 (2005) 81–95, https://doi.org/10.1159/000086749.
[12] X. Zhu, J. Chen, L. Scheideler, T. Altebaeumer, J. Geis-Gerstorfer, D. Kern, Cellular reactions of osteoblasts to micron- and submicron-scale porous structures of titanium surfaces, Cells Tissues Organs 178 (2004) 13–22.
[13] A. Cheng, D.J. Cohen, A. Kahn, R.M. Clohessy, K. Sahingur, J.B. Newton, Z. Schwartz, Laser sintered porous Ti-6Al-4V implants stimulate vertical bone growth, Ann. Biomed. Eng. 45 (8) (2017) 2025–2035.
[14] J.J. de Damborenea, M.A. Larosa, M.A. Arenas, J.M. Hernández-López, A.L. Jardini, M.C.F. Ierardi, C.A.C. Zavaglia, R.M. Filho, A. Conde, Functionalization ofTi6Al4V scaffolds produced by direct metal laser for biomedical applications, Mater. Des. 83 (2015) 6–13.
[15] S. Ghosh, S. Abanteriba, S. Wong, S. Houshyar, Selective laser melted titanium alloys for hip implant applications: surface modification with new method of polymer grafting, J. Mech. Behav. Biomed. Mater. 87 (2018) 312–324.
[16] S. Spriano, S. Yamaguchi, F. Baino, S. Ferraris, A critical review of multifunctional titanium surfaces: new frontiers for improving osseointegration and host response, avoiding bacteria contamination, Acta Biomater. 79 (2018) 1–22.
[17] M. Wang, Y. Wu, S. Lu, T. Chen, Y. Zhao, H. Chen, Z. Tang, Fabrication and characterization of selective laser melting printed Ti-6Al-4V alloys subjected to heat treatment for customized implants design, Progr. Nat. Sci.: Mater. Int. 26 (6) (2016) 671–677.
[18] T.J. Webster, J.U. Ejiofor, Increased osteoblast adhesion on nanophase metals: Ti, Ti6Al4V, and CoCrMo, Biomaterials 25 (2004) 4731–4739.
[19] N. Gui, W. Xu, D.E. Myers, R. Shukla, H.P. Tang, M. Qian, The effect of ordered and partially ordered surface topography on bone cell responses: a review, Biomater. Sci. 6 (2018) 250–264.
[20] A. du Plessis, S.G. le Roux, Standardized X-ray tomography testing of additively manufactured parts: a round robin test, Addit. Manuf. 24 (2018) 125–136.

[21] A. du Plessis, P. Sperling, A. Beerlink, O. Kruger, L. Tshabalala, S. Hoosain, S.G. le Roux, Standard method for microCT-based additive manufacturing quality control 3: surface roughness, MethodsX 5 (2018) 1111–1116.
[22] I. Yadroitsev, P. Krakhmalev, I. Yadroitsava, A. du Plessis, Qualification of Ti6Al4V ELI alloy produced by laser powder bed fusion for biomedical applications, J. Miner. Met. Mater. Soc. (TMS) 70 (3) (2018) 372–377.
[23] A. du Plessis, I. Yadroitsava, D. Kouprianoff, I. Yadroitsev, Numerical and experimental study of the effect of artificial porosity in a lattice structure, in: SSF 2018—The 29th Annual International Solid Freeform Fabrication Symposium—An Additive Manufacturing Conference, Austin, Texas, August 13–15, 2018, 2018.
[24] J. Xu, X. Chen, C. Zhang, Y. Liu, J. Wang, F. Deng, Improved bioactivity of selective laser melting titanium: surface modification with micro-/nano-textured hierarchical topography and bone regeneration performance evaluation, Mater. Sci. Eng. C 68 (2016) 229–240.
[25] M. Tsukanaka, S. Fujibayashi, M. Takemoto, T. Matsushita, T. Kokubo, T. Nakamura, K. Sasaki, S. Matsuda, Bioactive treatment promotes osteoblast differentiation on titanium materials fabricated by selective laser melting technology, Dent. Mater. J. 35 (1) (2016) 118–125.
[26] S.J. Park, S.B. Bae, S.Q. Kim, T.G. Eom, S.I. Song, Effect of implant surface microtopography by hydroxyapatite grit-blasting on adhesion, proliferation, and differentiation of osteoblast-like cell line, MG-63, J. Korean Assoc. Oral Maxillofac. Surg. 37 (2011) 214–224.
[27] B. Kasemo, J. Gold, Implant surfaces and interface processes, Adv. Dent. Res. 13 (1999) 8–20.
[28] C. Chen, S.-M. Zhang, I.P. Lee, Immobilizing bioactive molecules onto titanium implants to improve osseointegration, Surf. Coat. Technol. 228 (1) (2013) S312–S317.
[29] D.A. Puleo, M.V. Thomas, Implant surfaces, Dent. Clin. N. Am. 50 (2006) 323–338.
[30] J.Y. Li, L.W. Zheng, L. Ma, D.L.W. Kwong, L.K. Cheung, E.H.N. Pow, Effect of fluoride-modified titanium surface on early adhesion of irradiated osteoblasts, BioMed Res. Int. (2015) 219752. 9 pp. https://doi.org/10.1155/2015/219752.
[31] Y.L. Cai, J.J. Zhang, S. Zhang, S.S. Venkatraman, X.T. Zeng, H.J. Du, D. Mondal, Osteoblastic cell response on fluoridated hydroxyapatite coatings: the effect of magnesium incorporation, Biomed. Mater. 5 (2010) 054114 (7 pp). https://doi.org/10.1088/1748-6041/5/5/054114.
[31a] I.A.J. van Hengel, M. Riool, L.E. Fratila-Apachitei, J. Witte-Bouma, E. Farrell, A.A. Zadpoor, S.A.J. Zaat, I. Apachitei, Selective laser melting porous metallic implants with immobilized silver nanoparticles kill and prevent biofilm formation by methicillin-resistant *Staphylococcus aureus*, Biomaterials 140 (2017) 1–15.
[32] V. Karageorgiou, D. Kaplan, Porosity of 3D biomaterial scaffolds and osteogenesis, Biomaterials 26 (27) (2005) 5474–5491.
[33] A.H.P.D. Nouri, C. Wen, Biomimetic porous titanium scaffolds for orthopedic and dental applications, in: A. Mukherjee (Ed.), Biomimetics, Learning From Nature, Intech Open Access Publisher, 2010, pp. 415–450 (Chapter 21).
[34] M. Fousová, D. Vojtěch, J. Kubásek, E. Jablonská, J. Fojt, Promising characteristics of gradient porosity Ti-6Al-4V alloy prepared by SLM process, J. Mech. Behav. Biomed. Mater. 69 (2017) 368–376.
[35] M. Elbadawi, J. Meredith, L. Hopkins, I. Reaney, Progress in bioactive metal and ceramic implants for load- bearing application. in: Advanced Techniques in Bone Regeneration, IntechOpen, 2016, https://doi.org/10.5772/62598 (Chapter 10).
[36] D. Carpenter, B.S. Klosterhoff, F.B. Torstrick, K.T. Foley, J.K. Burkus, S.C.D. Leeh, K. Gall, R.E. Guldberg, D.L. Safransk, Effect of porous orthopaedic implant material and structure on load sharing with simulated bone ingrowth: a finite element analysis comparing titanium and PEEK, J. Mech. Behav. Biomed. Mater. 80 (2018) 68–76.

[37] M. Mounir, M. Atef, A. Abou-Elfetouh, M. Hakam, Titanium and polyether ether ketone (PEEK) patient-specific sub-periosteal implants: two novel approaches for rehabilitation of the severely atrophic anterior maxillary ridge, Int. J. Oral Maxillofac. Surg. 47 (2018) 658–664.
[38] I. Yadroitsev, A. Gusarov, I. Yadroitsava, I. Smurov, Single track formation in selective laser melting of metal powders, J. Mater. Process. Technol. 210 (12) (2010) 1624–1631.
[39] A. du Plessis, D.-P. Kouprianoff, I. Yadroitsava, I. Yadroitsev, Mechanical properties and in situ deformation imaging of microlattices manufactured by laser based powder bed fusion, Materials 11 (2018) 1663.
[40] H. Korn, P. Koch, R. Kordaß, S. Holtzhausen, C. Schöne, B. Müller, R. Stelzer, Adapted scan strategy and slicing tool for improvement of strut precision in lattice structures, in: J.S. Taylor (Ed.), American Society for Precision Engineering—ASPE; European Society for Precision Engineering and Nanotechnology—EUSPEN: ASPE and euspen Summer Topical Meeting "Advancing Precision in Additive Manufacturing" 2018: July 22–25, 2018, Berkeley, California, USA, 2018, pp. 50–54.
[41] H. Korn, P. Koch, R. Kordaß, C. Schöne, B. Müller, S. Holtzhausen, R. Stelzer, Influences of scan strategy and exposure parameters on diameter and surface quality of struts in lattice structures, in: Fraunhofer Direct Digital Manufacturing Conference, March 14–15, 2018, Berlin, Germany, 2018.
[42] S. Ghouse, S. Babu, A.J. Van Arkel, K. Naic, P.A. Hooper, J.R.T. Jeffers, The influence of laser parameters and scanning strategies on the mechanical properties of a stochastic porous material, Mater. Des. 131 (2017) 498–508.
[43] A. du Plessis, I. Yadroitsev, I. Yadroitsava, S.G. le Roux, X-ray micro computed tomography in additive manufacturing: a review of the current technology and applications. 3D Print. Addit. Manuf. (2018) https://doi.org/10.1089/3dp.2018.0060.
[44] T. Hildebrand, A. Laib, R. Müller, J. Dequeker, P. Rüegsegger, Direct three-dimensional morphometric analysis of human cancellous bone: microstructural data from spine, femur, iliac crest, and calcaneus, J. Bone Miner. Res. 14 (7) (1999) 1167–1174.
[45] T. Hildebrand, P. Rüegsegger, A new method for the model-independent assessment of thickness in three-dimensional images, J. Microsc. 185 (1) (1997) 67–75.
[46] T. McNulty, D. Bhate, A. Zhang, M.A. Kiser, L. Ferry, A. Suder, S. Bhattacharya, P. Boradkar, A Framework for the design of biomimetic cellular materials for additive manufacturing. Solid freeform fabrication, in: Proceedings of the 28th Annual International Solid Freeform Fabrication Symposium—An Additive Manufacturing Conference, Austin, Texas, 2017, pp. 2188–2200.
[47] O. Al-Ketan, R. Rowshan, R.K. Abu Al-Rub, Topology-mechanical property relationship of 3D printed strut, skeletal, and sheet based periodic metallic cellular materials, Addit. Manuf. 19 (2018) 167–183.
[48] P. Heinl, L. Muller, C. Korner, R.F. Singer, F.A. Muller, Cellular Ti-6Al-4V structures with interconnected macro porosity for bone implants fabricated by selective electron beam melting, Acta Biomater. 4 (5) (2008) 1536–1544.
[49] Q. Wang, C. Han, T. Choma, Q. Wei, C. Yan, B. Song, Y. Shi, Effect of Nb content on microstructure, property and in vitro apatite-forming capability of Ti-Nb alloys fabricated via selective laser melting, Mater. Des. 126 (2017) 268–277.
[50] Z. Wang, C. Wang, C. Li, Y. Qin, L. Zhong, B. Chen, Z. Li, H. Liu, F. Chang, J. Wang, Analysis of factors influencing bone ingrowth into three-dimensional printed porous metal scaffolds: a review, J. Alloys Compd. 717 (2017) 271–285.
[51] J. Parthasarathy, B. Starly, S. Raman, A. Christensen, Mechanical evaluation of porous titanium (Ti6Al4V) structures with electron beam melting (EBM), J. Mech. Behav. Biomed. Mater. 3 (3) (2010) 249–259.

[52] S.L. Sing, F.E. Wiria, W.Y. Yeong, Selective laser melting of lattice structures: a statistical approach to manufacturability and mechanical behaviour, Robot. Comput. Integr. Manuf. 49 (2018) 170–180.
[53] I. Yadroitsev, Selective Laser Melting: Direct Manufacturing of 3D-Objects by Selective Laser Melting of Metal Powders, Lap Lambert Acad. Publ, 2009. 309 pp.
[54] M.F. Ashby, T. Evans, N.A. Fleck, J.W. Hutchinson, H.N.G. Wadley, L.J. Gibson, Metal Foams: A Design Guide, Elsevier, 2000.
[55] S.Y. Chen, J.C. Huang, C.T. Pan, C.H. Lin, T.L. Yang, Y.S. Huang, C.H. Ou, L.Y. Chen, D.Y. Lin, H.K. Lin, T.H. Li, J.S.C. Jang, C.C. Yang, Microstructure and mechanical properties of open-cell porous Ti6Al4V fabricated by selective laser melting, J. Alloys Compd. 713 (2017) 248–254.
[56] L.J. Gibson, M.F. Ashby, Cellular Solids: Structure and Properties, Cambridge University Press, 1999.
[57] J. Wieding, A. Wolf, R. Bader, Numerical optimization of open-porous bone scaffold structures to match the elastic properties of human cortical bone, J. Mech. Behav. Biomed. Mater. 37 (37) (2014) 56–68.
[58] S.C. Kapfer, S.T. Hyde, K. Mecke, C.H. Arns, G.E. Schröder-Turk, Minimal surface scaffold designs for tissue engineering, Biomaterials 32 (29) (2011) 6875–6882.
[59] F.S.L. Bobbert, K. Lietaert, A.A. Eftekhari, B. Pouran, S.M. Ahmadi, H. Weinans, A.A. Zadpoor, Additively manufactured metallic porous biomaterials based on minimal surfaces: a unique combination of topological, mechanical, and mass transport properties, Acta Biomater. 53 (2017) 572–584.
[60] J. Kadkhodapour, H. Montazerian, A.C. Darabi, A. Zargarian, S. Schmauder, The relationships between deformation mechanisms and mechanical properties of additively manufactured porous biomaterials, J. Mech. Behav. Biomed. Mater. 70 (2017) 28–42.
[61] C. Yan, L. Hao, A. Hussein, P. Young, Ti-6Al-4V triply periodic minimal surface structures for bone implants fabricated via selective laser melting, J. Mech. Behav. Biomed. Mater. 51 (2015) 61–73.
[62] A. du Plessis, I. Yadroitsava, I. Yadroitsev, S.G. le Roux, D.C. Blaine, Numerical comparison of lattice unit cell designs for medical implants by additive manufacturing, Virt. Phys. Prototyp. 13 (4) (2018) 266–281.
[63] A. Panesar, M. Abdi, D. Hickman, I. Ashcroft, Strategies for functionally graded lattice structures derived using topology optimisation for additive manufacturing, Addit. Manuf. 19 (2018) 81–94.
[64] H. Heo, J. Jub, D.M. Kim, Compliant cellular structures: application to a passive morphing airfoil, Compos. Struct. 106 (2013) 560–569.
[65] I. Maskery, L. Sturm, A.O. Aremu, A. Panesar, C.B. Williams, C.J. Tuck, R.D. Wildman, I.A. Ashcroft, R.J.M. Hague, Insights into the mechanical properties of several triply periodic minimal surface lattice structures made by polymer additive manufacturing, Polymer 152 (2017) 62–71.
[66] A. du Plessis, I. Yadroitsava, I. Yadroitsev, Ti6Al4V lightweight lattice structures manufactured by laser powder bed fusion for load-bearing applications, Opt. Laser Technol. 108 (2018) 521–528.
[67] S.Y. Choy, C.N. Sun, K.F. Leong, J. Wei, Compressive properties of Ti-6Al-4V lattice structures fabricated by selective laser melting: design, orientation and density, Additive Manuf. 16 (2017) 213–224.
[68] B. Zhang, X. Pei, C. Zhou, Y. Fan, Q. Jiang, A. Ronca, U. D'Amora, Y. Chen, H. Li, Y. Sun, X. Zhang, The biomimetic design and 3D printing of customized mechanical properties porous Ti6Al4V scaffold for load-bearing bone reconstruction, Mater. Des. 152 (2018) 30–39.
[69] F. Liu, Z. Mao, P. Zhang, D.Z. Zhang, J. Jiang, Z. Ma, Functionally graded porous scaffolds in multiple patterns: new design method, physical and mechanical properties, Mater. Des. 160 (2018) 849–860.

[70] L. Xiao, W. Song, Additively-manufactured functionally graded Ti-6Al-4V lattice structures with high strength under static and dynamic loading: experiments, Int. J. Impact Eng. 111 (2018) 255–272.
[71] F. Liu, D.Z. Zhang, P. Zhang, M. Zhao, S. Jafar, Mechanical properties of optimized diamond lattice structure for bone scaffolds fabricated via selective laser melting, Materials 11 (2018) 374.
[72] C. Han, Y. Li, Q. Wang, S. Wen, Q. Wei, C. Yan, L. Hao, J. Liu, Y. Shi, Continuous functionally graded porous titanium scaffolds manufactured by selective laser melting for bone implants, J. Mech. Behav. Biomed. Mater. 80 (2018) 119–127.
[73] Q. Feng, Q. Tang, Y. Liu, R. Setchi, S. Soe, S. Ma, L. Bai, Quasi-static analysis of mechanical properties of Ti6Al4V lattice structures manufactured using selective laser melting, Int. J. Adv. Manuf. Technol. 94 (2018) 2301.
[74] Y.J. Liu, S.J. Li, L.C. Zhang, Y.L. Hao, T.B. Sercombe, Early plastic deformation behaviour and energy absorption in porous β-type biomedical titanium produced by selective laser melting, Scr. Mater. 153 (2018) 99–103.
[75] L. Yan, Y. Yuan, L. Ouyang, H. Li, A. Mirzasadeghi, L. Li, Improved mechanical properties of the new Ti-15Ta-xZr alloys fabricated by selective laser melting for biomedical application, J. Alloys Compd. 688 (2016) 156–162.
[76] L.C. Zhang, D. Klemm, J. Eckert, Y.L. Hao, T.B. Sercombe, Manufacture by selective laser melting and mechanical behavior of a biomedical Ti-24Nb-4Zr-8Sn alloy, Scr. Mater. 65 (1) (2011) 21–24.
[77] D. Ren, S. Li, H. Wang, W. Hou, S. Zhao, W. Jin, R. Yang, R.D.K. Misra, L.E. Murr, Fatigue behavior of Ti-6Al-4V cellular structures fabricated by additive manufacturing technique, J. Mater. Sci. Technol. (2018), https://doi.org/10.1016/j.jmst.2018.09.066.
[78] A. Khalili, M.R. Ahmad, A review of cell adhesion studies for biomedical and biological applications, Int. J. Mol. Sci. 16 (2015) 18149–18184.
[79] M. Benedetti, E. Torresani, M. Leoni, V. Fontanari, M. Bandini, C. Pederzolli, C. Potrich, The effect of post-sintering treatments on the fatigue and biological behavior of Ti-6Al-4V ELI parts made by selective laser melting, J. Mech. Behav. Biomed. Mater. 71 (2017) 295–306.
[80] C. Eriksson, H. Nygren, K. Ohlson, Implantation of hydrophilic and hydrophobic titanium discs in rat tibia: cellular reactions on the surfaces during the first 3 weeks in bone, Biomaterials 25 (19) (2004) 4759–4766.
[81] D. Zhao, Y. Huang, Y. Ao, C. Han, Q. Wang, Y. Li, J. Liu, Q. Wei, Z. Zhang, Effect of pore geometry on the fatigue properties and cell affinity of porous titanium scaffolds fabricated by selective laser melting, J. Mech. Behav. Biomed. Mater. 88 (2018) 478–487.
[82] S. Van Bael, Y.C. Chai, S. Truscello, M. Moesen, G. Kerckhofs, H. Van Oosterwyck, J.-P. Kruth, J. Schrooten, The effect of pore geometry on the in vitro biological behavior of human periosteum-derived cells seeded on selective laser-melted Ti6Al4V bone scaffolds, Acta Biomater. 8 (7) (2012) 2824–2834.
[83] M. Rumpler, J.W.D. Woesz, J.T. van Dongen, P. Fratzl, The effect of geometry on three-dimensional tissue growth, J. R. Soc. Interface 5 (2008) 1173–1180.
[84] H. Wang, K. Su, L. Su, P. Liang, P. Ji, C. Wang, The effect of 3D-printed Ti6Al4V scaffolds with various macropore structures on osteointegration and osteogenesis: a biomechanical evaluation, J. Mech. Behav. Biomed. Mater. 88 (2018) 488–496.
[85] X.P. Tan, Y.J. Tan, C.S.L. Chow, S.B. Tor, W.Y. Yeong, Metallic powder-bed based 3D printing of cellular scaffolds for orthopaedic implants: a state-of-the-art review on manufacturing, topological design, mechanical properties and biocompatibility, Mater. Sci. Eng. C 76 (2017) 1328–1343.
[86] Z.J. Wally, A.M. Haque, A. Feteira, F. Claeyssens, R. Goodall, G.C. Reilly, Selective laser melting processed Ti6Al4V lattices with graded porosities for dental applications, J. Mech. Behav. Biomed. Mater. 90 (2018) 20–29.

[87] K.-H. Frosch, F. Barvencik, V. Viereck, C.H. Lohmann, K. Dresing, J. Breme, E. Brunner, K.M. Stürmer, Growth behavior, matrix production, and gene expression of human osteoblasts in defined cylindrical titanium channels, J. Biomed. Mater. Res. A 68 (2004) 325–334.
[88] J. Markhoff, J. Wieding, V. Weissmann, J. Pasold, A.J. Heincke, R. Bader, Influence of different three-dimensional open porous titanium scaffold designs on human osteoblasts behavior in static and dynamic cell investigations, Materials 8 (2015) 5490–5507.
[88a] B. Wysocki, J. Idaszek, J. Buhagiar, K. Szlązak, T. Brynk, K.J. Kurzydłowski, W. Święszkowski, The influence of chemical polishing of titanium scaffolds on their mechanical strength and in-vitro cell response, Mater. Sci. Eng. C 95 (2019) 428–439.
[89] W. Xue, B.V. Krishna, A. Bandyopadhyay, S. Bose, Processing and biocompatibility evaluation of laser processed porous titanium, Acta Biomater. 3 (2007) 1007–1018.
[90] Y. Zhang, O. Andrukhova, S. Berner, M. Matejka, M. Wielandd, X. Rausch-Fana, A. Schedle, Osteogenic properties of hydrophilic and hydrophobic titanium surfaces evaluated with osteoblast-like cells (MG63) in coculture with human umbilical vein endothelial cells (HUVEC), Dent. Mater. 26 (2010) 1043–1051.
[91] T. Kokubo, S. Yamaguchi, Novel bioactive materials developed by simulated body fluid evaluation: surface-modified Ti metal and its alloys, Acta Biomater. 44 (2016) 16–30.
[92] M. Todea, A. Vulpoi, C. Popa, P. Berce, S. Simon, Effect of different surface treatments on bioactivity of porous titanium implants. J. Mater. Sci. Technol. (2018), https://doi.org/10.1016/j.jmst.2018.10.004.
[93] N. Kohli, S. Ho, S.J. Brown, P. Sawadkar, V. Sharma, M. Snow, E. García-Gareta, Bone remodelling in vitro: where are we headed?—a review on the current understanding of physiological bone remodelling and inflammation and the strategies for testing biomaterials in vitro, Bone 110 (2018) 38–46.
[94] M. Fukuda, T. Takemoto, S. Saito, M. Fujibayashi, D.K. Neo, T. Pattanayak, K. Matsushita, N. Sasaki, T. Nishida, T.N. Kokubo, Osteoinduction of porous Ti implants with a channel structure fabricated by selective laser melting, Acta Biomater. 7 (2011) 2327–2336.
[95] N. Taniguchi, S. Fujibayashi, M. Takemoto, K. Sasaki, B. Otsuki, T. Nakamura, T. Matsushita, T. Kokubo, S. Matsuda, Effect of pore size on bone ingrowth into porous titanium implants fabricated by additive manufacturing: an in vivo experiment, Mater. Sci. Eng. C 59 (2016) 690–701.
[96] S. Arabnejad, R.B. Johnston, J.N. Pura, B. Singh, M. Tanzer, D. Pasini, High-strength porous biomaterials for bone replacement: a strategy to assess the interplay between cell morphology, mechanical properties, bone ingrowth and manufacturing constraints, Acta Biomater. 30 (2016) 345–356.
[97] A.-F. Obaton, J. Fain, M. Djemaï, D. Meinel, F. Léonard, E. Mahé, B. Lécuelle, J.-J. Fouchet, G. Bruno, In vivo XCT bone characterization of lattice structured implants fabricated by additive manufacturing, Heliyon 3 (8) (2017) e00374.
[98] A. Cheng, A. Humayun, B.D. Boyan, Z. Schwart, Enhanced osteoblast response to porosity and resolution of additively manufactured Ti-6Al-4V constructs with trabeculae-inspired porosity, 3D Print. Addit. Manuf. 3 (1) (2016), https://doi.org/10.1089/3dp.2015.0038.
[99] D.J. Cohen, A. Cheng, K. Sahingur, R.M. Clohessy, L.B. Hopkins, B.D. Boyan, Z. Schwartz, Performance of laser sintered Ti-6Al-4V implants with bone-inspired porosity and micro/nanoscale surface roughness in the rabbit femur, Biomed. Mater. (2017) 025021. 12 pp.
[100] B. Otsuki, M. Takemoto, S. Fujibayashi, M. Neo, T. Kokubo, T. Nakamura, Pore throat size and connectivity determine bone and tissue ingrowth into porous implants:

three-dimensional micro-CT based structural analyses of porous bioactive titanium implants, Biomaterials 27 (35) (2006) 5892–5900.
[101] F. Bai, Z. Wang, J. Lu, J. Liu, G. Chen, R. Lv, J. Wang, K. Lin, J. Zhang, X. Huang, The correlation between the internal structure and vascularization of controllable porous bioceramic materials in vivo: a quantitative study, Tissue Eng. A 16 (2010) 3791–3803.
[102] B. Chang, W. Song, T. Han, J. Yan, F. Li, L. Zhao, H. Kou, Y. Zhang, Influence of pore size of porous titanium fabricated by vacuum diffusion bonding of titanium meshes on cell penetration and bone ingrowth, Acta Biomater. 33 (2016) 311–321.
[103] J. Liu, G. Han, S. Pan, Y. Ge, H. Feng, Z. Shen, Biomineralization stimulated peri-titanium implants prepared by selective laser melting, J. Mater. 1 (2015) 253–261.
[104] M. Figliuzzi, F. Mangano, C. Mangano, A novel root analogue dental implant using CT scan and CAD/CAM: selective laser melting technology, Int. J. Oral Maxillofac. Surg. 41 (2012) 858–862.
[105] A. Chug, S. Shukla, L. Mahesh, S. Jadwani, Osseointegration—molecular events at the bone-implant interface: a review, J. Oral Maxillofac. Surg. Med. Pathol. 25 (2013) 1–4.
[106] J.E. Davies, Understanding peri-implant endosseous healing, J. Dent. Educ. 67 (2003) 931–949.
[107] R.C. Dutta, A.K. Dutta, B. Basu, Engineering implants for fractured bones; metals to tissue constructs, Int. J. Mater. Eng. Appl. 1 (1) (2017) 9–13.
[108] I.V. Shishkovsky, Laser-assisted fabrication of tissue engineering scaffolds from titanium alloys, in: J. Lawrence, et al. (Eds.), Laser Surface Engineering. Processes and applications, Woodhead Publishing Series in Electronic and Optical Materials, Elsevier Science & Technology, 2015, pp. 603–653.
[109] G.F. de Grado, L. Keller, Y. Idoux-Gillet, Q. Wagner, A.M. Musset, N. Benkirane-Jessel, F. Bornert, D. Offner, Bone substitutes: a review of their characteristics, clinical use, and perspectives for large bone defects management, J. Tissue Eng. 9 (2018) 1–18.
[110] L.E. Murr, Frontiers of 3D printing/additive manufacturing: from human organs to aircraft fabrication, J. Mater. Sci. Technol. 32 (10) (2016) 987–995.

Further reading

[111] P.V. Giannoudis, H. Dinopoulos, E. Tsiridis, Bone substitutes: an update, Injury 36S (2005) S20–S27.

CHAPTER 13

Manufacturing technologies for titanium eyeglass frames

Hiroyuki Tada
Charmant Co. Ltd, Technology Development Section, Sabae, Japan

Contents

1 Components of the eyeglass frame and its manufacturing process

First, the components of the eyeglass frames and their required functions are described. It helps to understand why titanium was selected for eyeglass frames. Fig. 1 shows the standard names of the parts of current eyeglass frames.

1.1 Eyewire

The parts that cover the lens—called an eyewire—have been used for ages in the history of the eyeglass frames [1]. To secure the position of the lens firmly in front of the eyes and ensure stable vision, the eyewire must

Titanium for Consumer Applications
https://doi.org/10.1016/B978-0-12-815820-3.00020-4

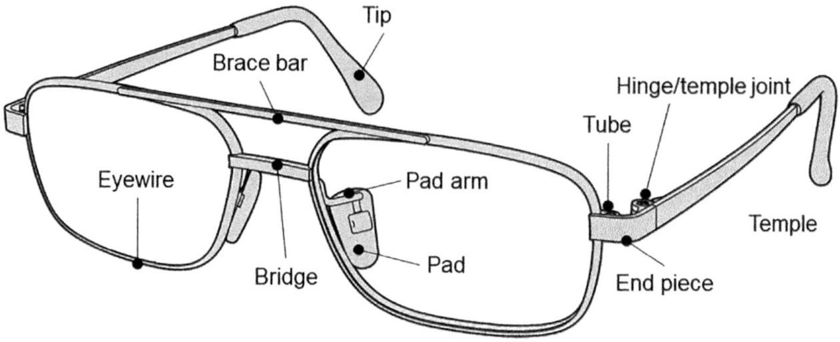

Fig. 1 Components of eyeglass frames and their official names.

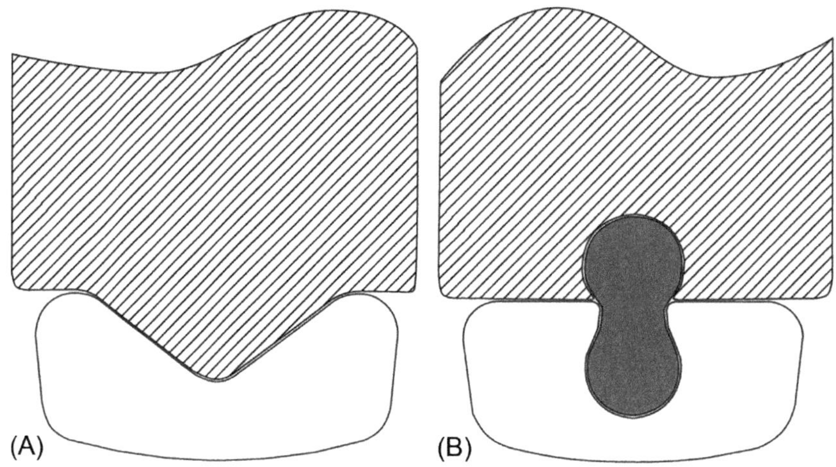

Fig. 2 Cross-sectional shape of two types of eyewire: (A) full eyewire type and (B) semi-rimless type.

have sufficient strength. Furthermore, to place the eyeglass frames in the consistently in the center of the face, the eyewire should follow the lens as much as possible. High fatigue strength is also required to endure stress, which is applied repeatedly from various directions by the wearer's movements. Fig. 2 shows the cross-sectional shapes of the eyewire. The eyewire commonly used is shown in Fig. 2A. The eyewire for a semi-rimless style eyeglass frames is shown in Fig. 2B. It surrounds half the periphery of the lens. The lenses are held with another wire. In the semi-rimless style eyeglass frame, a ditch is scraped into the lenses to fix the lens via a wire having a cross-sectional shape such as a combination of two circles. The unique cross-sectional shape of the eyewire part is manufactured from a square rod

Fig. 3 Three-dimensional curves of eyewire parts.

by cold rolling. This shape helps the lens to not drop off from the eyeglass frames. After rolling, the eyewire part is formed three-dimensionally with a bending machine as shown in Fig. 3.

1.2 Bridge and end-piece

The bridge is responsible for connecting the two lenses. It has a role in holding the lenses at a suitable position for the wearer by keeping the distance between the lenses constant. The end-piece connects the eyewire on the opposite side of the bridge with the temple. The end-piece must suppress the vibration transmitted from the temple. It is effective for maintaining proper positional relationship between the lens and eyes. It is desirable for the bridge and the end-piece to suppress the movement of the lens in the direction of opening and twisting when an external force is applied to the eyeglass frame. From this viewpoint, it is thought that high strength is required for materials used for the bridge and end-piece. These parts are usually formed from coiled or rod-shaped material. These materials are cut into an appropriate length by calculating from the part volume. After cutting, as shown in Fig. 4, a wired or rod-shaped material is placed sideways and cold worked with dies. For parts of eyeglass frames, cold working is generally used to improve strength. The results of cold forging analysis are shown in Fig. 5. Die forging complicatedly deforms the parts of eyeglass frames.

Fig. 6 shows simplified processing steps of manufacturing the bridge and end-piece. Because the bridge requires a three-dimensional shape, bending processing is applied. The end-piece has a two-dimensional shape, so it is formed only by forging and cutting.

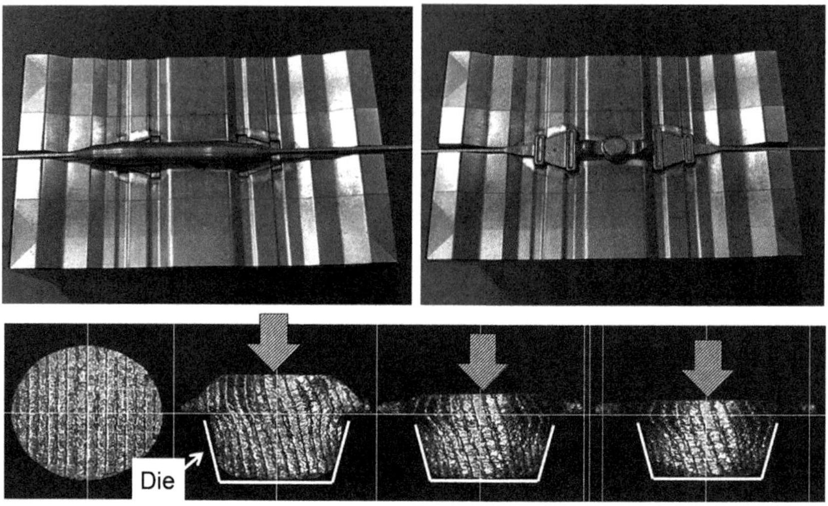

Fig. 4 Die forging for eyeglass frames (end-piece parts).

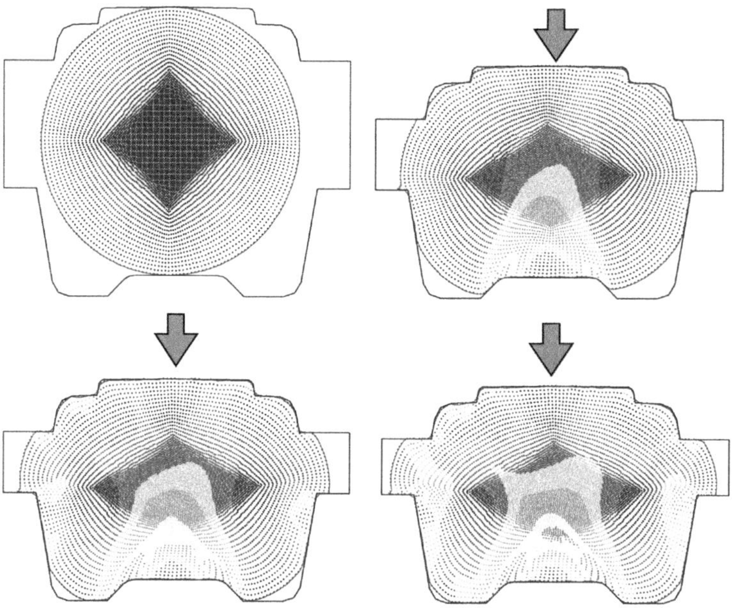

Fig. 5 Plastic flow analysis of die forging for eyeglass frame parts.

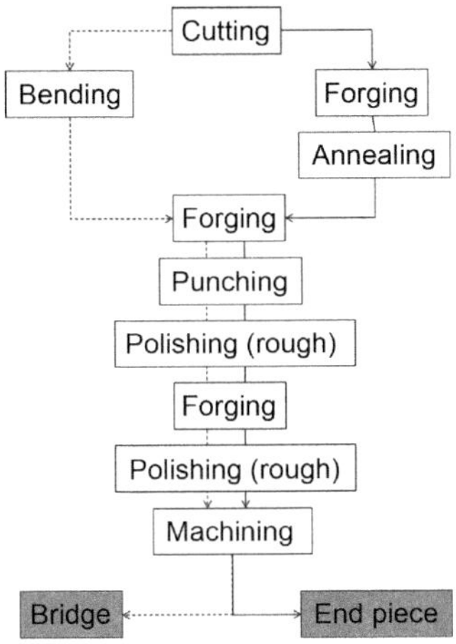

Fig. 6 Manufacturing process of eyeglass frames parts.

1.3 Temple

An unstable field of view arising from eyeglass frames and facial misalignment increases eye strain. To maintain a good view, it is desirable that the lens and eyeball positions are always kept constant. The temple part is required for this function. This part is fixed above the auricle of the wearer and fixes the lens in front of the face through the temporal region. Fig. 7 shows the relationship between positions of the temple and temporal part.

The gap between the curves created by the entire eyeglass frame and the head deforms the temple parts, and a load is generated as a reaction force.

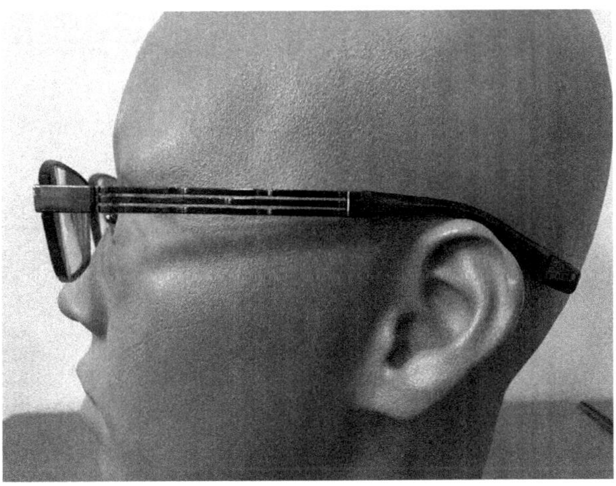

Fig. 7 Position of temple part and auricle.

The lens is kept in the proper position due to the temple and provides a constant load to the temporal part while absorbing external forces. If this load is too small, the lens would be displaced and become an obstacle to securing good visibility. If it is too much, the wearer would feel pain. Because it is preferable that the distance and angle between the lens and the eyes do not change, the front portion of the eyeglass frame is required to have high strength against bending and twisting. On the other hand, the temple parts can distribute the load by being bent along the wearer's head shape. Therefore, it is considered that the flexibility of the temple parts has a strong influence on the feeling of wearing the eyeglass frame. Many people use eyeglass frames around the world today, and there are many variations in the shape of the head. Therefore, the material constituting the temple part is required to have the following functions.

1. Change in load cannot be too large with respect to bending.
2. Will not break easily when shocking force is applied.

The temple is the part developed to fix the lens to the face without using the hands, and its cross-sectional shape varies greatly depending on the place where it fulfills its function. Therefore, swage processing is used to control the diameter of its wire material. In recent years, a method of freely working a wire's diameter by computer control, known as computerized numerical control (CNC) swage processing, is also generally performed. Fig. 8 shows the difference between CNC swage machining and conventional swage machining. The material is stretched while the wire diameter is narrowed by SW processing. As a result, a long part is obtained from a short material, thereby contributing to an efficient use of materials. With regard to the optimum diameter for each location, it is also possible to suppress the consumption of excessive machining energy in subsequent forging.

In eyeglass frames, the shape of the lens and temple parts are the most visible, not just in functionality but also the design required for the part.

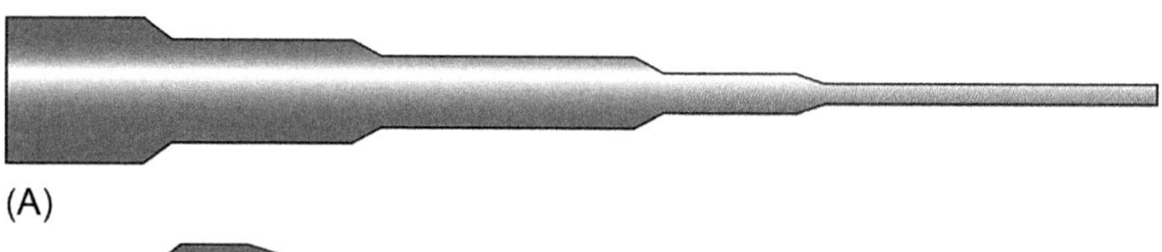

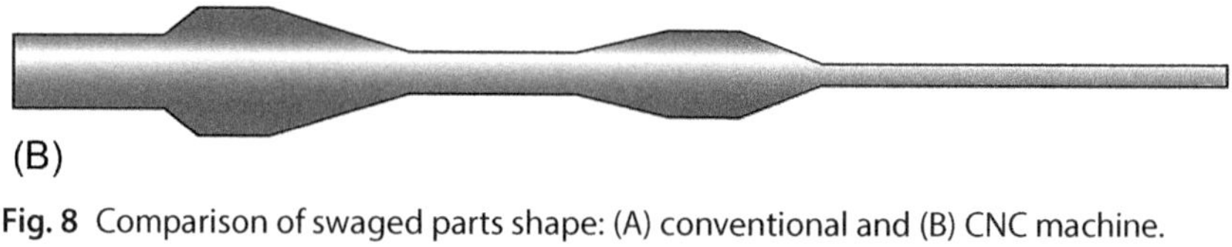

Fig. 8 Comparison of swaged parts shape: (A) conventional and (B) CNC machine.

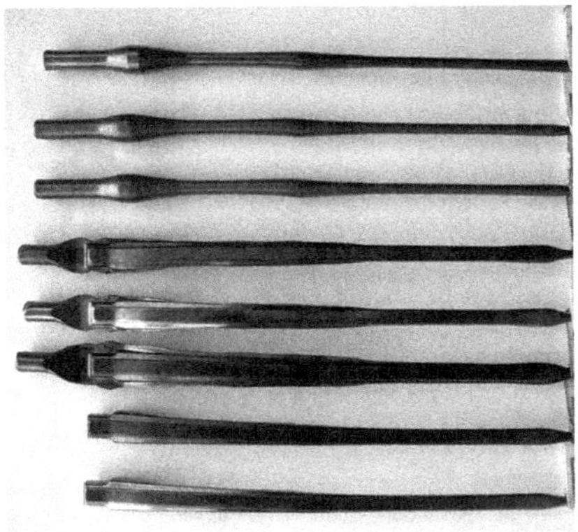

Fig. 9 Manufacturing process of temple parts.

Therefore, the temple parts are designed in a delicate shape, and many steps of processing are executed. The steps for the temple part and its shape changes are shown in Fig. 9.

1.4 Others (tube, hinge, pad arm, temple tip)

If the lens is placed directly on the nose, the eyeball and the lens are too close to each other and the eyelashes come into contact with the lens. This is not only unpleasant but also makes it difficult to keep the lens in a proper position. The positional relationship between the eyes and the nose or the shape of the nose is different for each person, and the eyeglass frames are required to have functions corresponding to these variations. The parts called pads and pad arms were developed to make the eyeglass frames have an adjustment function. It is possible to control the distance between the eyeball and the lens and the vertical position by deforming the pad arm. This component makes it possible to apply eyeglass frames to many people. Pads contact the skin directly and always transmits the load of the lens to the face, so the function to disperse the load and reduce pain is required. Meanwhile, ductility is required for a pad arm to protect the face from damage even if there is an impact due to an accident. The part covering the end of the temple is called a tip, which reduces pain caused by wearing eyeglass frames for a long time. Currently, plastics and elastomers are commonly used as material for pads or tips. There is also a part called a tube for tightening the eyewire covering the lens. This part connects the parts snugly, thus high strength is required. The appearance of these parts before assembly is shown in Fig. 10. In manufacturing eyeglass frames, materials that satisfy these functions are selected. The reason why titanium and its alloy started to be used for eyeglass frames will be explained in the next section.

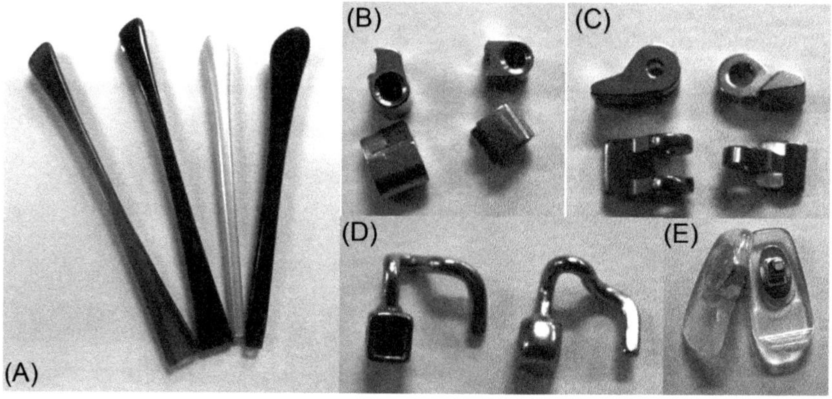

Fig. 10 Small parts for eyeglass frames: (A) tips, (B) tubes, (C) hinges/temple joints, (D) pad arms, and (E) pads.

2 History of materials transition for eyeglass frames—Until titanium is used

2.1 Transition of materials for eyeglass frames

There is a record that eyeglasses were invented for the correction of vision in the northern part of Italy at the end of the 13th century, and this is considered the prototype of the modern eyeglass frame [1]. The eyeglass frame at that time was one in which a convex lens was placed on the page so that small letters could be magnified and read. After that, the lens was surrounded by materials such as wood to make it easy to hold in the hands, and the precious lens was protected from damage. The parts making it easier to hold the lens could be considered the beginning of current eyeglass frames. Natural materials derived from animals and plants were used as the material for the initial eyeglass frames. The reason for this is believed that it was easier to process into a shape conforming to the lens' shape, and it was relatively durable. Since then, eyeglass frames' shape and materials have changed with the times. At first, improvements were made to the handpiece section to make it easy to carry, and monocular glasses like the current magnifying glasses were developed. After a while, the concave lens was invented, and a method of fixing the lens in front of the face was developed for a person with myopia, correcting their vision. The length of time that people use eyeglass frames within the day steadily increased, and the demand for use in various situations had also increased.

A design that fixed two lenses in front of the face that would lead to the current eyeglass frame was invented in the 14th century. Such eyeglass

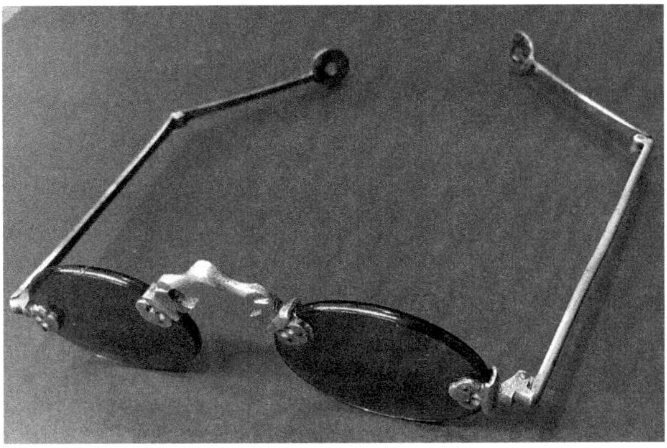

Fig. 11 Eyeglass frames close to modern products.

frames are called "Rivet spectacles." Until the 17th century, eyeglass frames in which the left and right lenses were fixed with a wire-shaped eyewire were used. It was clamped on the nose and fixed the lenses in front of eyes. In the 17th century, eyeglass frames were developed in which the temple extended from the side of the lens toward the ears. As shown in Fig. 11, this product is close to the design of current glasses frames. The frames in those days also had the aspect of showing off one's wealth and education, so expensive and valuable materials such as ivory, tortoiseshell, gold, and silver were preferably used. It seems that these frames were produced by craftsmen working with various techniques, so each frame was suitable for expensive materials. The material of the eyeglass frame was delicately patterned by the technique of engraving, which was not a tool of everyday life but a work of art.

In the 18th century, eyeglass frames of steel materials with high strength were developed using the steelmaking technology that had evolved greatly in the Industrial Revolution. In the 19th century, high quality stainless steel was able to be manufactured inexpensively and in large quantities. The equipment and technology for mass production of high quality products was also established, so that eyeglass frames permeated through the commoner class. In addition to having high corrosion resistance, stainless steel has low specific gravity compared to nickel alloy, and it has been actively used as a material for eyeglass frames, especially in Europe. It can be presumed that Germany had a technology for producing high-quality stainless steel material. In the case of Japan, nickel, silver, or monel metal was used actively as a spectacle frame material. These materials are used instead of silver alloys, showing a well-balanced property with excellent corrosion resistance and strength. However, as indicated in the EU directive of 1999, metallic allergies were regarded as a problem in the late 20th century. Under the circumstances, development of eyeglass frames using titanium as an alternative to nickel alloy and nickel-containing stainless steel was begun. Development of processing technology was necessary to produce titanium eyeglass frame. It was accompanied by difficulties, but in 1981, titanium eyeglass frames were commercialized in Japan for the first time in the world. As a matter of fact, the history of titanium eyeglass frames is less than 40 years old.

2.2 Titanium eyeglass frame parts

As previously described, the parts constituting the eyeglass frame are required to have different properties. For this reason, it is realized by using different forms of titanium and its alloy for each part in titanium eyeglass frames. Fig. 12 shows the representative metallic materials used for eyeglass frames. Among these materials, how the titanium and its alloys are used for each part of eyeglass frames is described in this section.

2.2.1 Durability

Eyewire is always subjected to a tensile force to hold the lens securely. Furthermore, as forces to the twisting direction are applied from the parts joined to eyewire, failures of eyewire for eyeglass frames is presently still a big problem. Titanium eyewire initially used pure titanium, but due to problems such as eyewire breakage caused by long-term use, it is not currently used very much. Fig. 13 shows the load directions applied to eyewires. The eyewire part needs to be fit along the lens shape to prevent the lens from coming off. Thus, the material for the parts needs to have strength and suppleness.

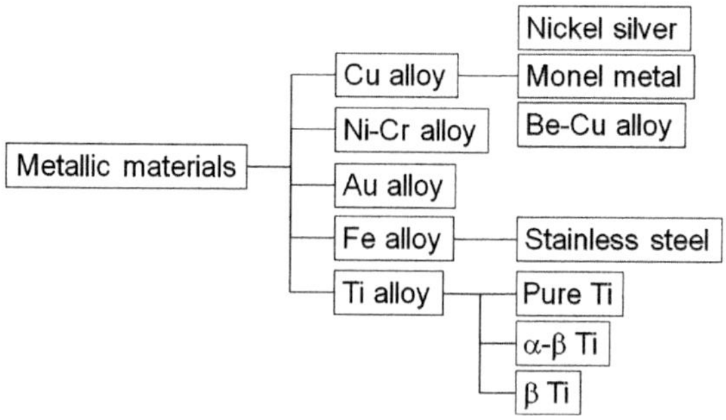

Fig. 12 Representative metallic material for the eyeglass frames.

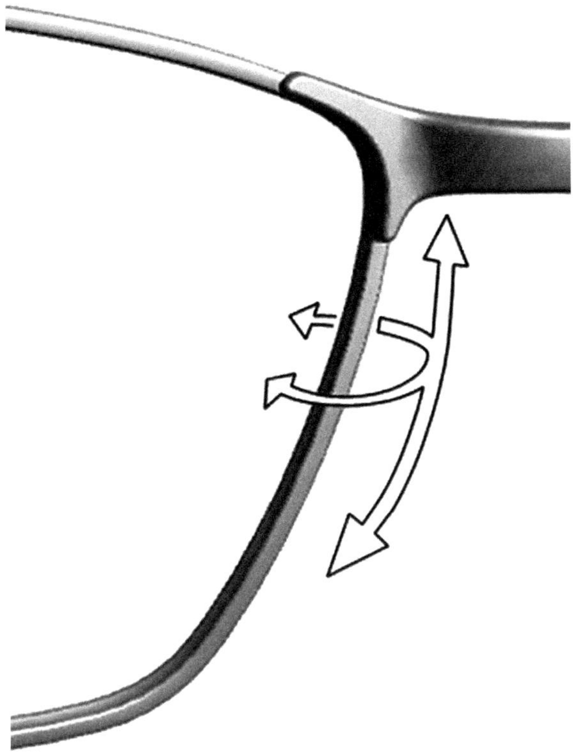

Fig. 13 Stress directions for eyewire parts.

Titanium alloys were examined as alternative parts material from pure titanium to satisfy these requirements. α-type titanium alloys typified by Ti-10Zr [2] and β-type titanium alloys such as Ti-3A-2.5V, and in recent years eyewires of β-type titanium alloys such as Ti-4Al-12V or Ti-15V-3Al-3Sn-3Cr are also commonly used [3]. In general, the β-type titanium alloy has high strength and low Young's modulus, and it is suitable as a material of eyewire parts. However, due to its wide elastic deformation range, there is a problem that the shape after processing is not stabilized even if bending deformation is added. A sufficient annealing must be applied before processing to obtain eyewire of a shape without variation. On the other hand, if the grain size is enlarged by excessive annealing, surface wrinkle failure and fatigue strength reduction could occur. Appropriate management of processing conditions is important. The relationship between surface wrinkle and average grain size after bending is shown in Fig. 14. Compared with the part with small grain size, a rough curved surface is obtained after the bending deformation in the part with large grain size.

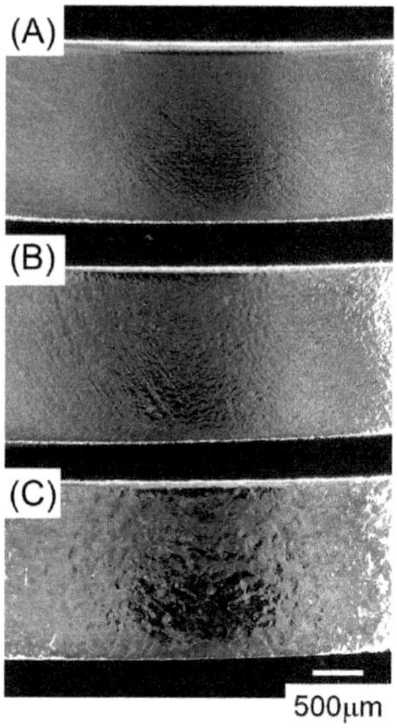

Fig. 14 Relationship between surface wrinkles and average grain size after bending. Average grain size: (A) 7.8 μm, (B) 14.5 μm, (C) 34.5 μm.

2.2.2 Strength and rigidity

Pure titanium has high flexural rigidity, and its price is reasonable compared to other titanium alloys. Although its maximum strength is relatively low, it is the first choice candidate for parts that have volume that won't be deformed by external forces. For these reasons, pure titanium is often used for the bridge and end-piece of titanium eyeglass frames. These parts do not easily deform, even when receiving a load from the temple due to the movement of the wearer, and will not change the positional relationship of the lens and eyes. Pure titanium also shows better processability than other titanium alloys. Therefore, it is possible to manufacture delicate patterns and parts with complicated shapes. β-Type titanium alloy is more suitable for small or elongated parts that require high elasticity.

2.2.3 Comfortability

In recent years, some eyeglass frames have been proposed that emphasize fit. As a feature of these eyeglass frames, many modifications have been made on temple parts to fix the eyeglass frame to the head without discomfort. This is because the temple part is thought to play an important role as an interface connecting the lens and user's head. The fit of the eyeglass frame is thought to depend on the magnitude and direction of pressure received from the temple part. It is effective to increase the contact area with the face by widening the temple part to reduce pressure. However, considering that it is a product to be worn on the face, the decorativeness loss due to an increase in size of parts is concerning. Therefore, applying various kinds of titanium alloys for the temple parts and distributing the load by utilizing the bending properties of the material have been proposed. As an applied material, a low Young's modulus titanium alloy developed for a biomedical material and a shape-memory alloy showing super elasticity at room temperature were also considered. In the eyeglass frame industry, TiNi alloy was adopted as a material for temple parts after pure titanium. This product retains its original shape even if it greatly deforms, so it has big impact as a tough eyeglass frame. This material combines superelastic properties and high strength; strong support is still obtained as a material for spectacle frames [4]. Compared to pure titanium, TiNi alloy has poor cold workability, so it is difficult to make parts with complicated shapes. Furthermore, even if a precise patterned or complicated shaped part is produced, the notch sensitivity of the material is high, thus the gap becomes a risk of failure due to stress concentration. Under the circumstances, β-type titanium alloy was adopted to eyeglass frames. This material was evaluated as very suitable for a temple part because of its high strength and toughness. Even now, β-type titanium alloy is the main material for titanium eyeglass frames. Research and development of new alloys with emphasized properties of β-type titanium alloy are actively carried out. For example, "BS-Ti" is designed to reduce the Young's modulus while considering its biocompatibility, such as not using heavy metal elements [5]. Similarly, "GUM metal" is an alloy with a low Young's modulus, but many unique properties at the time of deformation have been reported, and it is a promising material for eyeglass frames as well as for medical use [6–8]. "Biotitanium" also has high workability and attracts attention as a material that can obtain optimum Young's modulus by appropriate processing and heat treatment [9, 10]. Fig. 15 is a comparison of stress-strain curves obtained from titanium and titanium alloys under the condition of processed eyeglass frame parts.

Here, a mechanism in which the low Young's modulus material contributes to the fit of the eyeglass frames will be described. Human head shape varies according to age, sex, etc., as well as foot size or body shape. However, the eyeglass frame does not have many size variations like shoes or clothes. Even for the same eyeglass frame, some people feel they are too loose and the others feel they are too tight. It depends on the size of wearer. The load on the temple caused by deformation of the eyeglass frame according to the

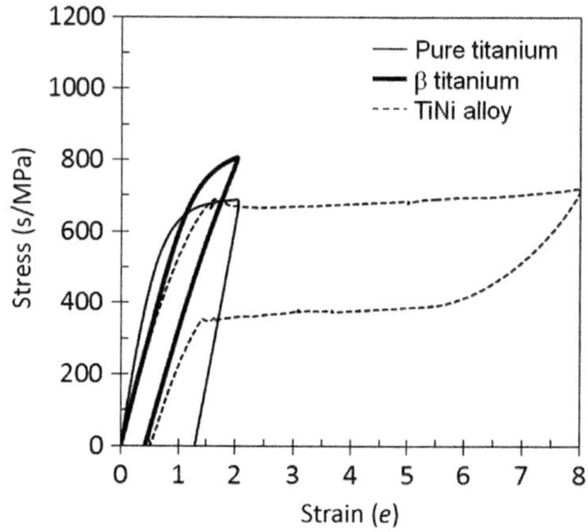

Fig. 15 Stress-strain curves of titanium alloys for temple parts.

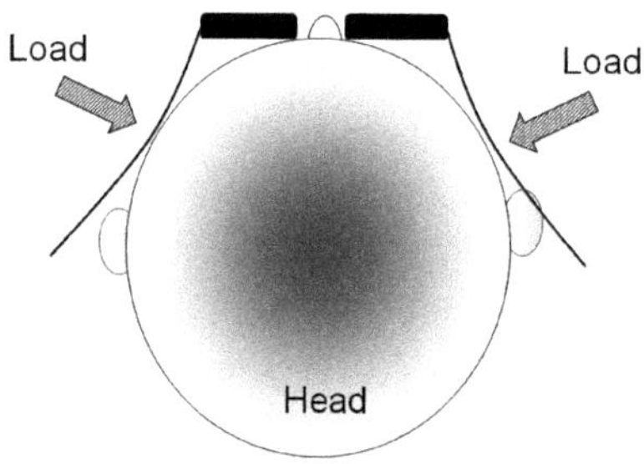

Fig. 16 Load given from temple parts to head.

wearer is shown in Fig. 16. The relationship between the deformation of the temple obtained with conventional materials and the load on the wearer is shown in Fig. 17A. As shown in Fig. 17B, the deformation range to generate a proper fit of the eyeglass frame is wider for the low Young's modulus material. In the case of superelastic material such as TiNi alloys, the range is further widened as shown in Fig. 17C. By using a material exhibiting low Young's modulus or superelasticity for the temple parts, the range of suitable fit can be further expanded, and eyeglasses frames with a comfortable feeling are expected.

In recent years, many reports on β-type titanium alloys exhibiting superelasticity at room temperature have been made. Active development has also been done to apply titanium alloys having both superelasticity and cold workability to eyeglass frames. "Excellence Titanium" is one of the materials that succeeded in practical use for spectacle frames [11]. Fig. 18 shows the results of comparing superelasticity of "Excellence Titanium" and

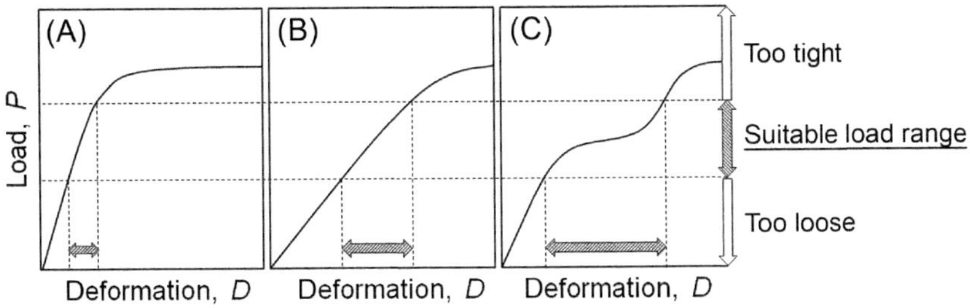

Fig. 17 Relationship between the amount of temple deformation and the load: (A) pure titanium, (B) β-type titanium (Low Young's modulus), and (C) superelasticity (TiNi alloy).

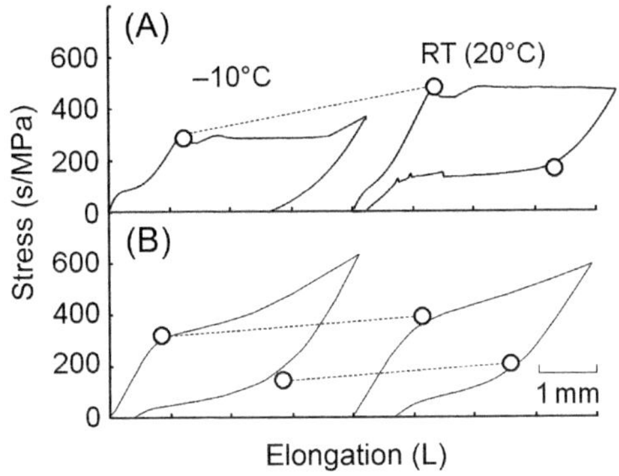

Fig. 18 Temperature dependence on superelasticity: (A) TiNi alloy and (B) Excellence Titanium.

TiNi alloy at room temperature and −10°C. Compared with the TiNi alloy, the titanium-based superelastic alloy has less change in properties due to temperature change in general. Thus it can be expected to exhibit stable properties in a wide temperature environment. In the next section, the manufacturing technique for titanium eyeglass frame will be described.

3 Manufacturing technology of titanium eyeglass frames

In the middle ages, selected craftsmen made spectacle frames. They made the frames while hitting or bending a metal. Therefore, it was often made from a wire or rod-shaped material. Even in current eyeglass frames, materials are generally supplied in the form of wires and bars, and formed to the parts. The original shapes of general material for eyeglass frames are shown in Fig. 19. It was the Sabae area (Fukui prefecture in Japan) that established the manufacturing technology of titanium eyeglass frames for the first time in the world. Before that, the primary materials of Japanese eyeglass frames

Fig. 19 Original shapes of materials for eyeglass frames.

were copper or nickel alloy. Processing for titanium required different technology from processing for copper and nickel alloy, and many difficulties were involved in practical application. In this section, the function of eyeglass frames for vision correction, the processing technology for realizing that function, and each processing method will be explained.

3.1 Machine processing

From the history of eyeglass frames manufacturing, the wire or bar-shaped material was mainly used. The parts are mass-produced by forging. Because surface quality is also important, cold forging is performed on pure titanium and β titanium alloy. The Ti-6Al-4V alloy has poor cold workability, and incremental processing such as cold rolling and swaging or cutting is performed.

3.1.1 Cold forging

In the cold forging of titanium, releasability from a die and forging scratch due to fine deposits on a die become a big problem. It is thought that such a fine deposit is generated by the strong oxide layer of the titanium surface being destroyed by plastic deformation and the newly formed surface coming into contact with the die. The results of X-ray fluorescence analyis of a die surface after titanium forging is shown in Fig. 20. Titanium particles were confirmed on the die where materials made contact. For this reason, die forging of titanium needs extra process to improve releasability. As a technique for improving releasability, anodic oxide film processing is generally performed. It is aimed at reducing the contact between the newly formed surface and the die by forming an oxide layer on the titanium surface. Further, as a die lubricant, boron nitride powders are also commonly used. However, at the current state, adhesion cannot be completely inhibited, and further technical development is necessary.

Here, the die forging workability of titanium in manufacturing eyeglass frames will be described. Fig. 21 shows the relationship between the reduction of area and the forging load of pure titanium and β titanium alloy.

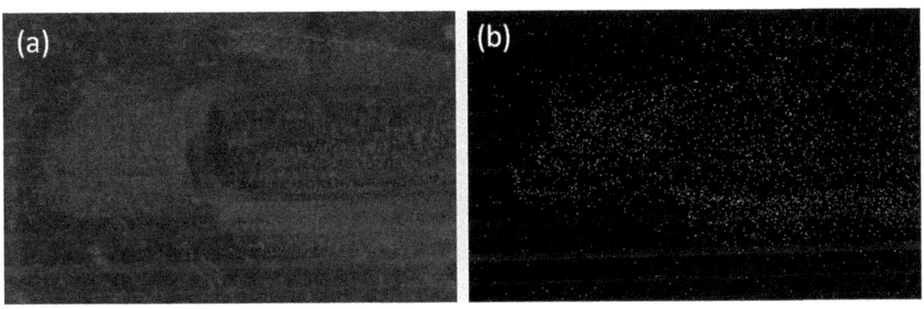

Fig. 20 X-ray fluorescence analysis of die surface after titanium forging. (A) Microscopic observation and (B) distribution of titanium particles.

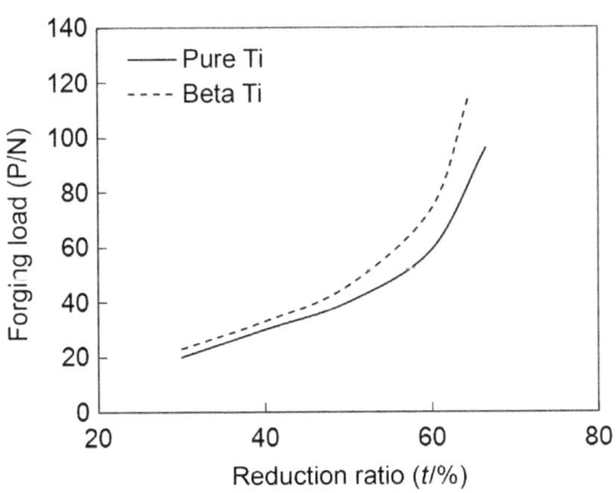

Fig. 21 Relationship between forging load and reduction ratio after forging.

Both materials increase the reduction of the area with an increasing forging ratio. As the reduction of area increases, the forging load of the β titanium alloy becomes larger than that of the pure titanium. For example, in forging with a reduction of area to 40%, a forging load of both materials is almost same. However, when the reduction of the area reaches 60%, the forging load of β titanium is about 1.25 times that of pure titanium. In addition, the cold forging of titanium has a larger springback compared to nickel silver, and it is difficult to suppress variations in component dimensions. The relationship between material springback and reduction of area is shown in Fig. 22. The vertical axis shows the difference in thickness between nickel silver and each material forged under the same condition. The β titanium alloy shows a larger springback when the reduction of area is less than 20% or greater than 50%. Because the processability of titanium varies depending on its type, it is necessary to study optimum processing conditions for each material for parts with stable dimensional accuracy.

3.1.2 Cutting

The bridge parts are milled to improve the joining strength with eyewire. This is a process of making a ditch along the shape of the eyewire at the end of the bridge. An example of a milled part is shown in Fig. 23. Because this processing greatly affects the joining quality of bridge and eyewire, high dimensional accuracy is required. However, variation in dimensional accuracy due to abrasion of the cutting blade has become a problem in this process. Compared to copper alloy, titanium has low thermal conductivity, high heat is generated locally at the cutting point, and the tool easily wears out. This problem was solved by optimizing the cutting amount, cutting

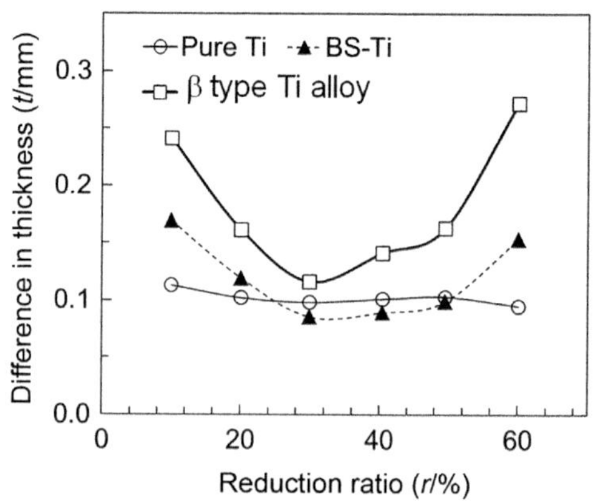

Fig. 22 Comparison of springbacks for titanium and its alloys by forging.

Fig. 23 Shape change before and after milling when making the ditch on a bridge part.

speed, cutting fluid, etc. to reduce heat generation accompanying titanium cutting. During the forging and cutting process, hardness adjustment by annealing, fine scratches, and burr removal by acid treatment or buff polishing are performed several times.

3.2 Joining process

The eyeglass frames are composed of several distinct parts, and these are joined together. Representative joining points of eyeglass frames are shown in Fig. 24. Typical eyeglass frames have about 10–16 joining points.

As a joining technique for eyeglass frames, brazing has been performed for a long period of time, and it is a method that is still commonly used today. The high frequency heating was used first as the heat source of brazing

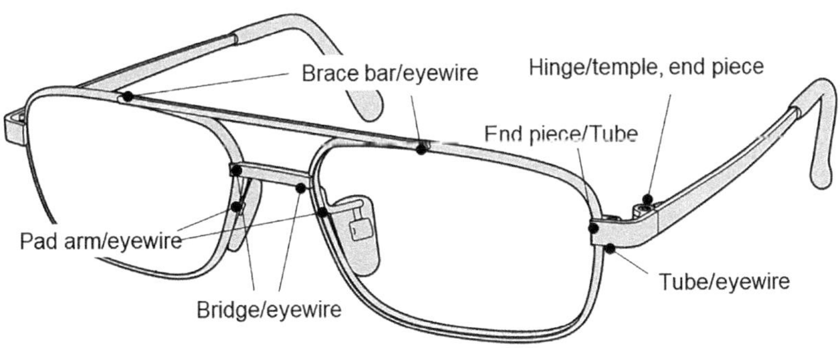

Fig. 24 Representative joining points of eyeglass frames.

the eyeglass frames. It was changed to resistance heating to adapt mass production. Fig. 25 shows a schematic diagram of high frequency brazing and resistance brazing.

Brazing is a method in which the filler metal melts at a temperature lower than the melting point of the base material, and it flows into the gap of the joining point by capillary phenomenon to join the parts. Fig. 26 is an enlarged view of the brazed part of the eyeglass frames. Because the filler metal smoothly fills the gap between the parts, the shape of the joint point becomes rounded. The rounded shape disperses the stress concentration at the joint point and is effective in increasing the fatigue strength of the joint.

Ni alloys and copper alloys were mainly joined by high frequency brazing, and enough strength was obtained by using an antioxidant and flux in combination with silver solder. However, titanium easily reacts with elements in the air such as oxygen, nitrogen, hydrogen, etc. to form a brittle

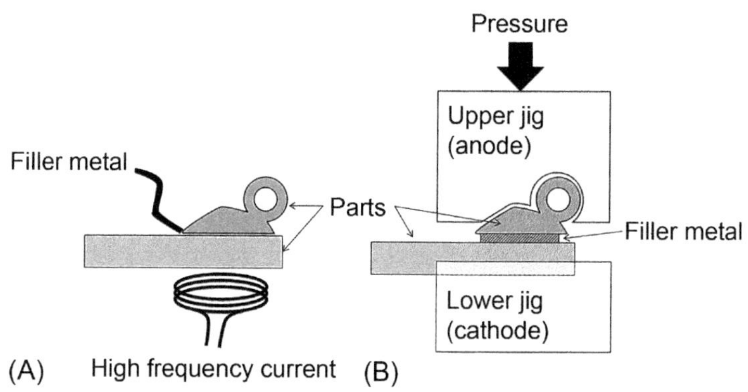

Fig. 25 Schematic diagram of two kind of brazing methods: (A) high frequency brazing and (B) resistance brazing.

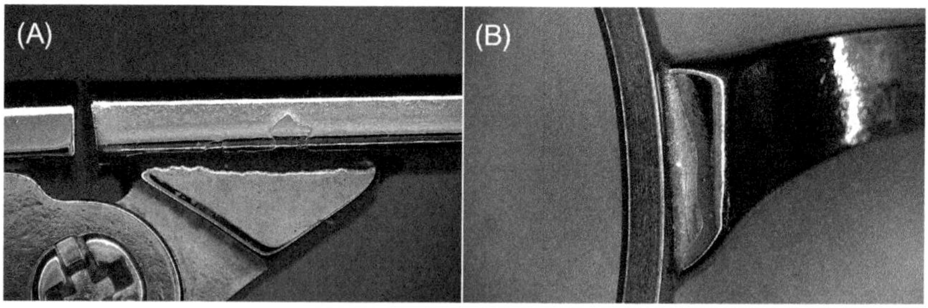

Fig. 26 Appearance of brazed part of eyeglass frames. (A) Hinge/temple and end-piece and (B) bridge/eyewire.

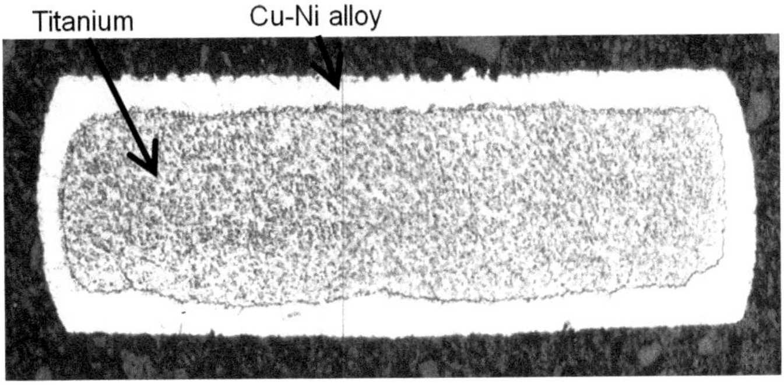

Fig. 27 Cross-section of titanium filler metal (Ti-Cu-Ni).

layer. This layer inhibits the stable joining strength. Therefore, Ni plating method was adopted for joining titanium parts with high frequency heating with silver braze at first. High frequency brazing requires only a simple jig. By using silver braze, it was possible to reheat after joining to correct the part's position. After that, a titanium filler metal capable of obtaining a high joining strength was developed, but the melting temperature was 950°C or more. Delicate temperature control is required to join the titanium parts using developed filler metal, and mass production with high frequency brazing is difficult [12, 13]. A cross-section of titanium filler metal (TiCuNi) is shown in Fig. 27. The inside is titanium, and it's covered with Cu-Ni alloy.

Table 1 shows the silver-based or titanium-based brazing filler materials and their melting points used for manufacturing eyeglass frames.

Titanium is a material suitable for resistance brazing because it has low thermal conductivity and high electrical resistance compared to the material that has been used for eyeglass frames until now. In resistance brazing, the contact resistance is stabilized by applying an appropriate load to the parts. This load on the heating part causes deterioration of appearance quality such as scratches and parts deformation. There are various setting conditions to control the appearance quality and joining strength, such as current, current period, pressure direction, and timing. Furthermore, it is not necessary

Table 1 Brazing filler metals and their melting points

Elements	Melting temp. (°C)	Comment
Ag, Cu, Zn, Sn	650	For Cu or Ni alloy
Ag, Cu, Zn, Ni	750	For Cu or Ni alloy
Ti, Cu, Ni	950	High strength
Ti, Zr, Cu, Ni	850	For strength supplement

to monitor the joining process allowing a worker to perform other work concurrently. This method fits well, especially for line production.

3.3 Surface modification

3.3.1 Polishing

As one of the surface modifications, buffing or wet barrel grinding is carried out to remove scratches and burrs. For relatively long parts of eyeglass frame parts, an automatic buffing machine as shown in Fig. 28A is used. Wet barrel grinding is carried out for small or short parts with a ceramic grindstone and water in a sealed vessel such as Fig. 28B and C.

On the other hand, the appearance of the eyeglass frames is important because of it's located in the center of the face. Therefore, a technique of finishing processing aimed at surface gloss is also indispensable. As a final polishing technique to obtain surface gloss, dry barrel polishing is used in eyeglass frame manufacturing. This polishing requires a long time such as several hours to 24 h. Fig. 29 shows the chips for (A) dry barrel polishing or (B) wet barrel grinding. Among the various types of media used for dry barrel polishing, crushed walnuts and similar compounds are commonly used in manufacturing eyeglass frames.

Fig. 28 Buff polishing and barrel grinding machine: (A) automatic buffing machine, (B) wet barrel machine, and (C) vessel for wet barrel machine.

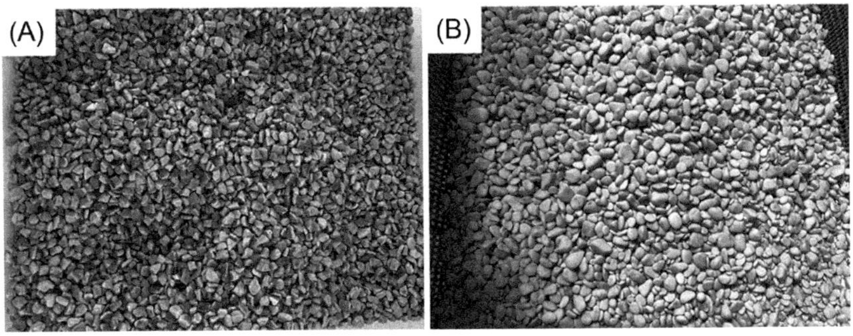

Fig. 29 Chips of barrel process for eyeglass frames: (A) dry polishing and (B) wet grinding.

3.3.2 Surface treatment

Brilliant colors or unique patterns are also essential elements for spectacle frames. Therefore, coating or plating is also applied to titanium eyeglass frames for decoration. The technique of plating applied to eyeglass frames is electroplating in general. The parts are immersed in a chemical solution with the ionic form of the metal, and a nonionic metal layer is produced by supplying an electron. Gold or chromium, nickel, and palladium are used for plating film. Plating on titanium was poor in adhesion with other metal elements, and many experiments were necessary to find a combination with good compatibility. After that, a technique to apply gold plating directly on titanium surface was established; it created a spectacle frame composed entirely of titanium parts. In recent years, besides gold, a technique of Pd plating has been developed contributing to an increase in color choices [14].

A coating is a method of covering with a plastic layer on eyeglass frames. Techniques of spray coating or electrochemical coating are usually used. Various methods such as multicolor or gradation color have been developed. The painting of eyeglass frames consists of generally three or more layers as shown in Fig. 30. It is a layer called primer, which directly applies to the substrate and has a role of enhancing adhesion with the base material. Even in coating, primers that have been used for copper or nickel alloy could not be used for titanium. Therefore, in coating for titanium, Ni plating had to be done on the substrate initially. Primers excellent in adhesion with titanium were developed, and direct coating to titanium became possible. A second layer for coloring and a third layer (top coat) for protecting the coating film are formed on the primer.

Although color tone is an important factor in these surface treatments, durability in various environments is also focused. These surface treatments are required to have fastness that does not easily peel off or crack even when

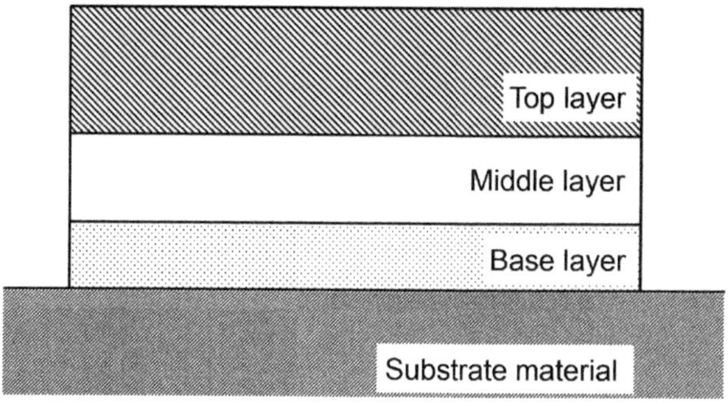

Fig. 30 General layer structure of the coating for eyeglass frames.

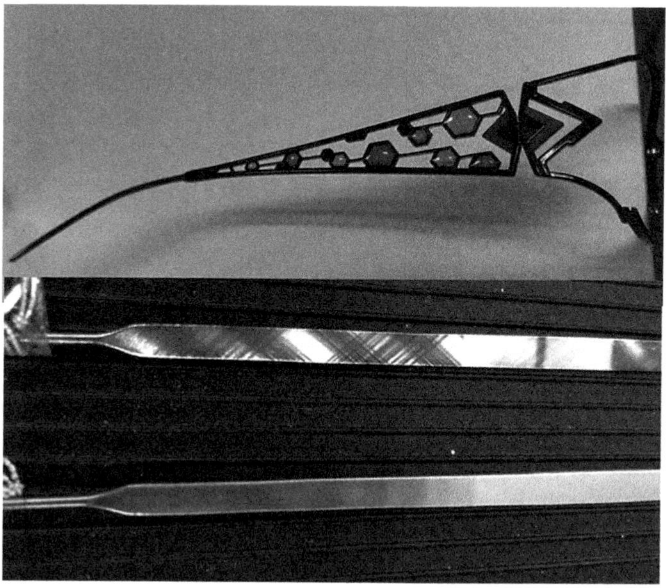

Fig. 31 Various color specifications for eyeglass frames.

exposed to hot and iced water or salt water. The eyeglass frames can be used during activities such as housework, sports, and outdoors by satisfying this requirement. When wearing or removing the eyeglass frame, the parts repeatedly undergo elastic or plastic deformation to fit the shape of the wearer's head. Of course, even if the surface treated film of the part is deformed, it requires proper adhesion.

Fig. 31 shows titanium eyeglass frames subjected to various decorations by surface treatment.

4 Latest processing technology and future challenges

A lot of information is shared all over the world in real time, and the changing speed of function, color, and shape required for the eyeglass frames are getting faster. The way of using eyeglass frames like an information terminal, which nobody has ever before imagined, has been proposed lately. Responding to such a market, not only a technology for producing a small number of various parts but also a technology for manufacturing a product in a short period, are required. Over the last decade, the eyeglass frames manufacturing industry has been actively incorporating new processing technologies to meet market needs. In this section, these recently adopted processing technologies are introduced.

4.1 Servo press machine

Most titanium eyeglass frame parts are manufactured by deforming the material plastically with a forging machine. Conventional forging machines work automatically, and it was impossible to control the working speed. Deforming the material generates heat according to the given energy. When deforming the material to strain ε with deformation resistance Y, the heat generated by forging is calculated by the following equation [15]:

$$\Delta T = \frac{\eta}{C\rho} \int Y d\varepsilon$$

Here, η is work efficiency, C is specific heat, and ρ is density. Normally it is said that 90% of the work given is converted to heat. When the deformation resistance of pure titanium and β titanium is 750 and 1000 MPa, respectively, and the materials are forged until strain 1, then calorific value of pure titanium is about 280°C and the value of β titanium is about 370°C. Heat generated by cold forging largely differs depending on the material. Titanium has low thermal conductivity, and processing heat generated locally by cold forging is hard to diffuse. This uneven heat distribution is considered to be one of the factors that lower the part's strength or the uniformity of its dimensions. In addition, as compared with pure titanium, because the titanium alloy has high deformation resistance, the forging load becomes excessive and the risk of damaging the die increases. It is a natural idea to perform forging processes multiple times to prevent die breakage. However, forging involves processes such as annealing, acid treatment, cleaning, etc., and multiple forging processes has a risk of increasing manufacturing cost and time. Therefore, utilization of a servo press machine

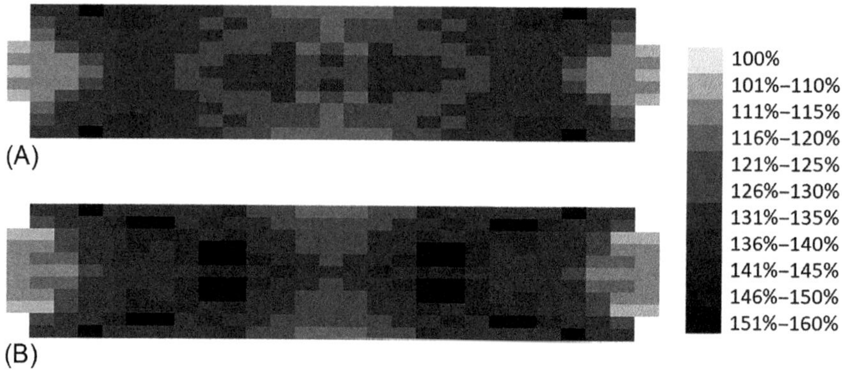

Fig. 32 Relationship between increasing ratio of hardness and forging speed: (A) 84 mm/s and (B) 670 mm/s.

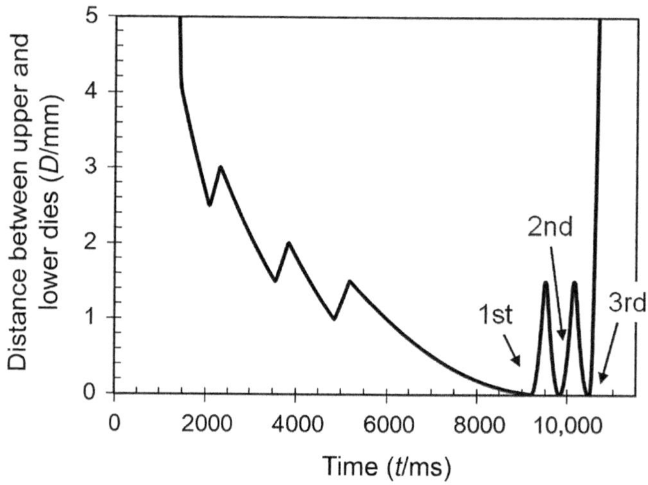

Fig. 33 Tested pattern of die moving.

that can generate high machining energy in the low speed region and can control the operating speed of the die was considered. Fig. 32 shows the relationship between the hardness increase ratio of the part cross-section in cold forged β titanium and the forging speed. When the material was forged at 84 mm/s die speed, the hardness is lower and more uniform.

With this result, research that changed multiple forge processes to single process was tried. Fig. 33 is one of the tested patterns of die operation in this research; the vertical axis shows the distance between the dies, and the horizontal axis shows the time. In this example, the part is forged three times in one process.

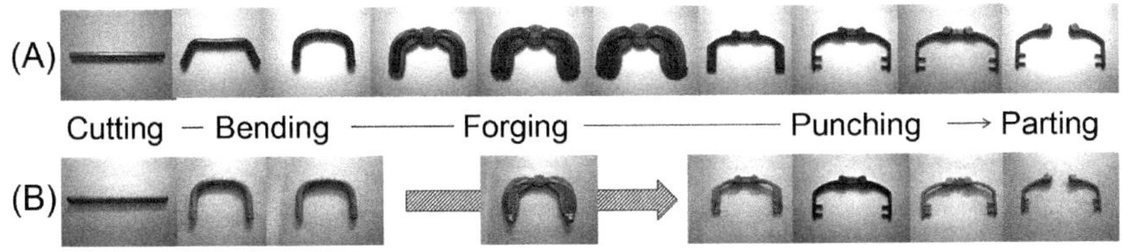

Fig. 34 Effect of servo forming methods. (A) Conventional process and (B) servo press method.

The die moving speed is set to be slower than the conventional one, and an increasing locally forging hardness is suppressed. A usage example of this technology is shown in Fig. 34. A conventional manufacturing process of β-type titanium alloy parts is shown in Fig. 34A, and a developed manufacturing process obtained in this research is shown in Fig. 34B. Due to the development of forging technology with a servo press machine, the number of manufacturing processes steps is greatly reduced from 28 to 17, contributing to shortening the product manufacturing period.

4.2 Laser microwelding

Because the eyeglass frames is a product that affects the impression of the wearer, appearance quality is emphasized. Obviously, high quality of appearance is also required for the joining process. This requirement becomes stronger with the added value of the product becoming higher. On the other hand, not only the appearance but also joining strength is required for the joining process. Among high-value-added spectacle frames, products delicately composed with thin wire parts of 1 mm or less diameter have been commercialized and achieve a more sophisticated design than ever. It is necessary to develop control technique to achieve a joining quality for the products. With the existing technology, it was difficult to achieve compatibility between appearance and strength with respect to joining of fine titanium parts, particularly parts having a big difference in volume. This is because the following problems cannot be avoided with existing technology:

1. The wide heat affected zone reduces the material strength around the joined portion.
2. Weld heat concentrates on small mass parts and these parts are overheated.
3. The contact portion between the parts and the jig causes a welding flaw.

Therefore, the development of laser microwelding technology using a laser beam narrowed to less than 1 mm diameter as a heat source has begun. A schematic diagram of laser microwelding is shown in Fig. 35.

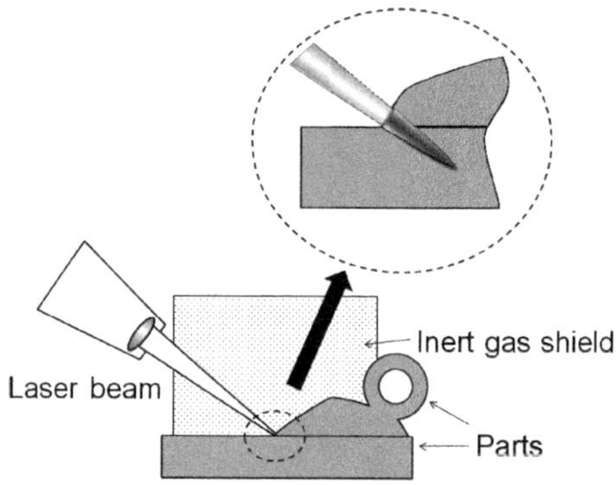

Fig. 35 Schematic diagram of laser microwelding.

A narrowed laser beam enables local and short-time welding of fine titanium parts. Welding in a short time does not require anything other than an inert gas shield to suppress deterioration of mechanical properties due to invasion of atmospheric impurities. It is also possible to reduce the thermal effect other than the molten portion compared to brazing. In addition, it is expected to suppress crystal grain size enlargement on titanium alloy.

Research on laser condition control to obtain optimum strength and appearance was required to adapt laser welding technology for eyeglass frames. For example, although the joining strength increases with increased laser output, scattering phenomenon of molten metal called spatter also increases. The appearance of general sputtering defect is shown in Fig. 36. This sputtering

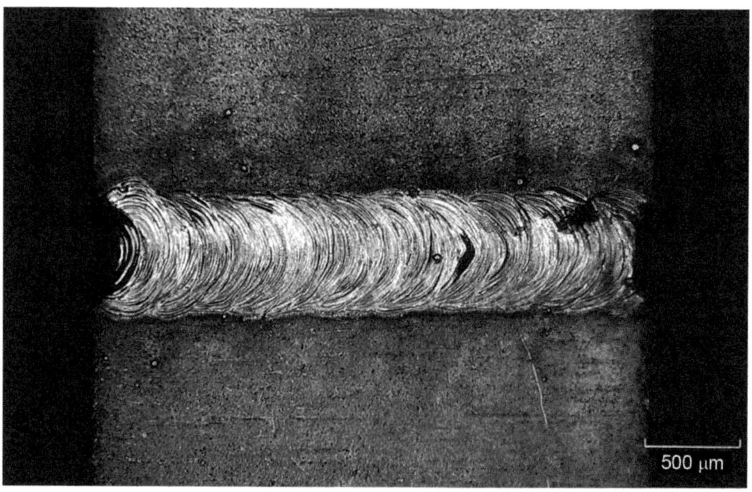

Fig. 36 Spatter generated by laser welding.

may require an additional process to remove or, in some worst cases, it may be an unrecoverable defect.

Kawahito et al. have reported suppressiung the weld defect by monitoring the state of the laser melted part in process and controlling the laser output in real time [16]. Nakamura et al. have reported that joining strength and appearance quality can be controlled by applying laser microwelding for eyeglass frames parts [17]. Fig. 37 is an optical micrograph of a cross-section of the parts joined by resistance brazing and laser microwelding. It can be confirmed that the heat-affected zone is small even when laser microwelding is performed to the parts having a large difference in volume.

Utilizing this technology, the complicated shaped products composed of delicate parts were put into practical use as shown in Fig. 38.

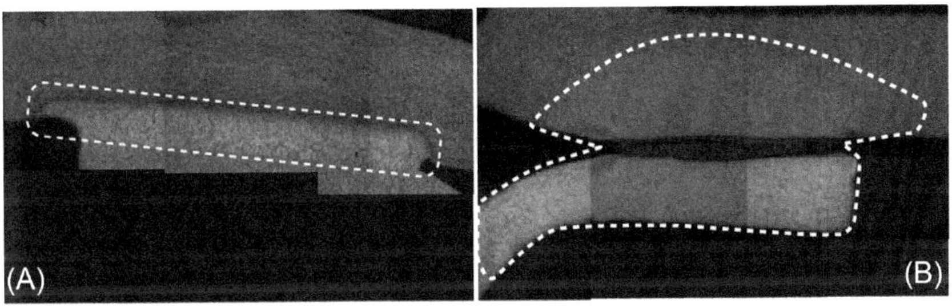

Fig. 37 Heat-affected zone of joined parts: (A) laser microwelding and (B) resistance brazing.

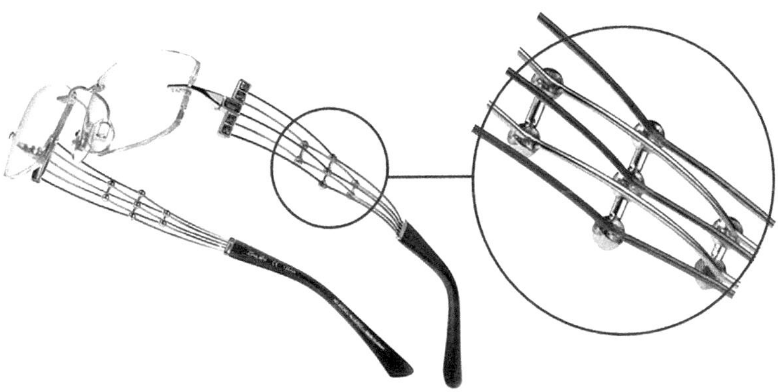

Fig. 38 Advantage product with laser microwelding.

4.3 Future challenges

4.3.1 Processing and heat treatment technology

The history of eyeglass frames evolution had a close relationship with the history of materials evolution. In the case of shape-memory alloy, which is a functional material used for eyeglass frames, it has a beneficial performance of changing the crystal structure by stress or heat, and recovering its original shape versus large deformation. Advanced control of processing and heat treatment makes it possible to show such functions properly. Many reports have been made on processing methods to obtain excellent properties for TiNi alloys, which are representative for shape-memory alloys [18–20]. Many reports also have been made about titanium-based superelastic alloys, which are being actively studied in recent years [21–24]. When applying these alloys to eyeglass frames, it is necessary to control age hardening at the temperature range from 100°C to 400°C [25–28]. This is because baking is performed at about 180°C to 350°C for improving the adhesion of the surface film of the eyeglass frames. As a result, the function of the eyeglass frames is changed before and after the surface decoration. To overcome this problem, development of a surface treatment with high adhesion, even at low temperature baking or a material with little properties change in the intermediate temperature range, are expected.

4.3.2 Adaptation to information and communication technology

Eyeglass frames have been used as a product that improves visual function, protection from external stimulus, and fashionability. However, owing to recent advances in Internet technology, eyeglass frames have attracted attention as information terminals [29–31]. Fig. 39 shows the conceptual image of this device. For these applications, it is presumed that a higher level of the fundamental function of the eyeglass frames previously described is required. This is because increasing the product weight due to the addition to

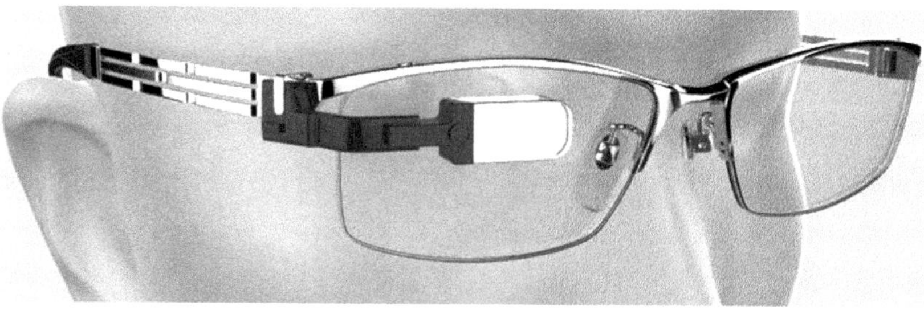

Fig. 39 Conceptual image of eyeglass frame as an information terminal device.

the function of holding a lens so far, a camera, various sensors and a power source, etc. are incorporated into the parts of the eyeglass frames, and adversely affects the fitting comfort. For eyeglass frames as electronic device, it is necessary to consider different kinds of safety as well. If both wearing performance as an eyeglass frames and evolution of information terminal function can be achieved at the same time, great expansion of the market of the product is expected.

4.3.3 Additive manufacturing technology

With the development of information and communication technology, greater attention is paid to new manufacturing technology. For example, eyeglass frame manufacturing technology utilizing a 3D printer called additive manufacturing has attracted attention as an advanced processing technique [32, 33]. With this technology, there is a possibility that products can be manufactured without using dies or other processing machines that have been commonly used before. Furthermore, it is easy to deal with individual production, thus it could be considered as a suitable manufacturing method for a small amount and a variety of products. Fig. 40 shows the prototype of eyeglass frames made with 3D printing technique. There are irregularities on the surface, and additional processing is necessary to obtain the same surface quality as conventional processing methods, but there is a great possibility to compensate for it. The current problems of 3D printing technology for eyeglass frames manufacturing are as follows.

1. Surface quality
2. Cost of powder materials and equipment
3. Corresponding to required part functions

Fig. 40 Prototype of eyeglass frame made with 3D printing technology and its surface.

Eyeglass frames made with 3D printing technology are currently used only for prototyping at present due to these problems. In the near future, the performance of this machine is expected to dramatically improve. Along with this evolution, this technology could become the mainstream of manufacturing technology. It is greatly expected to utilize 3D printing technique for eyeglass frames manufacturing method with variety composition alloys such as β-type titanium alloy, low Young's modulus alloy, shape memory, or superelastic alloy.

References

[1] R. Corson, Fashions in Eyeglasses (1967). Arranged With Peter Owen Ltd., UNI Agency Inc., Tokyo. ISBN 4-89694-431-3 C0039.
[2] K. Nakasuji, M. Okada, New high strength titanium alloy Ti10%Zr for spectacle frames, Mater. Sci. Eng. A 213 (1996) 162–165.
[3] M. Tubielewicz-Michalczuk, Application of titanium properties in civil engineering and architecture, Eng. Mater. 687 (2016) 220–227.
[4] H. Tobushi, K. Tanaka, H. Horikawa, M. Matsumoto, Shape Memory Materials and THEIR APPLICATIOns, CORONA Publishing CO., LTD, ISBN: 4-339-04572-1, 2004.
[5] A. Inoue, Advanced Technology and Application of Nanometals, CMC Publishing CO., LTD, ISBN: 978-4-7813-0033-7, 2003. C3043.
[6] S. Kuramoto, T. Furuta, J. Hwang, K. Nishino, T. Saito, Elastic properties of Gum Metal, Mater. Sci. Eng. A 442 (2006) 454–457.
[7] T. Furuta, S. Kuramoto, J.W. Morris Jr., N. Nagasako, E. Withey, D.C. Chrzan, The mechanism of strength and deformation in Gum Metal, Scripta Mater. 68 (2013) 767–772.
[8] D.M. Gordin, R. Ion, C. Vasilescu, S.I. Drob, A. Cimpean, T. Gloriant, Potentiality of the "Gum Metal" titanium-based alloy for biomedical applications, Mater. Sci. Eng. C 44 (2014) 362–370.
[9] T. Akahori, M. Niinomi, H. Toda, K. Yamauchi, H. Fukui, M. Ogawa, Mechanical properties and deformation behavior of β-type titanium alloy (TNTZ) drawn wires, J. Jpn. Inst. Metals 69 (2005) 530–537.
[10] D. Kuroda, M. Niinomi, H. Fukui, A. Suzuki, J. Hasegawa, Heat treatment processes and mechanical properties of new beta-type biomedical Ti-29Nb-13Ta-4.6Zr alloy, ISIJ Tetsu-to-Hagane 86 (2000) 610–616.
[11] H. Tada, T. Yamamoto, X. Wang, H. Kato, Non-monotonic aging temperature dependence of superelasticity of Ti72Nb15Zr10Al3 quaternary alloys, Mater. Trans. 54 (2013) 1502–1509.
[12] T. Onzawa, A. Suzumura, M.W. Ko, Brazing of titanium using low-melting-point Ti-based filler metals, Weld. Res. Suppl. 12 (1990) 462–467.
[13] O. Botstein, A. Rabinkin, Brazing of titanium-based alloys with amorphous 25wt.% Ti-25wt.%Zr-50wt.%Cu filler metal, Mater. Sci. Eng. A 188 (1994) 305–315.
[14] N.D. Tomashov, G.P. Chernova, T.A. Fedoseeva, Cathodic alloying of the surface of titanium, chromium, and stainless steels as a method of increasing their passivation and corrosion resistance, Corrosion 36 (1980) 201–207.
[15] H. Osakada, T. Ishikawa, M. Ono, K. Morishita, H. Andou, Seimitsu Tanzou, 2010, pp 36-40, Nikkankougyoushinbunsya, Japan, ISBN: 978-526-06554-5 C3053.
[16] Y. Kawahito, M. Kito, S. Katayama, In-process monitoring and adaptive control for gap in micro butt welding with pulsed YAG laser, J. Phys. D: Appl. Phys 40 (2007) 2972–2978.

[17] Y. Kawahito, M. Kito, S. Katayama, H. Nakamura, Fundamental research of laser micro welding for practical use of titanium eyeglass frames, Q.J.Jpn.Weld. Soc. 26 (2008) 203–209.

[18] T. Todoroki, H. Tamura, Effect of heat treatment after cold working on the phase transformation in TiNi alloy, Trans. Jpn. Inst. Met. 28 (1987) 83–94.

[19] S. Takaoka, 6—Overview of the development of shape memory and superelastic alloy applications, in: K. Yamauchi, I. Ohkata, K. Tsuchiya, S. Miyazaki (Eds.), Woodhead Publishing Series in Metals and Surface Eng., Shape Memory and Superelastic Alloys, Woodhead Publishing, ISBN: 9781845697075, 2011, pp. 77–84.

[20] K. Tsuchiya, M. Inuzuka, D. Tomus, A. Hosokawa, H. Nakayama, K. Morii, Y. Todaka, M. Umemoto, Martensitic transformation in nanostructured TiNi shape memory alloy formed via severe plastic deformation, Mater. Sci. Eng. A 438–440 (2006) 643–648.

[21] T.W. Duerig, J. Albrecht, D. Richter, P. Fisher, Formation and reversion of stress induced martensite in Ti-10V-2Fe-3Al, Acta Metall. 30 (1982) 2161–2172.

[22] C. Baker, The shape-memory effect in a titanium-35 wt.% niobium alloy, Metal. Sci. J. 5 (1971) 92–100.

[23] H. Hosoda, N. Hosoda, S. Miyazaki, Mechanical properties of Ti-Mo-Al biomedical shape memory alloys, Trans. MRS Jpn. 26 (2001) 243–246.

[24] P.J.S. Buenconsejo, H.Y. Kim, S. Miyazaki, Novel β-TiTaAl alloys with excellent cold workability and a stable high-temperature shape memory effect, Scr. Mater. 64 (2011) 1114–1117.

[25] S. Semboshi, T. Shirai, T.J. Konno, S. Hanada, In-situ transmission electron microscopy observation on the phase transformation of Ti-Nb-Sn shape memory alloys, Metall. Mater. Trans. A39 (2008) 2820–2829.

[26] B.S. Hickman, The formation of omega phase in titanium and zirconium alloys, J. Mater. Sci. 4 (1969) 554–563.

[27] H. Tada, T. Yamamoto, X. Wang, H. Kato, Non-monotonic aging temperature dependence of superelasticity of Ti72Nb15Zr10Al3 quaternary alloys, Mater. Trans. 54 (2013) 1502–1509.

[28] H. tada, H. Nakamura, T. Shobu, R. Kitai, K. Kikuta, Y. Kawahito, Effect of laser solution-treatment on a Ti-based shape memory alloy, Mater. Res. Bull. 90 (2017) 41–46.

[29] Y.H. Wang, C.C. Hsieh, Explore technology innovation and intelligence for IoT (Internet of Things) based eyewear technology, Technol. Forecast. Soc. Change 127 (2018) 281–290.

[30] D.M. Wilson, C. Chheng, L. Stimson, Safety Goggle compliance monitoring using hall effect sensors, IEEE Sens. Lett. 2 (2018) 1–4.

[31] M. Farooq, E. Sazonov, Real time monitoring and recognition of eating and physical activity with a wearable device connected to the eyeglass, in: Eleventh International Conference on Sensing Technology (ICST), Sydney, NSW, 2017, pp. 1–6.

[32] F.H. (Sam) Froes, M. Qian, 1.2—titanium background, alloying behavior and advanced fabrication techniques—an overview, in: F.H. Froes, M. Qian (Eds.), Woodhead Publishing Series in Biomaterials Titanium in Medical and Dental Applications, Woodhead Publishing, 2018, pp. 23–37.

[33] T.-C.T. Chen, Y.-C. Lin, A three-dimensional-printing-based agile and ubiquitous additive manufacturing system, Robotics Comput.-Integr. Manuf. 55 (2019) 88–95.

Further reading

[34] ISO/DIS 7998, Ophthalmic Optics—Spectacle Frames—Lists of Equivalent Terms and Vocabulary, 2005.

[35] ISO 12870, Ophthalmic Optics—Spectacle Frames—Requirements and Test Methods, 2016.

[36] C.E. West, D.G. Hunter, Displacement of optical centers in over-the-counter readers: a potential cause of diplopia, J. AAPOS 18 (2014) 293–294. 1091-8531.
[37] A. Vitols, Dictionaire des lunettes (Paris, Editions Christine Bonneton), Paris through Tuttle-Mori Agency Inc., Tokyo, ISBN: 4-938133-72-5, 1994. Japanese Translation Rights Arranged With les Editions Bonneton. C0030.

CHAPTER 14

Applications of titanium in the electronic industry

Shulong Ye[a,b], Yongyun Zhang[a], Peng Yu[a]
[a]Department of Materials Science and Engineering, Southern University of Science and Technology, Shenzhen, China
[b]Institute of Applied Physics and Materials Engineering, University of Macau, Macau, China

Contents

1 Introduction

Titanium is the fourth most-abundant metal existing in the upper crust of the earth, next only to aluminum, iron, and magnesium. The utilization of titanium and its alloys has been widely explored in many industries because they offer a unique combination of high specific strength, outstanding corrosion resistance, and good biocompatibility [1, 2]. However, until now, titanium and its alloys were mainly used in aerospace, chemical processing, marine engineering, defense, and biomedical areas, where their unique properties override affordability [1–6]. However, the requirements for low-cost raw materials and processing methods in other industries, such as electronic devices, automotive parts, architecture, sporting goods, and other consumer products, have become a major stumbling block for wider applications of titanium products. The applications of titanium and its alloys will continue to expand if affordability is improved.

With the rapid development of the electronics industry over the last decades, the demand for electronic products has rapidly grown. Consumers

Titanium for Consumer Applications
https://doi.org/10.1016/B978-0-12-815820-3.00019-8

have enjoyed the convenience brought by electronic technology. Excellent properties such as high specific strength and superior corrosion resistance have laid the firm groundwork for the applications of titanium and its alloys in high-quality consumer electronics. Hence, there is a trend in recent years that titanium and its alloys have become increasingly attractive for electronic applications such as computers, mobile phones, and wearable devices. However, due to the fast updating speed of electronic products, it is necessary to evaluate the relationship between performance, cost, and processing methods of titanium and its alloys, as compared to other, more affordable materials used in current electronic industries.

Traditionally, titanium parts are typically fabricated by subtractive manufacturing techniques, that is, primarily by machining forged blanks [7]. The insufficient formability and machinability of titanium and its alloys limit their applications and makes the products expensive. This problem becomes even more serious when titanium and its alloys are considered in electronic devices where many components have complex shapes in thin walls. New approaches that allow cost-effective production of high-performance intricate titanium products are expected to open new market opportunities for titanium and its alloys. Cui [3] compared the cost-saving potential of new technologies, such as powder metallurgy, hot isostatic press (HIP) casting, etc. Among them, powder metallurgy is considered most economically feasible to fabricate titanium and its alloys for electronic applications, and the demand for the technology is expected to increase.

Apart from the fabrication cost, the design of alloys using less costly alloying elements has also been investigated to improve the performance-to-cost ratio and to penetrate cost-sensitive markets such as the electronics industry. For example, a titanium alloy of Ti-0.5Fe-0.1N was reported to have the potential to be used as a structural material for consumer electronic goods [8]. The alloy was strengthened by low-cost elements, including O, N, and Fe. With grains refined by the addition of Fe, the alloy can achieve balanced strength and ductility. Furthermore, the alloy has an additional advantage in that the microstructures can be tailored during brazing when used in some practical application cases such as glasses frames. This alloy is therefore suitable for electronic applications that require high mechanical properties. Apart from this, titanium alloys with high interstitial element contents have also been tried for electronic applications, where the requirement for material ductility is not strict. Both commercially pure titanium (CP Ti) and Ti-6Al-4V alloy can be made from high-oxygen-content raw materials to reduce the material's cost if the ductility can meet requirements.

In the following sections, some electronic parts made from titanium will be introduced, which may help in understanding the recent development of titanium in the electronics industry.

2 Representative applications

2.1 The computer industry

With the living standards continuing to improve over the last decades, the demand for digital multimedia products has grown steadily. The laptop industry first catered to these trends. In 1999, Fujitsu Limited unveiled a new half-notebook-sized (A5) handheld PC, INTERTOP CX300, for the Japanese market (shown in Fig. 1A). It was the first reported commercial product in the computer industry that has used CP Ti (99.5% purity) for the top cover of the casing. The application demonstrates that titanium can be used as a computer casing material to achieve a thin-walled frame while offering sufficient strength at the same time. Compared with magnesium, which is a premier metal used for computer casings, the strength of titanium is much higher. Furthermore, the excellent corrosion resistance of titanium can benefit the subsequent coating treatment of the casing.

Another typical product is the Powerbook G4 designed by Apple Inc. (shown in Fig. 1B), in which Grade 1 CP Ti foil was used to make the outer skin. The titanium foil was wrapped around a very stiff carbon fiber lightweight support frame for the necessary structural rigidity. Thus, the high strength-to-weight ratio of titanium paved the way for the application of

(A) (B)

Fig. 1 (A) Fujitsu [9] and (B) Apple [10] laptops with titanium casing.

titanium on the Powerbook G4 and made this product one of the lightest laptops at that time.

Titanium has also found applications in some computer peripheral products [11]. The outer shell of the mouse is constructed of Grade 1 CP Ti to improve the touch feeling of fingertips during use. The mouse is designed to be an expression of identity and elegance. It is a unique product that demonstrates the elegance of titanium.

Apart from the previous examples, titanium is considered a promising candidate material for the substrate of hard disk drives [1]. As a nonmagnetic material, titanium can prevent detrimental interference with the data storage process. The most cost-effective choice among titanium alloys is Ti-3Al-2.5V (Grade 9), which is sometimes referred to as "half 6-4" and has high strength and ductility in the annealed condition [12]. The alloy can offer a mechanical strength that is 20%–50% higher than that of CP Ti at both room and elevated temperatures. More importantly, it has better cold formability than Ti-6Al-4V (Grade 5) and can be precisely cold-rolled into foil. Compared to 5086 aluminum alloy, which is commonly used for hard disk substrates, Ti-3Al-2.5V alloy provides significant advantages. In the annealed condition, the ultimate tensile strength-to-density ratio of Ti-3Al-2.5V is 1.49 times higher, and the yield strength-to-density ratio is 2.83 times higher than those of 5086 aluminum alloy [12]. Thus, for the disk substrate of the same weight, its thickness can be reduced with the use of the titanium alloy. Moreover, the melting temperature of the titanium alloy is substantially higher than that of the aluminum alloy, implying a much better heat-distortion resistance. The higher temperature stability permits the disk to be coated at high temperatures and thus improves the disc production rates. On the other hand, the lower coefficient of thermal expansion of titanium alloys contributes to the reduction of dislocation between the read/write head and disk. The permission of closer read/write head tolerances can increase disc storage capacity when using a titanium alloy substrate. Fig. 2 shows a thin titanium foil disk substrate working with recording heads.

2.2 The mobile phone industry

Mobile phones are representative items of communication products that have experienced the same trend as laptops. Some manufacturers tried to use titanium to produce upscale mobile phones for the demands of businessmen who require efficient and high-quality phones. For example, Nokia Inc. released a mobile phone named Nokia 8910 with a titanium casing,

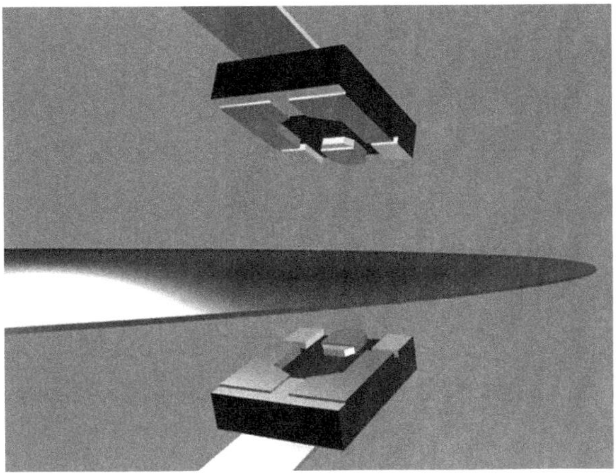

Fig. 2 Thin titanium foil disk substrate with recording heads [13].

which was welcomed by the business world (shown in Fig. 3). The use of titanium makes the mobile phone casing strong and significantly improves the feeling when held.

The miniaturization trend on mobile phones makes the phones more fragile when they drop to the ground. As a result, mobile phone shells have been developed recently as a remedy to protect the phones. Due to the high specific strength of titanium alloys, they have been used for phone shell fabrication. A typical example is the iPhone shells designed by Gray Inc., as shown in Fig. 4. These cases are made of machined Ti-6Al-4V alloy,

Fig. 3 Nokia mobile phone with titanium casing [14].

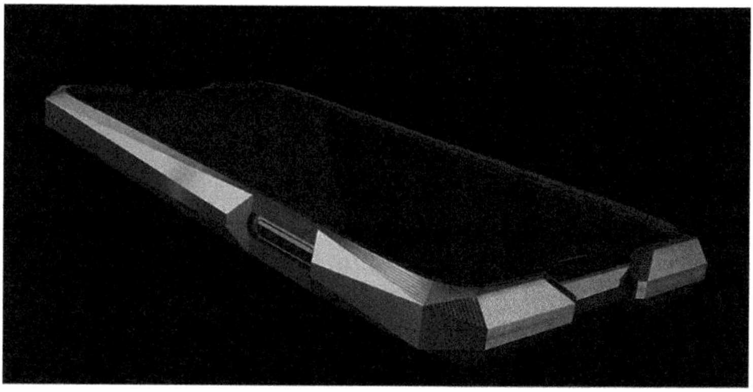

Fig. 4 Titanium cases provide persistent protection of phones [15].

which can fit the frame well. The high precision process during machining produces a series of textured lines across its body, which is impossible to be replicated by hand and adds a complication to the multifaceted profile of the cases. The cases are available in a variety of colors and are sold at a relatively high price. The protective effect of these cases can prolong lifespans of these mobile phones. Compared with the phone cases made of silica gel, titanium alloys phone cases would not be stained from being oxidized.

2.3 Wearable devices

Wearable devices are products controlled by electronic components and software that can be incorporated into clothing or worn on the body like accessories. Nowadays, a variety of wearable devices, such as smart glasses and smartwatches, have been invented. The sales of these consumer electronics have been growing steadily. In the fabrication of wearable products, the utilization of titanium can increase the accessional values of these devices. Materials with superior corrosion resistance and good biocompatibility are considered suitable for wearable devices frequently in contact with human skin. Titanium products are resistant to corrosion by perspiration, which makes titanium an ideal material for wearable devices used in the sports. Moreover, the high specific strength of titanium makes the frame of the devices strong enough to protect the electronic components inside and significantly decreases the overall weight of the devices.

Fig. 5 demonstrates a titanium glasses arm for augmented reality smart glasses fabricated by metal injection molding (MIM) [16]. MIM combines the merits of powder metallurgy and plastic injection molding, and offers design freedom in shape complexity and is suited for mass production [17, 18].

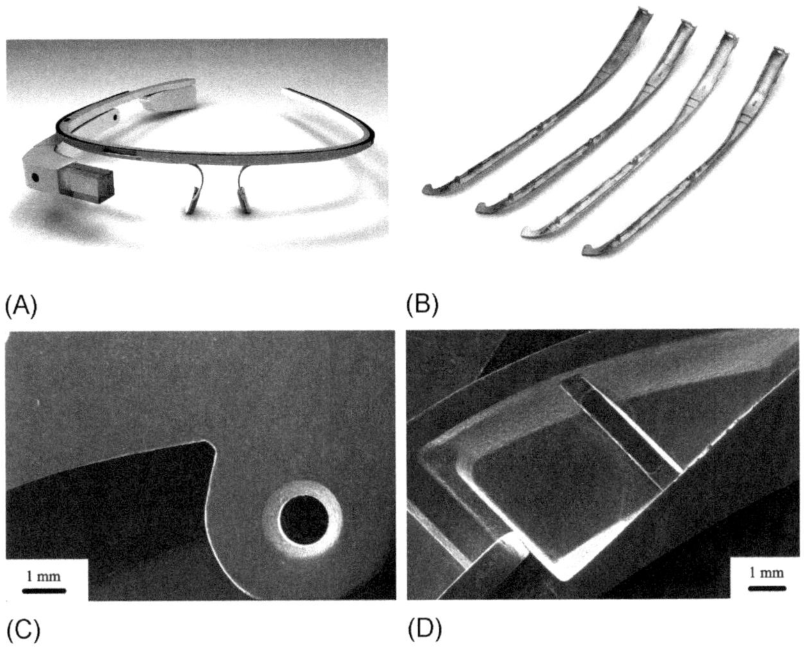

Fig. 5 (A) Product design of smart glasses, (B) a photo of the Ti glasses arms after coloring, and (C) and (D) SEM micrographs showing delicate structures of the MIMed Ti glasses arm. *(Courtesy of ElementPlus, Shenzhen, China.)*

The glasses arm can be made of CP Ti or Ti-6Al-4V alloy. Fig. 5A shows the product design of the augmented reality smart glasses whereas Fig. 5B is a photo of the MIM-fabricated titanium glasses arms. The glasses arms are anodized to obtain different colors, during which titanium reacts with oxygen to form an oxide layer. The thickness of the oxide layer can be adjusted by heat or electrochemical treatment. When the thickness of oxide varies, the reflected light waves vary resulting in different colors of the part. It is worth mentioning that the titanium glasses arm is more than 170 mm long but weighs only 6.8 g, featuring an extremely complex curved surface and a thin wall structure with smallest wall thickness ~0.5 mm. Fig. 5C and D shows SEM micrographs revealing the complex details in the glasses arm. Therefore, it is both difficult and expensive by machining and other processing methods, such as forging or casting, to mass produce the component. In fact, it is challenging to the MIM process as well. These MIM-fabricated glasses arms can achieve tolerances consistently controlled within ±0.5% of the nominal dimensions, as well as good densities and mechanical properties comparable to those of the cast counterparts.

There are other examples of titanium wearable parts made by MIM. Sintered to a high density, the watch frame weighs ~6 g and has a curved surface and internal complex structures. All the complex structures are net-shape-formed without machining. After being carefully designed to prevent dimensional distortion during sintering, tolerances controlled within ±0.5% of the nominal dimensions can be consistently achieved.

2.4 Other products

Apart from the previously mentioned sectors, titanium parts have been used in other areas of the electronics industry. One typical product is the Digital Single Lens Reflex (DSLR) camera. Titanium has been used for the fabrication of the shell and lens of DSLRs. Fig. 6 shows a photo of a DSLR camera (Leica M-P 240), in which the top plate, base plate, and control elements are all precisely machined from dense titanium. The application of titanium in shells of DSLRs not only reduces the weight of DSLRs and makes them more portable but also restrains stains produced on the bayonet connected to the lens.

Earphones can transform the electric signal received from the digital media receiver into sound waves. The market of earphones is expanding as the field of portable electronic products expands. Fig. 7 shows a photo of an earphone (ATH-CM2000Ti) in which titanium is used to produce its cavities. The use of titanium decreases the weight of the earphone. The machining process and carbon-like coating have ensured earphones are durable and portative. Moreover, titanium is nonmagnetic, which ensures that the electrical signal and sound quality are not affected by the material.

Fig. 6 Leica Digital Single Lens Reflex camera series "titanium" [19].

Fig. 7 Titanium used on the cavities of earphones [20].

3 Concluding remarks

In addition to their conventional markets in the aerospace, chemical processing, defense, and biomedical industries, titanium and its alloys nowadays are extending their applications into the electronics industry. Their exceptional attributes such as high specific strength, excellent corrosion resistance, significant biocompatibility, and nonmagnetic property make titanium and its alloys elegant materials for the fabrication of components in computers, mobile phones, digital cameras, and other wearable devices. Although most of the products are identified as upscale electronic devices, the market is expected to continue to grow. On the other hand, the rapid development of the electronics industry gives rise to increasing demand for cost-effective production methods and design of low-cost titanium alloys.

Acknowledgments

The authors thank the support from the Shenzhen Science and Technology Innovation Committee (grant number SGLH20161212155758670 and KQJSCX20180322152424539). In addition, the authors gratefully acknowledge the useful suggestions and critical read of this chapter by Professor Ma Qian of RMIT University, Melbourne, Australia.

References

[1] C. Leyens, M. Peters, Titanium and Titanium Alloys Fundamentals and Applications, Wiley-VCH, Weinheim, 2003.

[2] M. Qian, Titanium Powder Metallurgy: Science, Technology and Applications, Butterworth-Heinemann, Oxford, 2015.

[3] C. Cui, B. Hu, L. Zhao, S. Liu, Titanium alloy production technology, market prospects and industry development, Mater. Des. 32 (3) (2011) 1684–1691.
[4] S.R. Seagle, The state of the USA titanium industry in 1995, Mater. Sci. Eng. A 213 (1–2) (1996) 1–7.
[5] H.J. Rack, J.I. Qazi, Titanium alloys for biomedical applications, Mater. Sci. Eng. C 26 (8) (2006) 1269–1277.
[6] A.T. Sidambe, Biocompatibility of advanced manufactured titanium implants—a review, Materials (Basel) 7 (12) (2014) 8168–8188.
[7] F.H. Froes, M. Qian, Titanium in Medical and Dental Applications, Woodhead Publishing, Duxford, 2018.
[8] M. Yamada, An overview on the development of titanium alloys for non-aerospace application in Japan, Mater. Sci. Eng. A 213 (1–2) (1996) 8–15.
[9] Handheld PC With Titanium Casing, Fujitsu Limited, Japan, 1999. http://pr.fujitsu.com/jp/news/1999/May/18-e.html.
[10] Titanium PowerBook G4, Apple Inc., USA, 2001. https://en.wikipedia.org/wiki/PowerBook_G4.
[11] Computer Mouse With Titanium Shell, Intelligent Design B.V., The Netherlands, 2009. http://www.intelligent-design.nl/titanium_mouse.php.
[12] C.W. Nelson, R.D. Weir, R.S. Weir, Titanium or titanium-alloy substrate for magnetic-recording media, US Patent 5,707,705, 1998.
[13] J. White, The gas bearing Interface of opposed recording heads in a disk drive utilizing helium and thin titanium foil disks, J. Tribol. 136 (4) (2014). 041901-1-9.
[14] Mobile Phone With Titanium Casing, Nokia Corporation, Finland, 2002. https://en.wikipedia.org/wiki/Nokia_8910.
[15] Titanium Protecting Case for Mobile Phone, Gray Brand, Singapore, 2015. https://gray-international.com/.
[16] S. Ye, W. Mo, Y. Lv, X. Li, C.T. Kwok, P. Yu, Metal injection molding of thin-walled titanium glasses arms: a case study, JOM 70 (5) (2018) 616–620.
[17] R.M. German, Progress in titanium metal powder injection molding, Materials 6 (8) (2013) 3641–3662.
[18] A. Dehghan-Manshadi, M.J. Bermingham, M.S. Dargusch, D.H. StJohn, M. Qian, Metal injection moulding of titanium and titanium alloys: challenges and recent development, Powder Technol. 319 (2017) 289–301.
[19] Leica M-P 'TITANIUM' camera, Leica Camera AG, Germany, 2014. https://www.leicastore-uk.co.uk/products/leica-m-p-typ-240-titanium.
[20] Earphone With Titanium Cavities, Audio-Technica Corporation, Japan, 2018. https://www.audio-technica.co.jp/atj/show_model.php?modelId=3057.

CHAPTER 15

NbTi superconducting wires and applications

Pingxiang Zhang[a,b,c], Jianfeng Li[a,b], Qiang Guo[a,b], Yanmin Zhu[a,b], Kaijuan Yan[a,b], Ruilong Wang[a,b], Kailin Zhang[a,b,c], Xianghong Liu[a,b], Yong Feng[a,b]
[a]Western Superconducting Technologies Co., Ltd. (WST), Xi'an, China
[b]National Engineering Laboratory for Superconducting Materials Preparation, Xi'an, China
[c]Northwest Institute for Non-ferrous Metal Research (NIN), Xi'an, China

Contents

Titanium for Consumer Applications
https://doi.org/10.1016/B978-0-12-815820-3.00010-1

1 Introduction

Niobium-titanium (NbTi) is an alloy of niobium and titanium used industrially as a type II superconductor for superconducting magnets. Its critical temperature is about 9.7 K [1]. NbTi alloys are suitable for fabricating superconducting magnets, with magnetic fields up to about 10T [2].

In 1962, Berlincourt and Hake [3] discovered that the NbTi alloy showed superconductivity at 4.2 K. Two years later, Westhouse produced the first NbTi wire. In 1965, the critical current density of NbTi superconducting wire was significantly improved by low-temperature heat treatment and cold processing by Vetrano. After overcoming the instability of NbTi wire, the critical current density (J_c) was increased to 2000 A/mm^2 by applying an aging heat treatment process and multifilamentary technology in 1972 [4]. To reduce the loss of NbTi superconducting wire, the diameter of NbTi filaments must be small enough [3], which requires the NbTi alloy to have high ingredient uniformity. In the late 1970s, the high-homogeneity NbTi alloy was successfully produced by vacuum smelting techniques [5, 6], which greatly improved the critical current density of NbTi superconducting wire. In 1982, Wu [7] from Northwest Institute for Nonferrous Metal Research (NIN) improved the J_c of 61 filaments of NbTi superconducting wire to 3470 A/mm^2 (5T, 4.2 K).

In industrial applications, to obtain high critical current density of NbTi superconducting wire, the α-Ti precipitates must reach nanoscale during deformation process. Meanwhile, to reduce the superconducting hysteresis loss and eddy current loss, NbTi superconducting wires should be multifilamentary [8]. The NbTi alloy with Ti content of 46–50 wt% has proven to be the best compromise between precipitate yield for critical current density and alloy composition for the other superconducting properties [9–11].

There is one important question: How can NbTi superconducting wires realize engineering and mass production, and be widely used in magnets? The first reason is that all the raw materials of NbTi superconductors have easy deformation ability (i.e., NbTi alloy and high purity copper). Owing to the good deformation ability of these raw materials, long NbTi superconducting wires can be produced. The second reason is that all the raw materials of NbTi superconductors are readily available. This means that the materials cost is much lower than other superconductors. The third reason is that NbTi superconductors are much more reliable and stable than other superconductors, and its performance degradation is much slower during the fabrication and operation of magnets.

As a result, NbTi superconducting wires have been widely used in magnetic resonance imaging (MRI), nuclear magnetic resonance (NMR), high-energy particle accelerators, Tokamak fusion reactors (e.g., International Thermonuclear Experimental Reactor [ITER] project), magnetic separation systems, power systems, superconducting energy storage system (SMES), etc. [12, 13].

2 NbTi superconducting composition and microstructure

The composition of Nb and Ti not only determines the superconducting properties but also the amount of α-Ti pinning center. The higher the Ti content is, the more the pinning center is, and the lower cost of the alloy is. In addition, the higher Ti content increases the hardness of the NbTi filaments and may result in sausage filament and even filament breakage [14, 15]. As the filament diameter is usually below 10 μm and even below 1 μm, it requires that initial alloy composition must be homogeneous over the entire casting ingot. Thus, any Ti-rich region, hard inclusions, or unmelted Nb is not permitted in the initial Nb-Ti ingot. Therefore, the correct overall alloy composition to optimize H_{c2}, T_c, and precipitation is typically Nb-(47 ± 1) wt% Ti alloy. A fine and uniform grain size and grain boundaries in the Nb-Ti matrix are to control the distribution of precipitate nucleation sites.

Fig. 1 shows the standard flash radiograph of a slice from a NbTi billet with a diameter of ~200 mm. No Ti-rich regions, hard inclusions, or unmelted Nb particles and voids were detected. Fig. 2 shows the microstructures

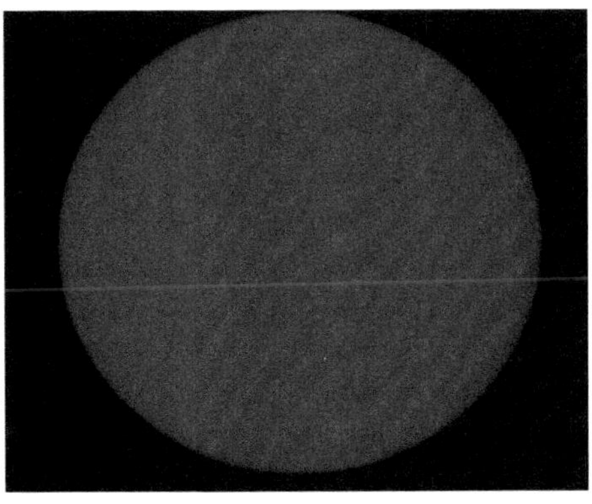

Fig. 1 Standard flash radiograph of a slice from a 200-mm diameter NbTi billet produced by Western Superconducting Technologies (WST), Xi'an, China.

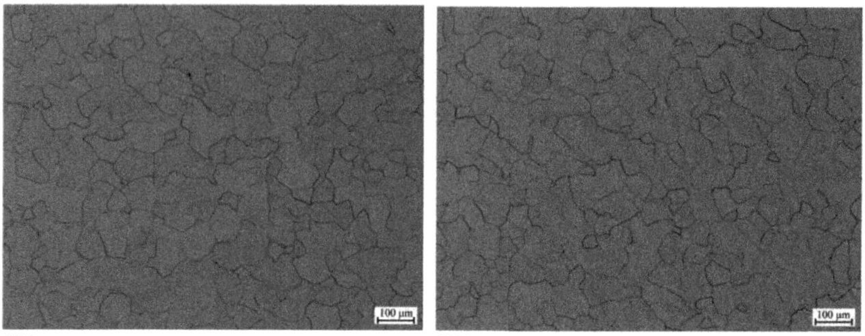

(A) (B)

Fig. 2 Microstructure of the cross-section of NbTi rod produced by WST. Center (A) and edge (B) of the NbTi rod.

at the center and edge regions of the cross-section of the NbTi rod. Grain size can satisfy the ASTM requirement, typically Grade 6 or smaller.

There are two stable phases in the Nb-Ti system: the BCC β phase and the HCP α-Ti phase that contains 1–2at.% Nb. Suitable heat treatments can make α-Ti precipitates occur at grain boundary intersections. High critical current density is obtained by restricting the movement of the superconducting vortices through the superconductor. At the same time, the barrier to vortex movement should not interfere with the overall transport of electrical current or significantly reduce the superconducting properties. An applied field of 5T and a temperature of 4.2K produce an equilibrium fluxoid radius of 5nm and spacing of 22nm. Heat treatment and artificial pinning center (APC) method can create vortex-pinning sites in Nb-Ti filament. The precipitate is deformed into an optimum nanostructure by deep cold drawing (see Fig. 3). The aging heat treatment limits the volume

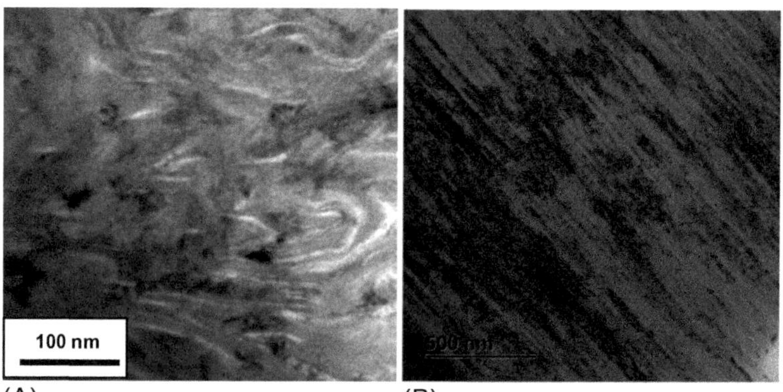

(A) (B)

Fig. 3 TEM images of the microstructure of NbTi filament in the final wire from WST. (A) Cross-section and (B) longitudinal section.

of α-Ti to 18%–28% of the filament [16, 17]. Thus, not only must the initial alloy composition be correct and homogeneous throughout the entire alloy ingot, but also must the microstructure be uniform and consistent on a submicron scale to ensure the wire ductility and the flux-pinning property.

3 NbTi superconducting wire types

To obtain a high critical current density NbTi wire, the microstructure of the NbTi filament must be a two-phase nanostructure after the final process. Meanwhile, to reduce the hysteresis loss and eddy current loss of the superconducting magnet, NbTi superconducting wires need to be multifilament and fine filament [11–14]. Because a NbTi superconductor has low processing cost, mature process routes, and stable performance, it has been widely used in various types of magnets. WST has successfully developed different types of NbTi superconducting wires, including monolith NbTi superconducting wire, wire-in-channel (WIC) NbTi superconducting wire, and superconducting cables. Fig. 4 shows the cross-sections of monolith NbTi superconducting wires with different filament numbers, different Cu/NbTi ratios, different shapes, and different low-temperature properties. At present, low-cost superconducting wire has been widely used in MCZ, MRI, NMR, and other applications.

To reduce the cost and increase the market share, WST independently developed WIC superconducting wire (Fig. 5). By soldering rod wire with low Cu/NbTi ratio into the U-shaped copper channel, WIC wire with a high Cu/NbTi ratio is fabricated. This type of wire has excellent stability and has been used in MRI magnets.

In addition, to meet the requirements of some special magnets, WST has also developed round cable-in-channel conductor and Rutherford Cable in Channel conductor with critical current higher than 10,000 A at 4T and 4.2K (see Fig. 6).

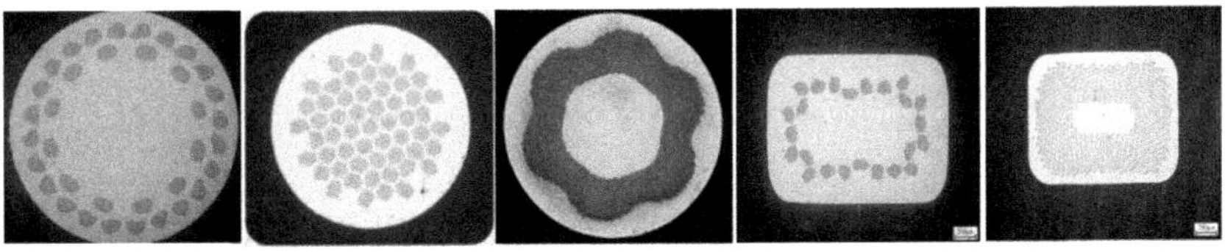

(A) (B) (C) (D) (E)

Fig. 4 Cross-section of monolith superconducting wires with different filament arrangement (from WST). (A) Cu/Sc ratio of 4.0 and filament number of 36, (B) Cu/Sc ratio of 1.0 and filament number of 55, (C) Cu/Sc ratio of 2.3 and filament number of 2616, (D) Cu/Sc ratio of 7 and filament number of 24, and (E) Cu/Sc ratio of 1.3 and filament number of 630.

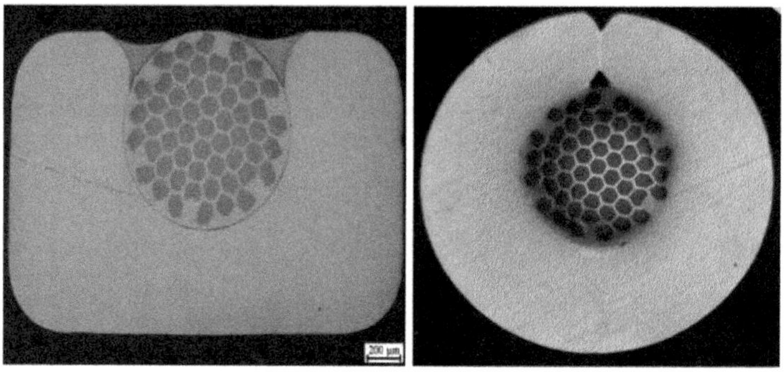

(A) (B)

Fig. 5 Cross-section of WIC superconducting wire (from WST). (A) Cu/Sc ratio of 7.0 and filament number of 55, and (B) Cu/Sc ratio of 12.0 and filament number of 55.

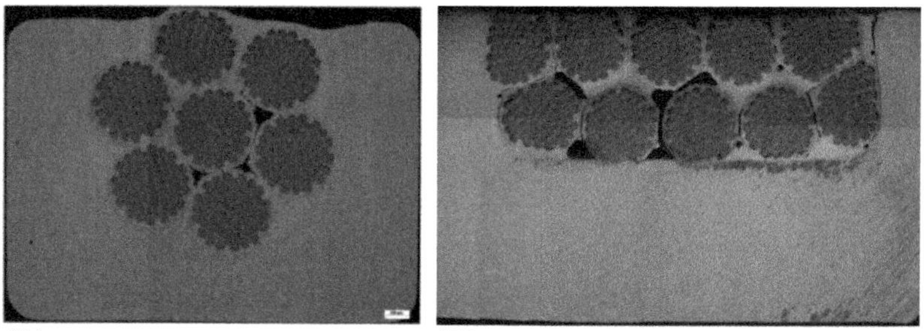

(A) (B)

Fig. 6 Cable in Channel conductor and metallographic section (from WST). (A) Cu/Sc ratio of 4.0 and filament number of 55 × 7, and (B) Cu/Sc ratio of 1.0 and filament number of 55 × 10.

4 Low-temperature properties and measurement method

The low-temperature performance measurement mainly includes critical current (I_c) measurement, Residual Resistance Ratio (RRR) measurement, and hysteresis loss (Q_h) measurement.

4.1 I_c measurement

The critical current is an important parameter to characterize the current carrying capacity of the superconductor, and it is also an important indicator to measure the quality of superconducting materials for commercial applications. For NbTi superconducting wires, the four-point probe method is generally used for critical current measurement. During the measurement, the sample is wound on the skeleton. The sample on the skeleton is together

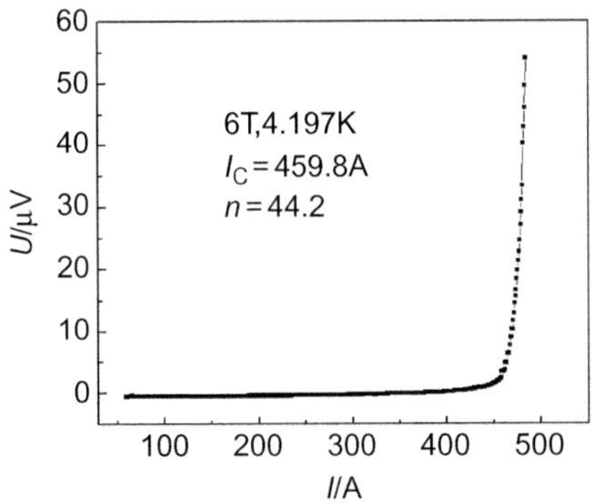

Fig. 7 A typical *U-I* curve [18].

placed in a magnet, and the magnetic field direction is vertical to the superconducting wire. The superconducting critical current measurement is to provide DC power for the superconductor sample, then measure the voltage of a short length of the sample. The current increases from zero at a certain rate, recording the *U-I* signal of the sample. The value of I_c should be determined by using an electric field criterion, E_c, or a resistivity criterion, ρ_c. Generally, the electric field criterion is 10 uV/m and the resistivity criterion is 10–14 Ω m. If the voltage shows an irreversible increase, then the conductor is considered to be quenched [18]. Fig. 7 shows a typical critical current measurement result.

The *n*-value is an important parameter that reflects the uniformity of the NbTi superconducting wires; it should be calculated as the slope of the Log$U$ versus Log$I$ in the region where the I_c is determined [19, 20], as shown by Eq. (1), and the typical calculation range is 10^{-14}~10^{-13} Ω m (resistance criterion) or 10~100 uV/m (electric field criterion) [19, 20].

$$U = U_0 \left(\frac{I}{I_0} \right)^n \tag{1}$$

where U represents the sample voltage, U_0 represents the reference voltage, I represents the sample current, and I_0 represents the reference current.

4.2 Residual resistance ratio (RRR) measurement

For measuring the RRR value, it is necessary to obtain the room temperature resistance, R_1, and the low-temperature resistance, R_2, of the sample.

For the room temperature (273 K or 300 K) resistance, R_1, the voltage (U_1) in the middle region of the sample is obtained by loading the current,

I_1, across the sample. Then the resistance at room temperature, R_1, is finally calculated. The formula is given as follows [21]:

$$R_1 = \frac{U_1}{I_1} \tag{2}$$

For measuring the low-temperature resistance, R_2, it is necessary to immerse the sample into a liquid helium environment and realize variable temperature control by raising the sample holder [21]. During the measurement, the sample is at a constant temperature, T_1 (10 K), and a certain current, I_2, is applied to the sample. To offset the effect of the thermoelectric power on the measured value, the current-reversal method was used to obtain two groups of voltage signals of the sample. One group is the voltage, U_{2+}, which is the sample voltage measured when the current passes through the sample in a positive direction, and the other group is the voltage, U_{2-}, which is the sample voltage measured after instantaneously changing the current direction. The average voltage U_2 is calculated according to Eq. (3) [21]:

$$U_2 = \frac{|U_{2+} - U_{2-}|}{2} \tag{3}$$

The low-temperature resistance R_2 can then be calculated by Eq. (4) [21]:

$$R_2 = \frac{U_2}{I_2} \tag{4}$$

Finally, the RRR of the NbTi superconducting wire is calculated according to Eq. (5) [21]:

$$\text{RRR} = \frac{R_1}{R_2} \tag{5}$$

4.3 Hysteresis loss (Q_h)

For the hysteresis loss measurement, it is generally obtained by using DC- or low-ramp-rate magnetometry. This will be performed using either a superconducting quantum interference device (SQUID magnetometer) or a vibrating-sample magnetometer (VSM).

For the VSM method [22], the sample to be measured is located in a uniform magnetic field, which causes it to become magnetized. The sample is mechanically oscillated near a set of pickup coils. The oscillating magnetic moment causes an oscillation in the magnetic field linking the pickup coils, thereby inducing an AC voltage that is then detected and converted into a magnetic moment value by electronic circuitry. The magnetometer is a "substitution" rather than "absolute" device, and its output signal requires calibration against a reference. Custom-made (hand-made) VSMs do exist, but increasingly, commercial versions of this machine are used. In general, they share the following characteristics. The sample to be measured is typically mounted on a vertical rod that vibrates longitudinally (vertically) with a position amplitude of about 1 mm and at a suitably low frequency [22].

The magnetic field may be supplied by either a horizontally mounted iron-core electromagnet (EM) or a vertically mounted superconducting solenoid (SCS). A horizontally mounted iron-core electromagnet causes the vibration direction of the sample to be perpendicular to the field direction. And a vertically mounted superconducting solenoid causes the vibration direction of the sample to be parallel to the field direction. The pickup coils are appropriately located and connected in pairs such that any external field oscillations (magnetic noise) are canceled and only the sample-generated field oscillations are detected.

The loss is determined from the numerically integrated area of the full *M-H* loop, which can be calculated with Eq. (6) [22].

$$Q_h = \frac{\oint m\mathrm{d}H}{\frac{M}{\rho} \times \frac{\pi D^2}{4}} \times \left(\frac{10^{-1}\,\mathrm{mJ}}{\mathrm{cm}^3} \right) \tag{6}$$

In Eq. (6), Q_h represents the hysteresis loss, M is the sample weight, ρ is the linear density, D is the sample diameter, m denotes the magnetic moment of sample, and H is the magnetization intensity.

5 NbTi superconducting wire fabrication process

The fabrication process from a NbTi alloy bar or billet to superconducting wires usually includes two main steps: monofilament bar fabrication and multifilament bar/wire fabrication.

NbTi bar, Nb barrier, and Cu can are cleaned and then assembled together (see Fig. 8). Nb barrier is warped around the NbTi bar to avoid the reaction of Ti and Cu during the precipitation heat treatment. The assembled billet is welded by electron beam in a vacuum condition. A subsequent warm extrusion is used to make the three components well bonded. By cold drawing, a monofilament bar is fabricated.

The monofilament bar is drawn into a hexagonal shape so that it can be restacked to create the final multifilament wire (see Fig. 9). The assembly process should be finished in a clean-room environment because any

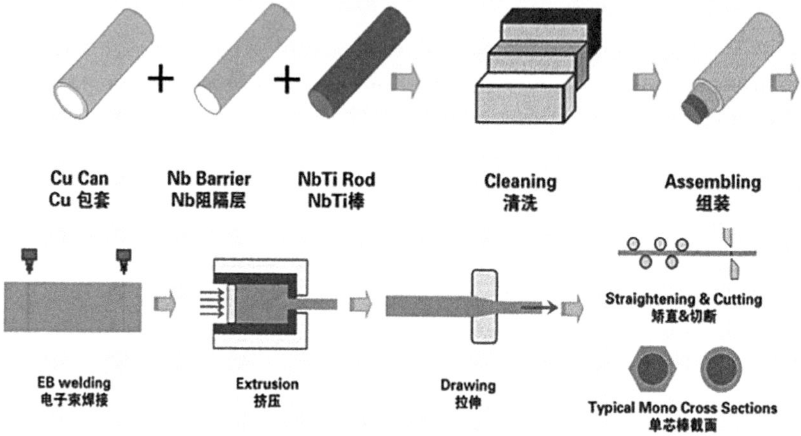

Fig. 8 Monofilament bar fabrication (this chart is from WST).

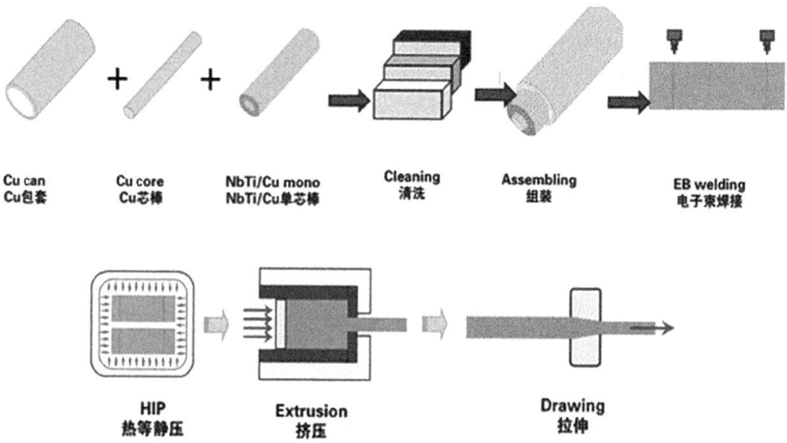

Fig. 9 Multifilament bar fabrication (this chart is from WST).

hard particulate incorporated into the billet could result in strand fracture at small size. After electron beam welding, hot isostatic pressing, and several times of drawing, the multifilament bar will be aged via heat treated and drawn alternately. Aging heat treatments are applied after sufficient cold work. For the Nb-47 wt% Ti alloy, the required prestrain is usually from 4% to 6%. Heat treatments are typically from 375 to 420°C for 20 to 80 h. The strain between heat treatments is usually from 0.8% to 1.5%, and at least three heat treatments are usually applied. α-Ti precipitates with the volume of ~20% of filaments are sufficient to obtain a critical current density of higher than 3000 A/mm^2 at 5T and 4.2 K.

6 Applications

6.1 International thermonuclear experimental reactor (ITER)

ITER project is one of the largest international scientific research cooperation projects in the world. The aim is to build a nuclear fusion reactor to verify the scientific and technological feasibility of peaceful use of nuclear fusion energy. To achieve this goal, two hydrogen isotopes—deuterium (D) and tritium (T)—need to be heated to over 100 million degrees centigrade.

ITER magnet system [23] is mainly composed of the following parts: 18 toroidal field (TF) coils, 1 center solenoid (CS), 6 poloidal field (PF) coils, 9 pairs correction field coil (CC), and compensating coil (Feeder). The magnet system with low-temperature superconducting wires is the core component of the ITER facility. The critical temperature (T_c) of the NbTi superconducting wire is 9.7 K, and its critical field (H) can reach 12T, which can be used to manufacture superconducting magnets with a magnetic field of 9T (4 K) or 11T (1.8 K). NbTi superconducting wire has good processing performance and high critical current density, and its conductor still has 1.5 K temperature allowance under the rated conditions [24]. Hence, it is selected as the conductor materials of poloidal field (PF) magnet coils for ITER [25]. WST has delivered about 174 tons of NbTi superconducting wires for PF coils and CC/Feeder coils.

6.2 Magnetic resonance imaging (MRI)

Magnetic resonance imaging (MRI) is a medical imaging technique used in radiology to form pictures of the anatomy and the physiological processes of the body in both health and disease. MRI scanners use strong magnetic fields, electric field gradients, and radio waves to generate images of the organs in the body (see Fig. 10). Energy from an oscillating magnetic field is temporarily applied to the patient at the appropriate resonance frequency. The excited hydrogen atoms emit a radiofrequency signal, which is measured by a receiving coil [26–28].

MRI requires a magnetic field that is both strong and uniform. Most clinical magnets are superconducting magnets, which require a liquid helium environment. MRI consumes about 2500 t of NbTi superconducting wires (WIC wire) per year. As this type of NbTi superconductor has high copper/NbTi ratio, high RRR, and high critical current, it is not only

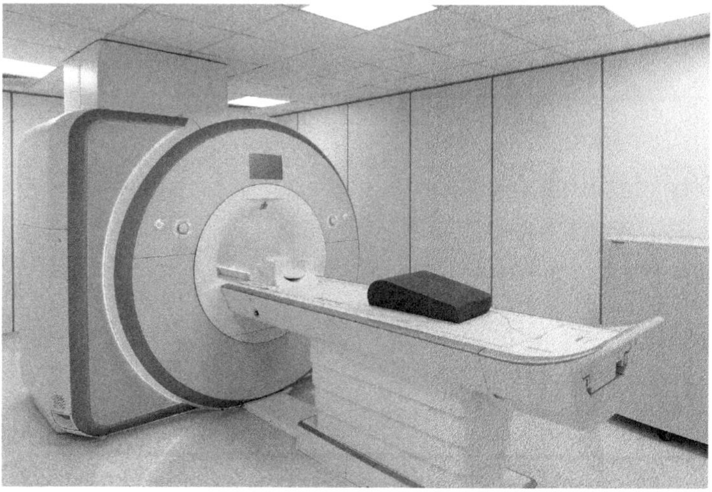

Fig. 10 Magnetic resonance imaging system. *(From https://baike.sogou.com/v432599.htm?fromTitle=MRI.)*

highly reliable for MRI but also helpful to lower the cost of MRI. Bruker EAS, Mitsubishi Luvata, and WST are three main suppliers of WIC wire for MRI manufacturers.

6.3 Nuclear magnetic resonance (NMR)

NMR is a physical phenomenon in which nuclei in a magnetic field absorb and re-emit electromagnetic radiation. This energy is at a specific resonance frequency that depends on the strength of the magnetic field and the magnetic properties of the isotope of the atoms. NMR allows the observation of specific quantum mechanical magnetic properties of the atomic nucleus. Many scientific techniques exploit NMR phenomena to study molecular physics, crystals, and noncrystalline materials through nuclear magnetic resonance spectroscopy [29]. NMR is also routinely used in advanced medical imaging techniques. Monolith NbTi wires with high n value, high diameter precision, low Copper/NbTi ratio of 1.3–4.0, and fine diameter of 0.4–1.5 mm are used in NMR magnets (Fig. 11).

6.4 Magnetic field-applied czochralski crystal growth system (MCZ)

In the MCZ system, a magnetic field is applied to the molten semiconductor material, thereby controlling the molten semiconductor material heat convection. High-quality monocrystalline silicon ingot with large diameter

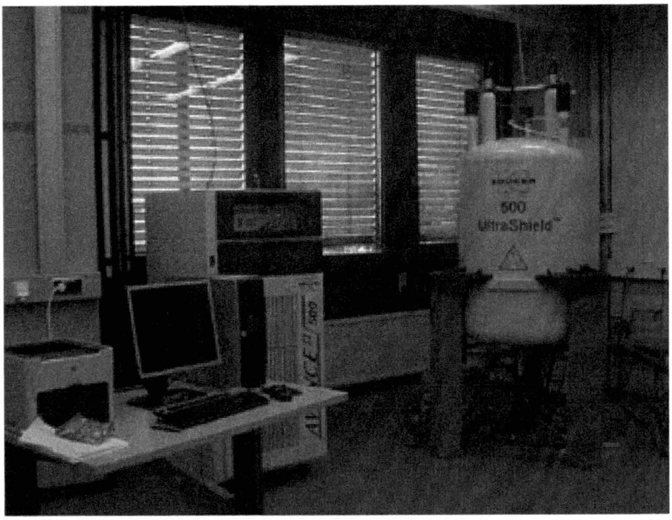

Fig. 11 A 500 MHz NMR system produced by Bruker.

can be manufactured by the MCZ system [30]. Fig. 12 shows a conventional MCZ system. This MCZ system is comprised of a magnetic field generator and a growing furnace. Semiconductor material in the growing furnace is melted, so that the crystal can be pulled upward from the molten semiconductor material. The magnetic field generator generates a magnetic field, which is applied to the molten semiconductor material in the growing

Fig. 12 MCZ magnet produced by WST. The height of the MCZ magnet is 2.2 m.

furnace. Growing furnace includes a furnace body, a crucible, a heater, a crucible support mechanism, and the lifter. The elevator is designed to melt from the crucible in the pulled semiconductor crystal material and the lifter mechanism rotating relative to each other, whereby the crystal and crucible rotating relative to each other. The magnetic field generator is mounted on the bracket, around the furnace body. The generator is comprised of a superconducting coil having a built-in cryostat. Monolith NbTi wires with high copper/NbTi ratio more than seven and with the shape of rectangle and round are used in MCZ magnets.

6.5 High energy particle accelerator

A particle accelerator is a machine that uses electromagnetic fields to propel charged particles to nearly light speed and to constrain them in well-defined beams [31]. Large accelerators are used for basic research in particle physics. The most powerful accelerator currently is the Large Hadron Collider (LHC) near Geneva, Switzerland, built by CERN. It is a collider accelerator, which can accelerate two beams of protons to an energy of 6.5TeV and cause them to collide head-on, creating center-of-mass energies of 13TeV. Other powerful accelerators are KEKB at KEK in Japan, RHIC at Brookhaven National Laboratory, and the Tevatron at Fermilab. Accelerators are also used as synchrotron light sources for the study of condensed matter physics. Smaller particle accelerators are used in a wide variety of applications, including particle therapy for oncological purposes, radioisotope production for medical diagnostics, ion implanters for manufacture of semiconductors, and accelerator mass spectrometers for measurements of rare isotopes such as radiocarbon. There are currently more than 30,000 accelerators in operation around the world [32]. Rutherford cable of NbTi wires with the total critical current higher than 10,000A (4T and 4.2K) is used in particle accelerator magnets.

6.6 High-intensity heavy-ion accelerator facility (HIAF)

The high-intensity heavy-ion Accelerator Facility (HIAF) is a new engineering project proposed by the Institute of Modern Physics (IMP) of China. Superconducting magnets will be used as part of the accelerator magnets. To obtain high beam intensities, fast-pulsed magnets are required for synchrotrons. When operating in a fast-pulsed cycled mode, the superconducting cable will dissipate power (so called AC loss), which affects the stability of the magnet [33]. The main contribution of the AC losses in superconducting wire are hysteresis loss and eddy current losses [34]. Fine filaments design can reduce the hysteresis loss, and a resistive (Cu-0.5% Mn or Cu-5-10% Ni)

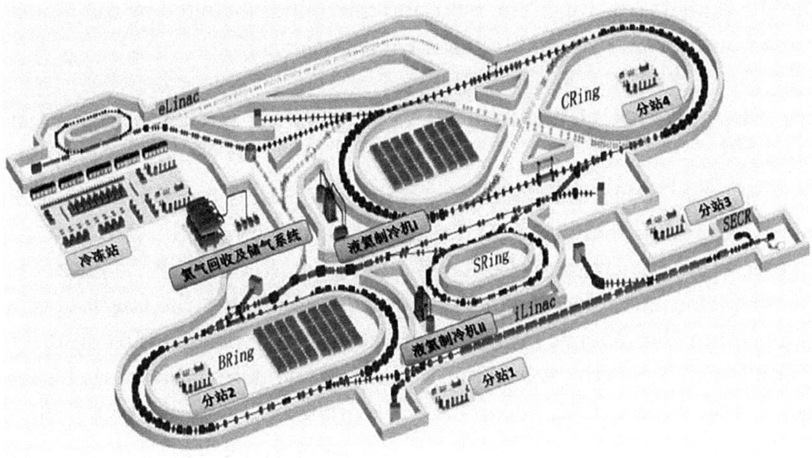

Fig. 13 The overall layout of HIAF [33].

interfilamentary matrix can reduce the interfilamentary eddy current losses. In addition, small twist pitch in superconducting wire is also required [35]. WST has developed two types of superconducting wires with the matrix of CuMn and CuNi, respectively, for the HIAF project (Fig. 13).

6.7 Magnetic separation

Magnetic separation is a process in which a magnetically susceptible material is extracted from a mixture using a magnetic force. This separation technique can be used for selecting iron as it will be attracted to the magnet. It is also used in electromagnetic cranes that separate magnetic material from scraps. Another application, not widely known but very important, is to use a magnet to remove metal contaminants from product streams [36]. Weifang Xinli Superconducting Technology Co., Ltd. focuses on developing superconducting magnetic separators and superconducting iron separators. Major breakthroughs have been achieved in superconducting technology and applied in industrial production. Monolith NbTi wire with Copper/NbTi ratio from 1.3 to 7 and with the shape of rectangle is used in magnetic separation magnets.

6.8 Magnetic levitation (Maglev)

Maglev is a system of train transportation that uses two sets of magnets, one set to repel and push the train up off the track as in levitation, and the other set to move the floating train ahead at great speed taking advantage of the lack of friction. In electrodynamic suspension (EDS), both the guideway and the train exert a magnetic field, and the train is levitated by the repulsive and attractive force between these magnetic fields [37].

6.9 Superconducting liquid level meter for superconducting magnet

To ensure its stable operation, the superconducting magnet must be immersed in liquid helium, and thus the liquid level must be detected. One type of liquid level meter is NbTi wire with a matrix of CuNi alloy. Because the magnet quench may occur in the excitation process, the fast flow of helium within the magnet may damage the liquid level meter. To ensure the reliability, it is necessary to arrange two superconducting liquid level meters. In this way, when a fault occurs, one is switched to the other one [38].

7 Summary

High-performance ductile NbTi superconducting wires are manufactured as a commodity product for applications where large and uniform magnetic fields are required. Western Superconducting Technologies Co., Ltd. (WST) has successfully developed different types of NbTi superconducting wires, including Monolith NbTi superconducting wire, wire-in-channel (WIC) NbTi superconducting wire, and superconducting cables. These wires and cables have found wide applications in international thermonuclear experimental reactors (ITER), magnetic resonance imaging (MRI), nuclear magnetic resonance (NMR), magnetic-field applied czochralski crystal growth systems (MCZ), high energy particle accelerator, high-intensity heavy-ion accelerator facility (HIAF), superconducting liquid level meter for superconducting magnets, magnetic separation, and magnetic levitation (Maglev). Just for MRI applications, the global annual demand is approximately 90,000 to 100,000 km. This means that NbTi superconducting wires still have a huge market in the future. The scientific studies of NbTi wires have largely come to an end, but the cost-control issue of NbTi superconducting wires still has a long way to go.

References

[1] Z. Charifoulline, Residual resistivity ratio (RRR) measurements of LHC superconducting NbTi cable strands. IEEE Trans. Appl. Supercond. 16 (May 2006) (2006) 1188–1191, https://doi.org/10.1109/TASC.2006.873322.

[2] T.G. Berlincourt, R.R. Hake, Pulsed-magnetic-field studies of superconducting transition metal alloys at high and low current densities, Bull. Am. Phys. Soc. II 7 (1962) 408. T.G. Berlincourt, Emergence of Nb-Ti as Supermagnet material, Cryogenics 27 (1987) 283.

[3] M.N. Wilson, IEEE Trans. Magn. MAG-13 (1977) 440.
[4] H. Hillmann, et al., Proceedings of 1972 Appl. Supercond. Conf., 1972, p. 429.
[5] T. Luhman, Treatise on materials science and technology, Metall. Supercond. Mater. 14 (1979) 99.
[6] N. Zhou, C.R. Li, X.Z. Wu, S.G. Xiong, New melting and casting process of NbTi50 superconducting alloy, Chin. J. Low Temp. Phys. 2 (1980) 159.
[7] X.Z. Wu, C.R. Li, N. Zhou, Observation of NbTi/Cu superconductor microstructure by SEM, Chin. J. Low Temp. Phys. 1 (1988) 35–39.
[8] P.J. Lee, J.C. McKinnell, D.C. Larbalestier, Microstructure control in high Ti NbTi alloys, IEEE Trans. Magn. 25 (2) (1989) 1918–1922.
[9] J.A. Parrel, P.J. Lee, D.C. Larbalestier, Cold work loss during heat treatment and extrusion of Nb-46.5wt.%Ti composites as measured by microhardness, IEEE Trans. Appl. Supercond. 3 (1) (1993) 734–737.
[10] L. Cooley, P.J. Lee, D.C. Larbalestier, Conductor Processing of Low Tc Materials: The Alloy Nb-Ti of the Handbook on Superconducting Materials, ResearchGate, 2003, pp. 603–639.
[11] P.J. Lee, C.M. Fischer, D.C. Larbalestier, M.T. Naus, A.A. Squitieri, W.L. Starch, R.J. Werner, P.J. Limon, G. Abbi, A. Zlobin, E. Gregory, Development of high performance multifilamentary Nb-Ti-Ta superconductor for LHC insertion quadrupoles, IEEE 9 (2) (1999) 1571–1574.
[12] N. Mitchell, D. Bessette, H. Fujieda, Y. Gribov, C. Jong, F. Simon, Improvements to the ITER magnets to extend the plasma operating window, Fusion Eng. Des. 84 (2009) 113–121.
[13] H. Rajainm, K. Humer, R. Prokopec, H.W. Weber, F. Rondeaux, P. Rossi, et al., Development of special resins and composite structures for the ITER toroidal field coils, in: Paper Presented at the 25th Symposium of Fusion Technology, Rostock, Germany, September, 2008.
[14] G.P. Vedernikov, A.K. Shikov, L.V. Potanina, E.I. Plashkin, E.V. Nikulenkov, I.N. Gubkin, N.I. Salunin, V.Y. Korpusov, Development of fine filament NbTi superconducting strands for magnet systems of fusion reactors, Physica C 354 (2001) 420–423.
[15] H.C. Kanithi, P. Valaris, B.A. Zeitlin, A novel approach to make fine filament superconductors, Supercollider 4 (1992) 41–47.
[16] P.J. Lee, D.C. Larbalestier, NbTi superconducting wires: nanostructures by extrusion and wire drawing, Wire J. Int. (2003) 61–66.
[17] V.V. Slezov, O.V. Chernyi, L.N. Davydov, Critical current density and magnetic pinning in Nb-Ti superconductor, Supercond. Sci. Technol. 18 (2005) 747–752.
[18] IEC 61788-2, Superconductivity—Part 1: Critical Current Measurement-DC Critical Current of Nb-Ti Composite Superconductors.
[19] D.M.J. Taylor, D.P. Hampshire, Relationship between the n-value and critical current in Nb3Sn superconducting wires exhibiting intrinsic and extrinsic behavior, Supercond. Sci. Technol. 18 (2005) S297–S302.
[20] D.M.J. Taylor, S.A. Keys, D.P. Hampshire, *E–J* characteristics and n-values of a niobium-tin superconducting wire as a function of magnetic field, temperature and strain, Physica C: Supercond. 372 (2002) 1291–1294.
[21] IEC 61788-2, Superconductivity—Part 4: Residual Resistance Ratio Measurement—Residual Resistance Ratio of Nb-Ti Composite Superconductors.
[22] IEC 61788-2, Superconductivity—Part 13: AC Loss Measurements-Magnetometer Methods for Hysteresis Loss in Cu/Nb-Ti Multifilamentary Composites.
[23] M. Huguet, The ITER magnets: preparation for full size construction based on the results of the model coil program, Nucl. Fusion 43 (2003) 352.
[24] Task Agreement #12TT08FR Development Programme for NbTi Conductors for the ITER PF Coils, Sept. 20, 1999.
[25] M. Huguet, The ITER magnets: preparation for full size construction based on the results of the model coil program, Nucl. Fusion 43 (2003) 352.

[26] T. Wong, M.K. Rudziak, J.M. Seuntjens, et al., Variable composition NbTi superconductors produced by artificial pinning center process, IEEE Trans. Appl. Supercond. 7 (2) (1997) 1126–1129.
[27] T. Hartikainen, A. Korpela, J. Lehtonen, et al., A comparative life-cycle assessment between NbTi and copper magnets, IEEE Trans. Appl. Supercond. 14 (2) (2004) 1882–1885.
[28] S. Nenonen, E. Friman, O. Ikkala, et al., Construction and test results of a compact 0.8 meter warm bore 1.5-tesla high-homogeneity superconducting magnet for MR-spectroscopy, IEEE Trans. Magn. 28 (1) (1992) 637–640.
[29] N.I. Salunin, E.I. Plashkin, E.V. Nikulenkov, et al., Superconducting wires for NMR-application, IEEE Trans. Magn. 32 (4) (1996) 2902–2904.
[30] Y. Nagai, S. Nakagawa, K. Kashima, Crystal growth of MCZ silicon with ultralow carbon concentration, J. Cryst. Growth 401 (2014) 737–739.
[31] S. Witman, Ten things you might not know about particle accelerators, in: Symmetry Magazine, Fermi National Accelerator Laboratory, 2014. (Retrieved 21 April 2014).
[32] E.D. Courant, M.S. Livingston, H.S. Snyder, The strong-focusing synchrotron—a new high energy accelerator, Phys. Rev. 88 (5) (1952) 1190–1196.
[33] Y.Q. Chen, R&D of Liquid Helium Inner Cooling Superconducting Cable for HIAF Fast-pulsed Magnet, University of Chinese Academy of Science, 2015.
[34] F. Borgnolutti, B. Auchmann, L. Bottura, et al., Construction of the CERN fast cycled superconducting dipole magnet prototype, IEEE Trans. Appl. Supercond. 22 (3) (2011) 4001604.
[35] Y.V. Karasev, L.V. Potanina, N.I. Salunin, Study of low loss experimental superconducting Nb-Ti wires to be used in FAIR magnets, IEEE Trans. Appl. Supercond. 25 (3) (2015). (June. 2015, Art. no. 6001405).
[36] Y. Zhao, B. Xi, Y. Li, M. Wang, Removal of phosphate from wastewater by using open gradient superconducting magnetic separation as pretreatment for high gradient superconducting magnetic separation, Sep. Purif. Technol. 86 (3) (2012) 255–261.
[37] Railway Technical Research Institute, http://www.rtri.or.jp/rd/division/rd79/yamanashi/english/principle_E.html. Retrieved 25 May 2012.
[38] Cheng Yu. Composite superconductive liquid level meter for superconducting magnet (Patent).

Further reading

[39] Z. Guo, W.H. Warnes, Mechanical behavior of fine filament Nb-Ti as a function of processing, IEEE Trans. Appl. Supercond. (1993) 1022–1025.

CHAPTER 16

Titanium springs and fasteners

N. Gui, Tingting Song, Ma Qian
Centre for Additive Manufacturing, School of Engineering, RMIT University, Melbourne, VIC, Australia

Contents

1 Titanium springs

Springs are typically manufactured from high-strength alloys. They can withstand considerable compression, extension, twisting, or bending, and then return to their original shapes without distortion when the force is removed. If the force applied exceeds the yield strength, permanent deformation will occur, and the spring will be distorted. Similar to steel springs, titanium (Ti) springs can be made into a variety of shapes, for example, compression springs (the most common type), torsion springs, clock springs, and simple clip devices, depending on applications.

The elastic potential energy (U_e) stored in a spring during deformation is defined by

$$U_e = \frac{1}{2}kx^2 \tag{1}$$

where k is the spring constant (N/m) and x is the length of extension or compression in relation to the original spring length (L). The relationship between k and elastic modulus (E) is defined according to Hooke's law ($F = -kx$; F: the force) and the definition of elastic modulus by

Titanium for Consumer Applications
https://doi.org/10.1016/B978-0-12-815820-3.00015-0

$$k = E\frac{A}{L} \tag{2}$$

in which A is the area over which the force (F) is applied.

Substituting Eq. (2) and $F=-kx$ into Eq. (1) gives

$$U_e = \frac{1}{2}\frac{F^2L}{EA} \tag{3}$$

Eq. (3) reveals that, for a given spring length (L) and a cross-section area (A), the spring alloy with low elastic modulus (E) offers high elastic potential energy (U_e), provided that the applied stress (F/A) is below the yield strength of the alloy (high yield strength is thus desired).

Compared with steel alloys, β-Ti alloys offer much lower elastic modulus (103 GPa vs. 200 GPa for steel, Table 1) but high yield strength up to 1600 MPa (e.g., Ti-1Al-8V-5Fe, in wt.% throughout this chapter) [1]. Together with their low density, excellent corrosion resistance, fracture toughness, fatigue strength, and good performance at moderate elevated temperatures [1], β-Ti alloys are thus ideal materials for spring applications. In addition, their excellent biocompatibility extends their usage as spring alloys to biomedical applications. Similarly, their unique compatibility with composites allows them to be used as ideal springs in contact with composites.

The conventional definition of β-Ti alloys is based on the molybdenum equivalence (Mo-Eq) value of a titanium alloy. In general, titanium

Table 1 Mechanical properties of titanium alloys and commonly used engineering alloys [1–3]

Material	Average specific gravity (g/cm^3)	Strength-to-weight ($kPa\ m^3/kg$)	Young's modulus (GPa)	Yield strength (MPa)
CP titanium (α)	4.51	55–100	106.3	250–450
Ti-6Al-4V (α-β)	4.43	200–250	110	900–1100
Ti-10V-2Fe-3Al (near-β)	4.65	252	110	1170
Ti-8V-6Cr-4Mo-4Zr-3Al (β-C; metastable β-Ti)	4.82	276	103	1330
Carbon steel	7.8	45–60	200	350–450
18–8 Stainless steel (SUS304)	7.9	51–101	200	400–800
18 Ni Maraging steel (250 Grade)	8.0	82	190	655
Aluminium alloy (7075-T76)	2.8	167	71.5	470
Magnesium alloy AZ91 (Mg-9Al-1Zn)	1.81	83	45	150

alloys can be classified into α-, (α-β)- and β-alloys, based on the presence of the principal alloying elements, which include α-stabilizers and β-stabilizers [1, 4]. Near-β-Ti alloys refer to those with the Mo-Eq in the range of $5 \leq$ Mo-Eq ≤ 10, whereas metastable β-Ti alloys are those with the Mo-Eq in the range of $8 \leq$ Mo-Eq ≤ 30 [1]. Stable β-Ti alloys have the Mo-Eq value of >30 [1]. A number of near-β-Ti alloys can offer low elastic modulus and ultrahigh yield strength up to 1600 MPa, which can effectively reduce the volume of the spring products [5]. For example, Ti alloy springs can achieve weight reductions by up to 70% without compromising mechanical properties compared with steel springs [6, 7]. Moreover, Ti alloy springs exhibit better lifecycle cost than steel springs due to their superior corrosion resistance [7]. Fig. 1 compares a Ti alloy spring and a steel spring of similar performance. Table 1 lists the properties of commercially pure titanium (CP Ti, α-alloy), Ti-6Al-4V (α-β), Ti-10V-2Fe-3Al (near-β; Mo-Eq = 9.5), and Ti-8V-6Cr-4Mo-4Zr-3Al (β-C, metastable β-Ti; Mo-Eq = 16.0) in relation to those of the commonly used engineering alloys [2].

Fig. 1 A carbon steel spring (left, 12.04 kg) and a titanium spring (right, 4.73 kg) designed for the same diameter and spring rate requirements. The titanium spring is 60% lighter and has a free length of 7.5 in., 28% shorter than that of the steel spring. *(Adapted with permission from MW Industries Inc. Photo licensed by ©2019 MW Industries, Inc. (MWI) All rights reserved. www.mw-ind.com.)*

According to Boyer and Briggs [7], the first titanium springs on commercial aircraft were introduced in about 1970 by McDonnell Douglas on the DC-10. They were fabricated from Ti-13V-11Cr-3Al (cold-worked and aged). The Ti-13V-11Cr-3Al alloy has a Mo-Eq value of 23 and was introduced in the mid-1950s by the Crucible Steel Corporation [8]. Due to its high Mo-Eq value, this early β-Ti alloy is unnecessarily stable. In fact, one of the tricks to using the alloy was to "not fully recrystallize the structure, thereby leaving residual hot/cold work within the bulk to assist aging response" [8].

The metastable Ti-1Al-8V-5Fe (Ti-1-8-5, Mo-Eq = 19) alloy containing 0.30%O-0.50%O was developed in the late 1950s [9]. It is capable of achieving very high strength/ductility combinations [10]. In that regard, it can still be regarded as a good candidate titanium alloy for spring applications.

Ti-3Al-8V-6Cr-4Mo-4Zr (referred to as β-C; Mo-Eq=15.96) is a premier metastable β-Ti alloy for spring applications (a standard titanium alloy for aircraft spring applications). It was first used on the Boeing 777 to make coil-type springs with a minimum ultimate tensile strength (UTS) of 1240 MPa, whereas the Ti-15V-3Cr-3Al-3Sn (Ti-15-3-3-3, Mo-Eq = 11.9) alloy was used to make clock-type springs with a minimum UTS of 1034 MPa for the same aircraft [7]. Other important β-Ti spring alloys include Ti-11.5Mo-6Zr-4.5Sn (Beta III, Mo-Eq=12), Ti-4.5Fe-6.8Mo-1.5Al (TIMETAL LCB, Mo-Eq=18) [11], and Ti-10V-2Fe-3Al (Ti-10-2-3, Mo-Eq=9.5) [7]. The Fe-containing TIMETAL LCB is a low-cost β-Ti alloy designed specifically for automotive coil spring applications [12]. It was first used in the 2000 Volkswagen Lupo FSI as suspension springs. From 2003, similar suspension springs were used for Ferrari Challenge Stradale [12]. A more detailed discussion of β-Ti alloy spring applications can be found in Refs. [7, 12].

2 Titanium fasteners

The term "fasteners" refers to tools that can mechanically join two or more components together [13]. The failure of a single fastener can risk causing failure or disfunction of the entire assembly [14]. In general, high yield strength (for tensile-loaded fastener assemblies) and high shear strength are essential for many structural applications with bolts and screws [4]. Because yielding can result in immediate loosening of the joint, it is always more important to have sufficiently high yield strengths for tensile-loaded fastener assemblies (more important than the ultimate tensile strength) [14]. On the other hand, for fasteners that are subject to cyclic loading, adequate fatigue performance will be an additional essential design criterion [14]. In addition, because failure often occurs in the threaded section due to the high stress concentration, sufficient notched fatigue resistance is particularly important.

Over the past 60 years, Ti-6Al-4V alloy has been the most commonly used Ti alloy for Ti fasteners to replace conventional steel and stainless steel fasteners [15]. However, the size of Ti-6Al-4V fasteners is generally limited to 0.625–0.75 in. (~16–19 mm) due to their inadequate hardenability (i.e., not possible to heat-treat them to full strength) and formability [15, 16]. As a result, larger-diameter parts of Ti-6Al-4V must be downgraded in strength due to the hardenability issue.

CP Ti Grade 7 (i.e., Grade 2 containing 0.12–0.25%Pd) and Ti-5111 (Ti-5Al-1Sn-1Zr-1V-0.8Mo) are ideally suited for applications in marine environments where corrosion resistance to seawater is essential. In particular, Ti-5111 offers excellent seawater stress corrosion, cracking resistance, and high dynamic toughness, and can be the material of choice for a variety of marine applications. American Society for Testing and Materials (ASTM) Grade 13 (Ti-0.5Ni-0.05Ru or known as Ticorex) is another titanium alloy that can offer excellent corrosion resistance in marine environments [17].

In an effort toward reducing the cost of manufacturing, Honda Motor Co. proposed to use a Ti-(0.6–1.4)Fe-(0.24–0.44)O alloy to make titanium bolts [18]. This Ti-Fe-O alloy offers tensile strength of ≥800 MPa, but it allows plastic working at ambient temperature, which can make a product close to its final shape and thereby enables the effective use of the material with a reduced cost of manufacturing [18].

Much effort has been made to develop high-strength β-Ti alloy fasteners with tensile strength of ≥1380 MPa and shear strength of ≥760 MPa, in addition to satisfying the fatigue criterion [7]. The goal was to replace high-strength steel fasteners. However, research has shown that it was challenging to meet the tensile strength, shear strength, and fatigue strength requirements at the same time. Unlike tensile strength, ASTM specifications offered no detailed information about the requirements for the minimum shear strength values. As an empirical rule, it has been suggested that, at least for carbon steel bolts, the shear strength should reach ~60% of the ultimate tensile strength. As can be seen from Tables 1 and 2, this requirement can be challenging for β-Ti alloys. In addition, it is challenging to manufacture high-strength β-Ti alloy bolts by cold forging because they can cause wear

Table 2 Representative titanium springs and fasteners, their microstructures, and mechanical properties at room temperature [1, 4]

Alloy composition	Commercial name	Microstructure	0.2% YS (MPa)	UTS (MPa)	El (%)	Condition
99.2Ti-≤0.25O-≤0.3 Fe	Grade 2/Ti50A	α	275	344	20	Mill-annealed 675°C
Ti-≤0.25O-≤0.3Fe-0.2 Pd	Grade 7/Ti75A	α	275	344	20	Mill-annealed 675°C
Ti-0.5Ni-0.05Ru	Grade 15	α	380	483	18	Mill-annealed ~675°C
Ti-6Al-4V ELI	Ti-6-4	α-β	790	860	15	Mill-annealed ~740°C
Ti-6Al-4V (IMI 318)	Ti-6-4	α-β	880	950	14	Mill-annealed ~740°C
			1100	1170	10	ST (α+β) aged 500°C
Ti-5Al-1Sn-1Zr-1V-0.8Mo	Ti-5111	Near-α	720	835	13	24.5 mm plate β roll 954°C 1 h air-cooled
Ti-(3.5–4.5)Al-(2–3)V-(1.2–1.8)Fe-(0.2–0.3)O-≤0.08C-≤0.03N-≤0.015H	ATI 425/Grade 38	α-β	1090	1123	14	Cold-rolled coil and sheet, mill-annealed
Ti-6V-6Mo-5.7Fe-2.7Al	Timetal 125	Near-β	1150	1439	3	As-cast
Ti-8V-5Fe-1Al	Ti-1-8-5	Near-β	1233	1263	15	Annealed 675°C
Ti-6.8Mo-4.5Fe-1.5Al	Timetal LCB	β	1241	1296	6	8.5–25 mm coils and bars STA
Ti-8V-6Cr-4Mo-4Zr-3Al	β-C	Near-β	1130	1225	10	ST(β) aged 540°C
Ti-10V-2Fe-3Al	Ti-10-2-3	Near-β	1000	1103	6	ST(β) aged at 510–535°C
Ti-13V-11Cr-3Al	B120VCA	Near-β	1200	1280	8	ST(β) aged 480°C 36 h
Ti-15V-3Cr-3Al-3Sn	Ti-15-3-3-3	β	1050	1160	11	790°C 15 min AC 538°C 4–16 h
Ti-15Mo-3Nb-3Al-0.2Si	β-21S	Near-β	1338	1427	6	ST 834–900°C aged at 538°C 8 h

ELI, Extra low interstitial; *YS*, yield stress; *UTS*, ultimate tensile strength; *EI*, elongation; *ST*, solution treated; *TIMET*, Titanium Metals Corporation; *LCB*, low-cost beta titanium alloy.

of the dies. To address this issue, since early 2000, Kobel Steel has developed a multistep cold-forging process for the manufacture of Ti-15-3-3-3 bolts (see Table 2) [19].

Fatigue strength is another important requirement for titanium fasteners. For aerospace fasteners, their fatigue strength is typically assessed under $R=0.1$ conditions for up to an average of 65,000 cycles, where R is the ratio of the minimum fatigue load to the maximum fatigue load [20]. For high-performance racing car fasteners, a much higher number of cycles (i.e., 2.5 million) is required when tested under high mean-load conditions [20].

Crevice corrosion occurs in small gaps and openings in many fasteners. In general, titanium alloys exhibit superior resistance to crevice corrosion versus stainless steels. This is another advantage of choosing titanium fasteners over stainless steel fasteners. However, crevice corrosion may become an issue when titanium fasteners are used in hot aqueous chloride, bromide, iodide, or sulfate solutions where tight crevices exist [21].

Galvanic corrosion has been frequently observed with 4340 steel (standard corrosion potential, $E_{corr}=-0.6V$) and aluminium alloy, Al-7075 ($E_{corr}=-0.78V$) fasteners that connect the carbon fiber/epoxy ($E_{corr}=0.12V$) skins of an aircraft to its inner substructure due to the noticeable difference in E_{corr}. Such structural damage is expensive to repair because it involves redrilling and installation of the oversized fasteners from an appropriate material [22]. The use of titanium ($E_{corr} = -0.42V$) fasteners can effectively mitigate this issue.

Strict microstructural control is essential to ensure reliable performance of titanium fasteners for critical applications. In a reported case study, a detailed analysis was made of the failure of a Ti-6Al-4V bolt used in aerospace applications [23]. This Ti-6Al-4V bolt was coated with MoS_2 as solid lubricant. During the trial assembly, the Ti-6Al-4V bolt failed such that the head of the bolt had sheared off the shank [23]. It was identified that there was process deviation during the processing of the bolt. As a result, the microstructure from head-to-shank area of the bolt became inconsistent, which deteriorates the properties of the Ti-6Al-4V bolt and hence the failure occurred [23].

In addition to microstructural variations, small changes in the frictional characteristics of the fastener system due to time-dependent degradation could result in substantial tensile overloading of the bolts at the prescribed torque level and therefore failures of the titanium bolts [24]. For this reason, in general, reuse of titanium fasteners that had previously been in service is not recommended. A detailed case study can be found in Ref. [24], which examined the failure of two Ti-6Al-4V wing attachment bolts from a commercial jetliner that occurred during the re-torque process when the wing was being reattached.

Long-term measurements have indicated that Ti-6Al-4V fasteners can exhibit noticeable creep behavior and accumulate large creep strain [25]. Consequently, precautions should be taken for preloaded Ti-6Al-4V bolts [25], including limiting the preload to ≤65% of the yield strength and minimizing preload uncertainty [26]. In addition, preload relaxation due to creep of solid-film lubricants applied onto the bolts can compound this issue [26].

Compared with A286 stainless steel aerospace fasteners, the tensile ductility of most titanium fasteners appears to be less sufficient. For example, the average tensile ductility index of the Ti-6Al-4V aerospace fasteners is 0.53, compared with 5.40 for the A286 aerospace fasteners [26]. As a result, the tensile fracture surfaces of Ti-6Al-4V aerospace fasteners are found to be perpendicular to the axis of the tensile loading (brittle fracture), while those of the A286 aerospace fasteners are at about 45 degrees to the axis of the tensile loading (ductile fracture) [26]. This could lead to uncertainty or potential catastrophic failures that can result in brittle materials. This may be regarded as a deficiency of most titanium alloys fasteners. In that regard, the interstitial content in the titanium fasteners needs to be strictly controlled.

In summary, utilization of Ti alloy springs and fasteners conforms to standard specifications, such as the ASTM specifications A 193/A 193M, A 194, A 540, B265, and B348. Table 2 summarizes representative examples of Ti springs and fasteners, and their mechanical properties at room temperature. The specific applications of Ti springs and fasteners are discussed in the following section.

3 Application of titanium springs and fasteners

3.1 Automobiles

The use of Ti springs and fasteners can help reduce the weight of automobiles for improved fuel consumption and reduced environmental pollution [27, 28]. However, the high cost of Ti springs and fasteners limits their usage mainly to racing automobiles such as in Formula One, Grand Touring, and NASCAR Winston Cup or high-end automobiles [28–30]. Fig. 2 illustrates the potential applications of Ti springs and fasteners in automobiles [29]. Table 3 lists Ti springs and fasteners that have been used in the racing and high-end automobiles [28].

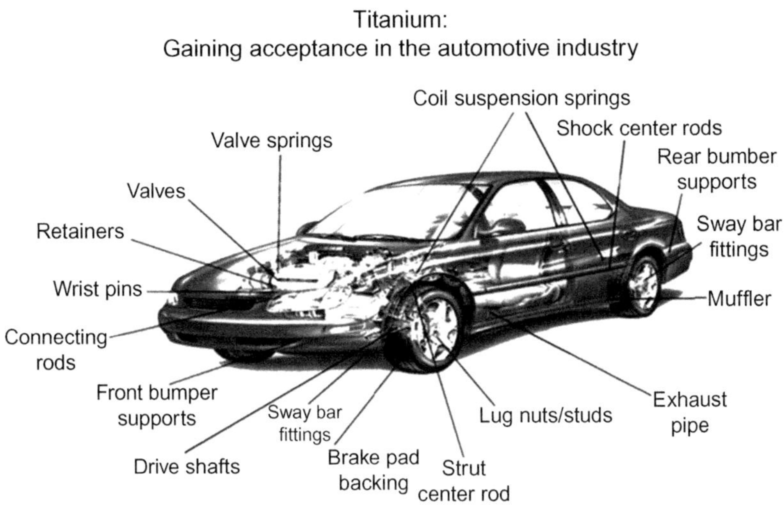

Fig. 2 Potential applications of titanium springs and fasteners and other potential titanium applications in automobiles [29]. *(Adapted with permission from Springer Nature.)*

Table 3 Ti springs and fasteners used in racing and high-end automobiles [28]

Material	Application	Manufacturer	Model
Ti-6Al-4V	Wheel rim screws	Porsche	Sport wheel option
Ti-6Al-4V	Wheel rim screws	BMW	M-Tech. option
Ti-6Al-4V	Wheel rim screws	Volkswagen	Sport package *GTI*
Ti-6Al-4V Grade 5	Valve springs and retainers	Audi	A4
Ti Grade 2	Brake pad guide pins	Daimler	S-Class
Ti Grade 1 s	Brake sealing washers	Volkswagen	All
Timetal LCB	Suspension springs	Volkswagen	*Lupo* FSI
Timetal LCB	Suspension springs	Ferrari	*360 Stradale*

To reduce the cost of production, Titanium Metals Corporation (TIMET) developed a low-cost β-Ti alloy Ti-4.5Fe-6.8Mo-1.5Al (Low-Cost Beta LCB®) for suspension spring applications [29]. Lower-cost elements iron (Fe) and molybdenum (Mo) were used to replace expensive vanadium (V) in Ti alloys [29]. The manufacturing method is based on the standard steel spring manufacturing line, which is composed of hot rolling and cold drawing [29]. Consequently, this alloy can reach high tensile strength of 1350 MPa and adequate ultimate strain after short aging [31]. It was reported that the fine grain size of this alloy (~3 μm) results in good tensile ductility [31]. The microstructure of this alloy is mainly composed of a primary α phase and supersaturated β phase [31]. After coiling and thermal treatment, precipitation of the secondary phase α (α_s) occurs, which is dependent on the aging time and temperature and the degree of deformation [31]. Table 4 lists the typical fatigue life data of Ti-4.5Fe-6.8Mo-1.5Al [32].

Similarly, Nippon Steel developed a series of low-cost Super-TIX™ Ti alloys as automotive fasteners, valves, connecting rods, etc. [33]. Inexpensive elements such as Fe, nitrogen (N), and oxygen (O) are used to replace expensive elements such as V, chromium (Cr), and nickel (Ni) [33]. The tensile strength of Super-TIX™ 51 AF (Ti-5Al-1Fe) is around 1000 MPa, which is similar to that of Ti-6Al-4V [33]. Super-TIX™ 800 (Ti-1Fe-0.35%O-0.01%N) exhibits high tensile strength of 800 MPa, and good hot and cold workabilities [33]. Fig. 3 compares the strength and ductility between Super-TIX™ series and commonly used titanium alloys.

SPS Technologies Inc. developed SPS Titan™ 200 fasteners including bolts, studs, and nuts with a minimum tensile strength of 1400 MPa, a minimum shear strength of 745 MPa, and comparable fatigue strength to that of steel [34]. The development was based on SPS Titan™ 761 (Ti-3Al-8V-6Cr-4Zr-4Mo) for high performance racing applications [34].

Pro-Bolt Co. Inc. manufactures a variety of Ti-6Al-4V fasteners for racing applications. Their fasteners can reach ultimate tensile strength of 1030 MPa, yield strength of 828 MPa, and ductility of 10% elongation at break. Moreover, both fatigue strength and antigalling properties have been enhanced. To improve the lifetime of these fasteners, prevent galling, and enhance corrosion resistance capability, a diamond-like coating can be applied to the surfaces of these fasteners. Fig. 4 shows the Ti fastener products with different colors developed by Pro-Bolt Co. Inc.

Table 4 Room temperature fatigue life for LCB coil/spring alloys in a solution-treated and aged condition [32]

Specimen type	Stress-concentration factor, K_t	Stress ratio	Cycles	Fatigue limit (MPa)
Unnotched	1	0.1	10^7	800–850
Notched	3	0.1	10^7	415

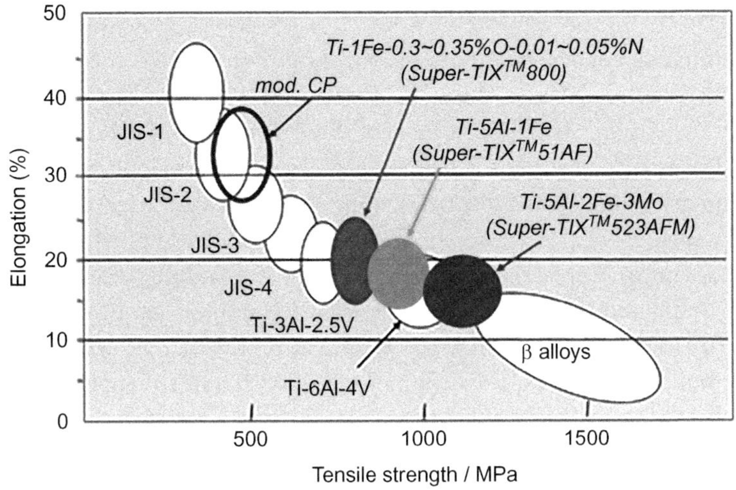

Fig. 3 Relationship between strength and ductility of Super-TIX series and commercially pure titanium and its alloys. *(Adapted with permission from Nippon Steel & Sumitomo Metal Corporation.)*

Fig. 4 Titanium fasteners with different colors for automotive applications. *(Adapted with permission from Pro-Bolt Co. Inc.)*

3.2 Aviation and aerospace

Titanium has been increasingly used in both commercial and military aircraft in recent decades, as shown in Fig. 5 [35]. Titanium alloys account for about 15% of the structural weight of the Airbus 350/380 and Boeing 787 [35]. Ti springs have been used as up-lock and down-lock springs in the landing gear, flight control springs, door counter-balance springs, hydraulic return springs, etc. [7]. Also, Ti fasteners have been applied to airframes, engine, wings, propellers, and some parts of the landing gear system [36]. It was estimated that 0.5 million pieces of Ti fasteners are installed in each new Boeing 787 [37]. Ti-6Al-4V fasteners represent the vast majority of structural fasteners used on airframes today, with annual usages in the millions by Boeing alone [38]. On average, it was estimated that about 2.4 million fasteners are used to assemble a Boeing 787 aircraft, in which 22% are structural bolts (mostly titanium) and the rest are rivets [38]. Boeing has developed a new titanium fastener alloy, designated as VST2F, which can replace Ti-6Al-4V by offering lower risks of producing metallurgical defects during thread rolling due to its higher ductility [38].

In general, Ti fasteners in aircraft have been performing reliably and safely. For example, the November 2001 accident of AA flight 587 is one of the worst aviation accidents on U.S. soil resulting in the death of all 260 people aboard the aircraft and five people on the ground [39]. The aircraft vertical stabilizer (consisting of 24 attachment lugs connected by 12 titanium

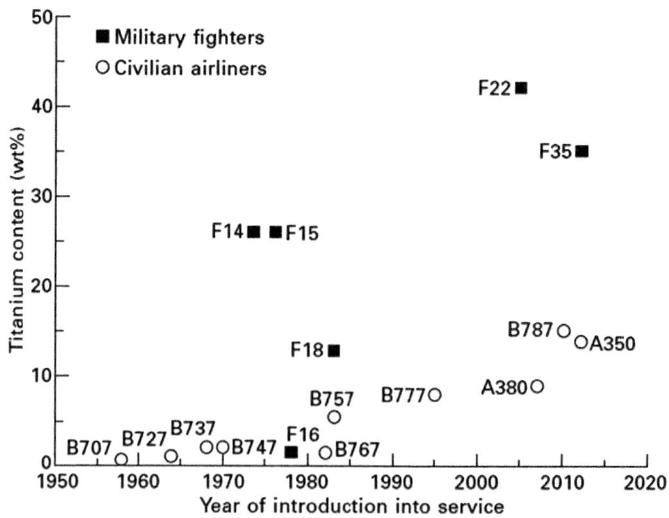

Fig. 5 Titanium content used in an aircraft [35]. *(Adapted with permission from Elsevier.)*

bolts) detached from the aircraft causing the aircraft to crash. Damage analysis showed that all 12 Ti bolts were intact but not the composite lugs [40].

It was reported that each of China's domestic C919 aircraft requires 200,000 titanium alloy fasteners, and it was estimated that nearly 30 million titanium alloy fasteners are needed to complete the planned orders every year [41]. Similarly, titanium alloy fasteners have been used in the wing and fuselage assemblies of the Airbus A380 jet liner to reduce weight. For example, nearly 9000 titanium MaxiBolt Plus blind bolts were used to replace stainless steel bolts in each carbon fiber vertical tail plane section, with a weight saving of 8.5 kg [1].

Commercial contracts for titanium fasteners can be significant. For example, in October 2015, Alcoa signed a contract worth \$1 billion from Airbus for the supply of titanium fasteners. In the same year, Alcoa signed another contract worth \$1.1 billion to supply titanium fasteners for Lockheed Martin's F-35 Joint Strike Fighter program, under which Alcoa would provide the titanium fasteners for the three variants of F-35 fighter aircraft, during 2016–2024 [42]. Stratview Research forecasts that there will be 6.4% compound annual growth rate in the global aircraft Ti fasteners market between 2018 and 2023 [43]. Fig. 5 illustrates the increased usage of titanium in both military and commercial aircraft since 1955 [35].

For aerospace applications, Ti-6Al-4V fasteners with a diameter of ≤0.625 in. are required to possess a tensile strength of 1103 MPa and a shear strength of 655 MPa based on National Aerospace Standards (NAS) [13]. However, for Ti-6Al-4V fasteners with a diameter of >0.625 in., their desired tensile strength and shear strength are 1034 MPa and 620 MPa, respectively [13]. Boeing recently filed a patent on an annealed Ti alloy with the composition of (5.50–6.75)% Al, (4.01–4.05)% V, (0.2–0.5)% O, and more than (0.551–0.80)% Fe to reach desirable mechanical properties [13]. The addition of O and Fe enabled the improvements in UTS and shear strength of the fastener, resulting in the elimination of the need for solution treatments and aging processes [13].

A new (α-β)-Ti alloy (ATI425) for fasteners has been developed with the composition of (3.9–4.5)% Al, (2.2–3.0)% V, (1.2–1.8)% Fe, (0.24–0.3)% O, ≤0.08% C, ≤0.05% N, and ≤0.3% other elements and titanium [44]. The alloy possesses a double shear strength of ≥710 MPa and an ultimate tensile strength of ≥1172 MPa for fasteners with diameters ranging from 0.18 in. (4.57 mm) to 1.25 in. (31.8 mm) [44]. Its tensile strain-to-fraction is ≥10% [44]. Hot rolling is needed for this alloy rather cold forging. Fig. 6 illustrates the manufacturing flow chart suggested specifically for this alloy [44].

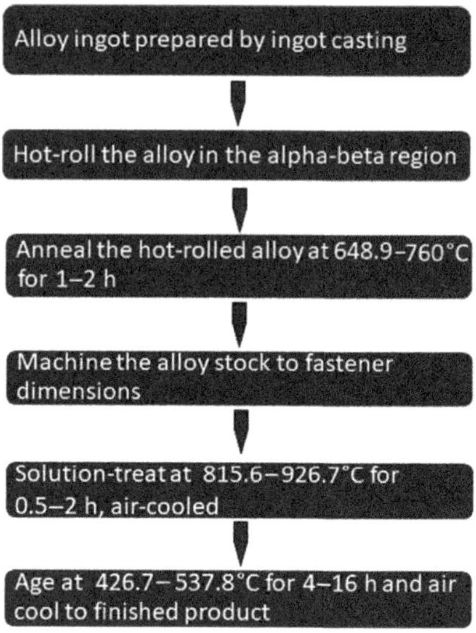

Fig. 6 A flow chart of producing a (α-β)-Ti (ATI 425) fastener by hot rolling. *(Drawn based on description of D.J. Bryan, High Strength Alpha/Beta Titanium Alloy Fasteners and Fastener Stock, ATI Properties LLC, 2012.)*

To reduce fastener installation costs and the weight of fasteners (e.g., to be 30% lighter), Boeing has filed a patent on a method for preparing highly deformed titanium and a one-piece titanium alloy fastener system incorporating rivets [45]. It features a head and shank [45]. These Ti and Ti-alloy fasteners with ultrafine grains, and cold-worked head and shank parts can exhibit a minimum ultimate shear strength of 655 MPa [45].

Ti-13V-11Cr-3Al (Ti-13-11-3) is the first β-Ti alloy used in the aerospace industry since the mid-1950s [1]. It was extensively used as rivets for the American SR-71 Blackbird reconnaissance aircraft due to its good thermal stability [46]. The alloy processed by forging can reach tensile strength of 1240 MPa and a minimum elongation of 2% [46]. Nowadays, a minimum elongation value of 2% is not considered to be sufficient for important structural applications.

Ti-3Al-8V-6Cr-4Zr-4Mo (Beta C) was a commonly used β-Ti alloy for aerospace fasteners and springs due to its low density (4.8 g/cm^3), high tensile strength of 1600 MPa, and good corrosion resistance [47]. Bombardier's regional jet has used this Beta C alloy as up-lock and down-lock coil springs with a minimum UTS of 1240 MPa [7]. After solution-treatment (850°C for 1.5 h), water quenching, and aging at 520°C for 10 h followed by air cooling, its tensile elongation and reduction of area increased by 10% and 45%, respectively [16]. Table 5 lists axial fatigue life data of Beta C alloy [32].

Table 5 Room temperature fatigue life for Beta C spring alloys in the solution-treated and aged condition [32]

Specimen type	Stress-concentration factor, K_t	Stress ratio	Cycles	Fatigue limit (MPa)
Unnotched	1	0.1	10^5	855
Unnotched	1	0.1	10^7	600
Notched	3	0.1	10^5	303
Notched	3	0.1	10^7	276

SPS Technologies Inc. provides optimized Beta C fasteners: SPS Titan™ 160 and SPS Titan™ 180. SPS Titan™ 160 Ti tension bolts display tensile strength of 1103 MPa, shear strength of 655 MPa, and good fatigue performance, whereas SPS Titan™ 180 Ti fasteners can offer a minimum tensile strength of 1241 MPa and a minimum shear strength of 745 MPa. They both meet mechanical property requirements of NAS 621, NAS 4004, AS 7460, and AS 7461.

High-strength β-Ti alloy Ti-5Al-5Mo-5V-3Cr (Ti-5553) has been introduced by TIMET Corporation as shear-type countersunk high-lock bolt [48]. Alcoa Fastening System Corporation subsequently enhanced the tensile strength of the Ti-5553 alloy to 1309 MPa with >10% elongation [49]. Moreover, the alloy exhibits good hardenability up to 12.24 cm (6 in.)-thick sections [50], good quenching capability, high fracture toughness, and low segregation sensitivity [16]. Fig. 7 shows the relationship between tensile mechanical properties and solution treatment and aging temperature [48].

Timetal®21S (Ti-15Mo-3Nb-3Al-0.2Si) [51] is featured by high temperature strength, enhanced oxidation resistance, thermal stability, and creep resistance. Its heat treatment includes solution-treatment at a temperature between 816 and 843°C for 3–30 min, followed by air cooling and aging at 538°C for 8 h [51]. Its UTS, yield strength, and elongation can reach 1427 MPa, 1338 MPa, and 6%, respectively, in both strip and 0.1875 in.-thick sheet [51]. Its working temperature can range from 228 to 593°C, which is suitable for usage as aerospace fasteners [51].

Cessna jet fleets use Ti-15V-3Cr-3Al-3Sn (Ti-15-3-3-3) alloy as clock-type springs. The advantages of this alloy include: (1) it can be strip-producible and has a lower production cost than Ti-6Al-4V sheet, (2) its forming process is easier than that of Ti-6Al-4V, and (3) it can reach a minimum UTS of 1034 MPa after heat treatment [7].

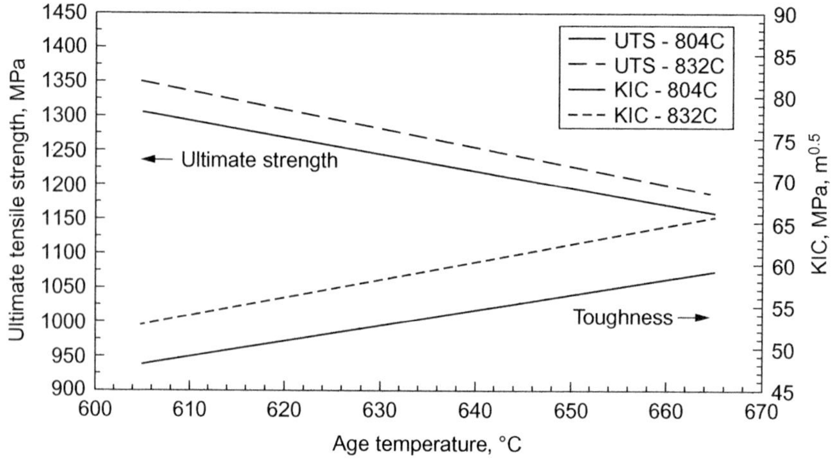

Fig. 7 Ultimate tensile strength and tensile elongation of Timetal Ti-5553 fastener stock with a diameter of 10.7 mm (0.42 in.) [48]. It was solution-treated and aged for 8 h, followed by air-cooling. *(Adapted with permission from Springer Nature.)*

3.3 Dental and bone implants

Thanks to their good biocompatibility, titanium alloys have been widely applied as bone fixation materials such as staples, screws, nails, nuts, and plates [52, 53]. Stryker Corp. has developed Ti-6Al-4V screws in various forms. AxSOS 3® titanium screws are featured with a round threaded screw head, one thread in a dual-threaded plate, and anodized surfaces, as shown in Fig. 8. It was reported that rounded threads on the locking screw head can facilitate improved tightening and releasing from the adjoined plate than sharp threads. In addition, anodized surfaces can increase fatigue strength by up to 17% and reduce friction and wear performance.

Nickel-titanium (NiTi) alloys (also named as nitinol) have received significant attention and interest in recent years due to their shape memory effect [54]. Moreover, their mechanical properties and density are closer to those of the natural bone than standard medical materials, as shown in Table 6 [55]. For example, the stress-strain relationship of nitinol can match well with that of human bones and tissues, as shown in Fig. 9 [56, 57]. For heat-active NiTi implants, the working principle is that they can deform to the desired shape in the martensite phase. After insertion into human body, they are heated to body temperature (35–37°C), which is above austenite transformation temperature. Consequently, they can return to their original shape where they produce force and tighten the bone or tooth fractures [55].

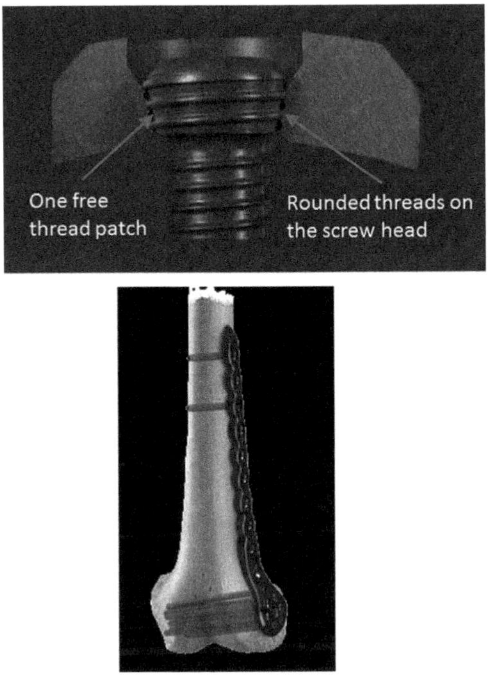

Fig. 8 AxSOS 3® Ti-6Al-4V screws. *(Adapted with permission from Styker Corp.)*

Table 6 Material density and Young's modulus of bone, stainless steel 316L, Ti-6Al-4V, and NiTi alloys [55]

Material	Density (g/cm^3)	Young's modulus (GPa)
Bone	1.7–2.0	0.2–19.4
Stainless steel 316L	7.95	193
Ti-6Al-4V	4.43	113.8
Nitinol	6.45	Austenite = 53.5 Martensite = 29.2

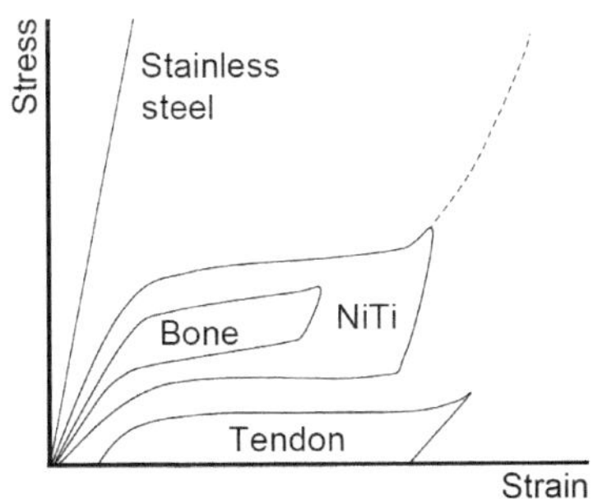

Fig. 9 The relationship between stress and strain for superelastic nitinol, stainless steel, bone, and tendon tissues [56]. *(Adapted with permission from Elsevier.)*

Nitinol bone staple is used to bond two bone fragments together [55]. One example is EasyClip®, devised by Stryker Corp. as osteosynthesis compression staples used in hand, foot, and ankle. This device can provide easy and fast fixation between bones and does not require heat activation. Its superelasticity effect responds only to mechanical stress. A newly developed device is EasyStep® for fixation in the foot (Stryker Corp.) to provide compression between the barbed legs. Fig. 10 shows nitinol bone staples developed by Stryker Corp.

Another application that has been exploited for shape memory effect is intramedullary nails. MedShape Inc. (Atlanta, GA) recently developed DynaNail® tibiotalocalcaneal (TTC) Fusion System to correct deformed joints in the hinderfoot. Immediately after surgery, internal nitinol compressive element stretches into its activated state. It can maintain active compression during the bone healing process.

Nitinol closed-coil springs are commonly used to replace elastomeric products in orthodontic space closure due to the faster rates of space closure provided [58]. Ormco Corp. developed a series of Copper Ni-Ti™ archwires, featured by different levels of loading force, as shown in Fig. 11. The loading force can be controlled via modifying the austenite transformation temperature ranging from 27°C (high force), 35°C (medium force), to 40°C (low force). In addition, it was reported that the addition of copper can increase thermal reactivity and keep tooth-moving forces consistent. It is noted that the release of Ni ion from the implant can cause a toxic effect in the human body [59]. Therefore, Ni-free superelastic alloys are being developed for future biomedical applications.

3.4 Marine applications

Been and Faller [14] have discussed the design considerations for marine fastener applications in detail along with the unique advantages that titanium alloys can offer for such applications. In a seawater environment, localized corrosion (e.g., crevice corrosion and pitting attack) can be a significant issue, which limits the applications of many otherwise corrosion-resistant bolting materials, such as stainless-steel fasteners, which are prone to crevice corrosion and chloride-related stress corrosion cracking [14]. Hence, localized corrosion demands regular maintenance and replacements when non-titanium fasteners are used. In general, Ti fasteners and springs offer excellent corrosion resistance in a seawater environment due to naturally formed titanium oxide (TiO_2) film [14].

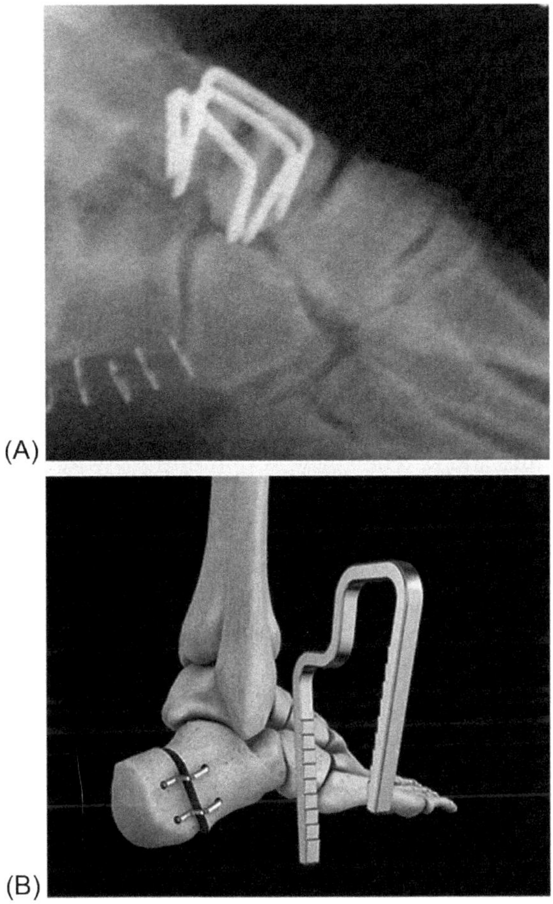

Fig. 10 NiTi bone staples. (A) EasyClip® and (B) EasyStep®. *(Adapted with permission from Stryker Corp.)*

Ti-5111 (Ti-5Al-1Sn-1Zr-1V-0.8Mo) is a near-α Ti alloy, which was designated as ASTM Grade 32 [14]. It was reported that 16 mm-diameter Ti-5111 fasteners with cold-rolled threads exhibit a 10-degree wedge tensile strength of around 930 MPa [14]. Moreover, as pointed out earlier, Ti-5111 alloy displays good resistance to stress-corrosion cracking in a marine environment at room temperature [14]. It can be manufactured at either the β or α/β region then postprocessed via either water quenching or air cooling [14].

A recent study showed that Ti-15V-3Cr-3Al-3Sn (Ti-15-3-3-3) alloy in a solution-annealed condition can be used as a medium strength fastener [15]. More importantly, it possesses higher resistance to degradation in

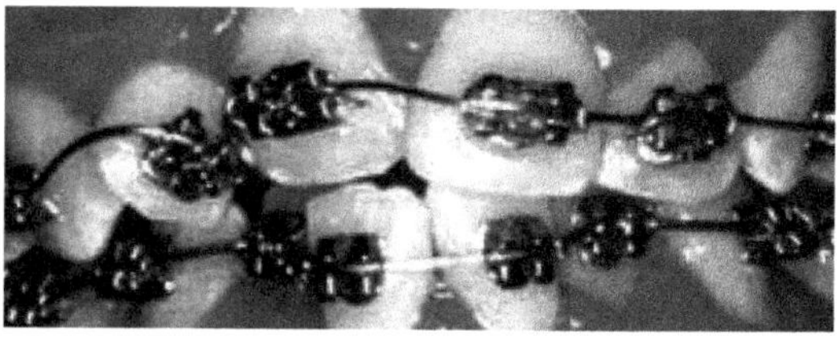

Fig. 11 Copper NiTi™ archwires. *(Adapted with permission from Ormco Corporation.)*

seawater than Ti-6Al-4V in mill-annealed condition [15]. In addition, the Ti-15-3-3-3 alloy exhibits good cold formability [15].

4 Concluding remarks

Titanium springs and fasteners have found important applications in aerospace, biomedical, marine engineering, racing automobiles, and other industrial sectors due to the combination of a broad range of the unique properties of titanium alloys. Along with the continuous development in new high-performance and more functional titanium alloys and improvement in their manufacturing processes, it is expected that titanium springs and fasteners will find broader and more extensive applications in the future.

Acknowledgments

The authors thank the Australian Research Council (ARC) for the financial support through LP140100608 and Chongqing Dien Technology China. Dr. N. Gui further acknowledges support from the Higher Degree by Research Publication Grant (HDRPG) of RMIT University.

References

[1] I. Polmear, D. StJohn, J.-F. Nie, M. Qian, Titanium alloys, in: I. Polmear, D. StJohn, J.-F. Nie, M. Qian (Eds.), Light Alloys, fifth ed., Butterworth-Heinemann, Boston, 2017, pp. 369–460.
[2] Titanium Products, Nippon Steel & Sumitomo Metal Corporation, 2014.
[3] A.P. Mouritz, Titanium Alloys for Aerospace Structures and Engines, Introduction to Aerospace Materials, Woodhead Publishing, Cambridge, 2012, pp. 202–223.
[4] C. Leyens, M. Peters, Titanium and Titanium Alloys: Fundamentals and Applications, John Wiley & Sons, Hoboken, NJ, 2003.
[5] G. Lütjering, J.C. Williams, Beta Alloys, Titanium, Springer Berlin Heidelberg, Berlin, Heidelberg, 2007, pp. 283–336.
[6] X. Zhao, S. Sun, L. Wang, Y. Liu, J. He, G. Tu, A new low-cost β-type high-strength titanium alloy with lower alloying percentage for spring applications, Mater. Trans. 55 (9) (2014) 1455–1459.

[7] R.R. Boyer, R.D. Briggs, The use of β titanium alloys in the aerospace industry, J. Mater. Eng. Perform. 14 (6) (2005) 681–685.

[8] P.J. Bania, Beta titanium alloys and their role in the titanium industry, JOM 46 (7) (1994) 16–19.

[9] A. Stanley, P.E. Moorhead, J.R. Gross, High-Strength Titanium Base Aluminumvanadium-Iron Alloys, U.S. Patent 2,884,323, RMI Co, 1959.

[10] A. Devaraj, V.V. Joshi, A. Srivastava, S. Manandhar, V. Moxson, V.A. Duz, C. Lavender, A low-cost hierarchical nanostructured beta-titanium alloy with high strength, Nat. Commun. 7 (2016) 11176.

[11] R. Santhosh, M. Geetha, M. Nageswara Rao, Recent developments in heat treatment of Beta titanium alloys for aerospace applications, Trans. Indian Inst. Metals 70 (7) (2017) 1681–1688.

[12] Y. Kosaka, S.P. Fox, K. Faller, S.H. Reichman, Properties and processing of TIMETAL LCB, J. Mater. Eng. Perform. 14 (6) (2005) 792–798.

[13] R.D. Briggs, Titanium Alloy for Fastener Applications, U.S. Patent 9,956,629, Boeing Company, 2018. May 1.

[14] J. Been, K. Faller, Using Ti-5111 for marine fastener applications, JOM 51 (6) (1999) 21–24.

[15] H.R. Tiyyagura, S. Kumari, M.K. Mohan, B. Pant, M. Nageswara Rao, Degradation behavior of metastable β Ti-15-3 alloy for fastener applications, J. Alloys Compd. 775 (2019) 518–523.

[16] S.R.C. Qing Yun Zhao, D.W. Li, M.D. Li, F.L. Liu, Effects of heat treatment on microstructures and mechanical properties of Ti-38644 alloy for aerospace fasteners, Mater. Sci. Forum 913 (2018) 109–117.

[17] M. Yamada, An overview on the development of titanium alloys for non-aerospace application in Japan, Mater. Sci. Eng. A 213 (1) (1996) 8–15.

[18] H. Horimura, K. Doi, H. Sumitomo, Titanium Alloy Bolt and its Manufacturing Process, U.S. Patent 2006/0234800 A1, Honda Motor Co Ltd and Meira Co Ltd, 2006.

[19] N. Matsukura, F. Kamiyama, T. Morikawa, Development of beta-titanium alloy bolts by cold forging, J. Jpn. Inst. Light Metals 55 (11) (2005) 599.

[20] S. Buzolits, Tough titanium fasteners, Adv. Mater. Process. 161 (11) (2003) 53.

[21] M. Ishii, M. Kaneko, T. Oda, Titanium and its alloys as key materials for corrosion protection engineering, Nippon Steel Tech. Rep. (Jpn.) 87 (87) (2003) 49–56.

[22] B.A. Miller, S.G. Lee, The Effect of Graphite-Epoxy Composites on the Galvanic Corrosion of Aerospace Alloys, Air Force Materials Lab Wright-Patterson AFB Ohio, 1976.

[23] A.K. Jha, S.K. Singh, M. Swathi Kiranmayee, K. Sreekumar, P.P. Sinha, Failure analysis of titanium alloy (Ti6Al4V) fastener used in aerospace application, Eng. Fail. Anal. 17 (6) (2010) 1457–1465.

[24] M.E. Stevenson, J.L. McDougall, K.G. Clin, Metallurgical failure analysis of titanium wing attachment bolts, Pract. Fail. Anal. 3 (4) (2003) 75–80.

[25] A. Takano, M. Matsubayashi, T. Matsuda, T. Obata, H. Morikawa, Long-term preload measurement and prediction for Ti-6Al-4V and A286 bolts, Trans. J. Soc. Mech. Eng. Ser. A 79 (804) (2013) 1201–1209.

[26] J.T. Whittaker, D.P. Hess, Ductility of titanium alloy and stainless steel aerospace fasteners, J. Fail. Anal. Prev. 15 (5) (2015) 571–575.

[27] D. Croccolo, M. De Agostinis, N. Vincenzi, Influence of tightening procedures and lubrication conditions on titanium screw joints for lightweight applications, Tribol. Int. 55 (2012) 68–76.

[28] F.H. Froes, H. Friedrich, J. Kiese, D. Bergoint, Titanium in the family automobile: the cost challenge, JOM 56 (2) (2004) 40–44.

[29] K. Faller, F.H. Froes, The use of titanium in family automobiles: current trends, JOM 53 (4) (2001) 27–28.

[30] M. Jackson, Automotive Applications for Titanium, Encyclopedia of Automotive Engineering, John Wiley & Sons, Ltd, Hoboken, NJ, 2014.
[31] O. Schauerte, Titanium in automotive production, Adv. Eng. Mater. 5 (6) (2003) 411–418.
[32] S.L. Nyakana, J.C. Fanning, R.R. Boyer, Quick reference guide for β titanium alloys in the 00s, J. Mater. Eng. Perform. 14 (6) (2005) 799–811.
[33] H. Fuji, K. Takahashi, Y. Yamashita, Application of Titanium and its Alloys for Automobile Parts, Nippon Steel Technical Report, 2003, pp. 70–75.
[34] SPS TITAN™ Titanium Fasteners for High Performance Racing, SPS Technologies Inc., 2003.
[35] A.P. Mouritz, Aerospace Materials: Past, Present and Future, Introduction to Aerospace Materials, Woodhead Publishing, Cambridge, 2012, pp. 15–38.
[36] Titanium Bolts in Miltary and Commerical Aviation, Aerospace Manufacturing, https://www.aero-space.us/2017/07/05/titanium-bolts-aviation/, 2017.
[37] F. Chen, An Overview of Aerospace Fasteners and Material Applications, http://www.fastener-world.com.tw/0_magazine/ebook/pdf_download/FW_166_E_306.pdf, 2017.
[38] The Quest for Stronger, Cheaper Titanium Alloys, https://www.boeing.com/features/innovation-quarterly/feb2018/feature-titanium.page, 2018.
[39] Aircraft Accident Report: In-Flight Separation of Vertical Stabilizer, American Airlines Flight 587, Airbus Industrie A300-605R, N14053, Belle Harbor, New York, November 12, 2001.
[40] Anon., American Airlines Flight 587, https://en.wikipedia.org/wiki/American_Airlines_Flight_587, 2014.
[41] Anon., What are the Applications of Titanium Materials in Modern Aerospace Industry and Engines? http://www.jzmm.com/en/news/359.html, 2017.
[42] Anon., Aerospace Fasteners Market Robust Expansion by Global Key Manufactures, Growth Opportunities, Driving Factors, Demand and Supply To 2023, Oct, https://www.reuters.com/brandfeatures/venture-capital/article?id=59539, 2018.
[43] Aircraft Titanium Fasteners Market Is Likely to Witness an Impressive CAGR of 6.(4% during 2018 to 2023, NASDAQ OMX's News Release).
[44] D.J. Bryan, High Strength Alpha/Beta Titanium Alloy Fasteners and Fastener Stock, ATI Properties LLC, 2012.
[45] S.G. Keener, Method for Preparing Highly-Deformable Titanium and Titanium-Alloy One-Piece Fasteners and Fasteners Prepared Thereby, U.S. Patent Application May 23, 2013.
[46] R.R. Boyer, An overview on the use of titanium in the aerospace industry, Mater. Sci. Eng. A 213 (1) (1996) 103–114.
[47] I. Polmear, D. StJohn, J.-F. Nie, M. Qian, Light Alloys: Metallurgy of the Light Metals, Elsevier, Butterworth-Heinemann, 2017.
[48] J.C. Fanning, Properties of TIMETAL 555 (Ti-5Al-5Mo-5V-3Cr-0.6Fe), J. Mater. Eng. Perform. 14 (6) (2005) 788–791.
[49] L.H.L. Zeng, A. Zenke, M. Murc, J. Fanning, S. Sweet, Evaluation of Newly Developed Ti-555 High Strength Titanium Fasteners, Aeromat, 2006.
[50] S.D. Bartus, Evaluation of Titanium-5Al-5Mo-5V-3Cr (Ti-5553) Alloy Against Fragment and Armor-Piercing Projectiles, U.S. Army Researchd Laboratory, September 2009.
[51] TIMETAL®: Ti-15Mo-3Nb-3Al-0.2Si, Titianium Metals Corporation, 2000.
[52] A. Barfeie, J. Wilson, J. Rees, Implant surface characteristics and their effect on osseointegration, Br. Dent. J. 218 (2015) E9.
[53] V.P. Mantripragada, B. Lecka-Czernik, N.A. Ebraheim, A.C. Jayasuriya, An overview of recent advances in designing orthopedic and craniofacial implants, J. Biomed. Mater. Res. A 101 (11) (2013) 3349–3364.

[54] V. Dunić, R. Slavković, E.A. Pieczyska, Properties and behavior of shape memory alloys in the scope of biomedical and engineering applications, in: F. Zivic, S. Affatato, M. Trajanovic, M. Schnabelrauch, N. Grujovic, K.L. Choy (Eds.), Biomaterials in Clinical Practice: Advances in Clinical Research and Medical Devices, Springer International Publishing, Cham, 2018, pp. 303–331.
[55] F. Auricchio, E. Boatti, M. Conti, SMA biomedical applications, in: L. Lecce, A. Concilio (Eds.), Shape Memory Alloy Engineering, Butterworth-Heinemann, Boston, 2015, pp. 307–341 (Chapter 11).
[56] N.B. Morgan, Medical shape memory alloy applications—the market and its products, Mater. Sci. Eng. A 378 (1) (2004) 16–23.
[57] J. Mohd Jani, M. Leary, A. Subic, M.A. Gibson, A review of shape memory alloy research, applications and opportunities, Mater. Des. (1980–2015) 56 (2014) 1078–1113.
[58] A.F. Magno, A.d.C. Monini, M.V. Capela, L.P. Martins, R.P. Martins, Effect of clinical use of nickel-titanium springs, Am. J. Orthod. Dentofac. Orthop. 148 (1) (2015) 76–82.
[59] B.A. Ramazanzadeh, F. Ahrari, B. Sabzevari, S. Habibi, Nickel ion release from three types of nickel-titanium-based orthodontic archwires in the as-received state and after Oral simulation, J. Dent. Res. Dent. Clin. Dent. Prosp. 8 (2) (2014) 71–76.

Further reading

[60] R.T. Cole, E.J. Bateh, J. Potter, Fasteners for composite structures, Composites 13 (3) (1982) 233–240.

CHAPTER 17

Titanium alloys for brass instruments

Mitsuo Niinomi[a,b,c,d,e]

[a]Institute for Materials Research, Tohoku University, Sendai, Japan
[b]Department of Materials and Manufacturing Science, Graduate School of Engineering, Osaka University, Osaka, Japan
[c]Department of Materials Science and Engineering, Graduate School of Science and Technology, Meijo University, Nagoya, Japan
[d]Institute of Materials and Systems for Sustainability, Nagoya University, Nagoya, Japan
[e]Faculty of Chemistry, Materials, and Bioengineering, Kansai University, Osaka, Japan

Contents

1 Introduction

Music instruments, such as trumpets, and their mouthpieces are either made of brass, which is a copper (Cu)-zinc (Zn) alloy, or nickel silver, which is a Cu-nickel (Ni)-Zn alloy. These exhibit good machinability. Cu, Ni, and Zn are also harmful elements from the viewpoint of cell viability as shown in Fig. 1 [1]. Ni is also a high-risk element from the viewpoint of metallic allergy as shown in Fig. 2 [2] and Fig. 3 [3]. Therefore, during long-term use of mouthpieces made of Cu-Zn alloys or Cu-Ni-Zn alloys, the constitutional elements dissolve into the body, as mouthpieces are in direct contact with saliva and the lip. As a result, inflammation or metallic allergy occurs.

Titanium for Consumer Applications
https://doi.org/10.1016/B978-0-12-815820-3.00008-3

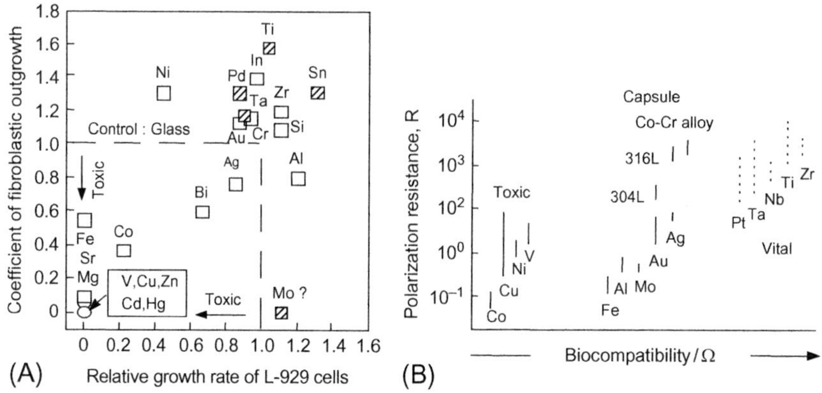

Fig. 1 (A) Cytotoxicity of pure metals; only relative growth rate of L-929 cells for Mo, and (B) relationship between polarization resistance and biocompatibility of pure metals, Co-Cr alloy, and stainless steel: SUS 316L and SUS 304L.

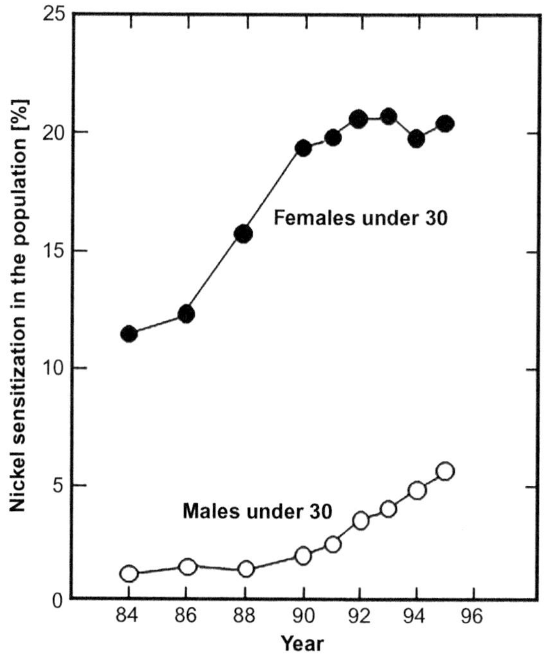

Fig. 2 Frequency of nickel sensitivity in the population.

Hence, most mouthpieces are subjected to various kinds of coatings such as silver (Ag), even though its cytotoxicity is not low, and lacquer. However, the surface of a mouthpiece itself is exposed after long-term use. Therefore, nontoxic and allergy-free metallic materials are required for mouthpieces

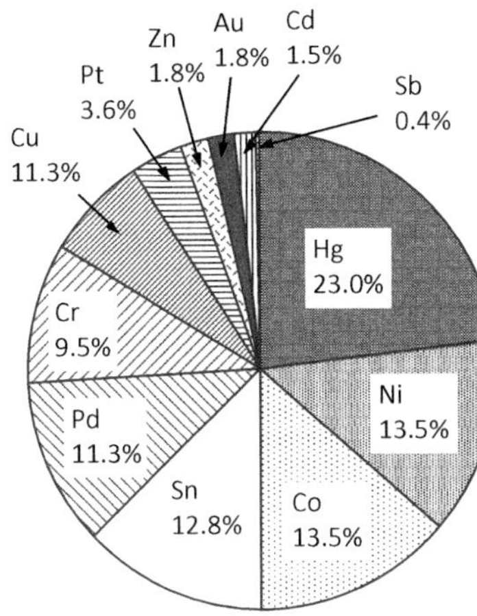

Fig. 3 Percentage of metal allergy caused by each alloy.

of musical instruments. The most expected candidate materials for this purpose are titanium (Ti) and its alloys.

Ti and its alloys are already used in various kinds of musical instruments such as drums, trumpets, electric guitars, and flutes as shown in Figs. 4–7 [4–7]. The acoustic characteristics of Ti and its alloys are expected to also be suitable for mouthpieces of brass instruments. The development of Ti alloys suitable for mouthpieces of brass instruments is necessary. This development, for use in mouthpieces of trumpets, is discussed in this chapter.

Fig. 4 Ti drum.

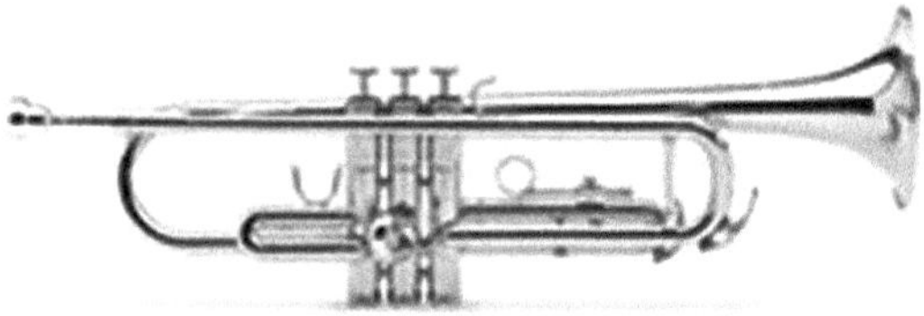

Fig. 5 Ti trumpet.

Fig. 6 Ti electric guitar.

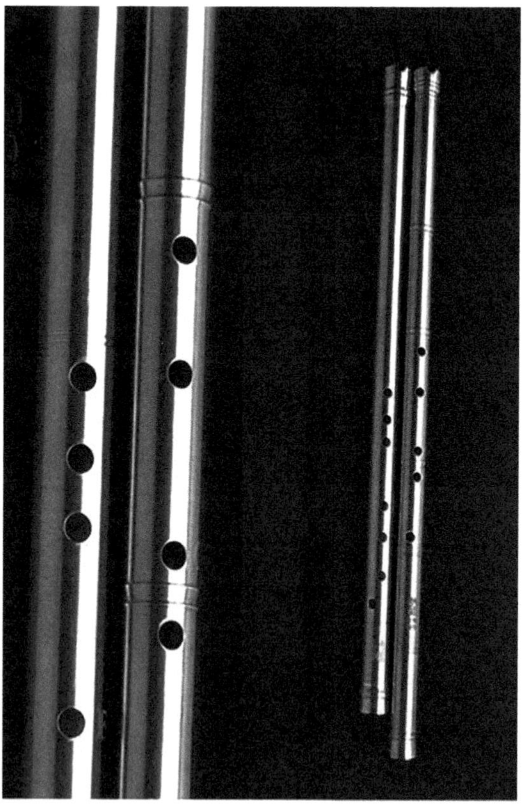

Fig. 7 Ti flute.

2 Brass instruments

Musical instruments that make a sound by feeding air into them are grouped into wind instruments. The wind instruments are further divided into woodwind and brass instruments. Woodwinds are classified as lip reed instruments because their sounds are made by the lip and the reed. Trumpet, trombones, horns, and tubas are, for example, brass instruments. The common structures comprising these brass instruments are the mouthpiece, mouth pipe, valve, tuning slide, and bell. Brass instruments have been made of Cu alloys including brass, stated earlier, for a long time. Cu alloys are used widely as materials for brass instruments, because they exhibit a good outside appearance, ductility, machinability, and tone color. However, there is a possibility for them to exhibit toxicity and allergic problems as stated earlier.

3 Acoustic characteristics

The sound of brass instruments is mainly related to the vibration of the air core between the mouthpiece and bell, but the siphon also vibrates and the sound is radiated from its surface. Therefore, the tone color is changed according to the materials. Instrument manufacturers once tried to make these instruments using stainless steel or aluminum instead of brass, but these failed because the tone color was wrong.

Materials with a high value of E (Young's modulus)/ρ (density) are conventionally suitable for acoustic materials. Namely, the materials with light weight and bending difficulty are suitable for music instruments. The propagation rate of the sound is theoretically higher by increasing the value of E/ρ. On the other hand, the materials, which do not absorb the vibrations are suitable. From this point of view, the materials with low internal damping rate are suitable.

The relationship between internal damping rate and E/ρ of currently used metallic materials for music instruments is shown in Fig. 8 [8]. The materials with low internal damping rate (internal friction: Q^{-1}) are found to be suitable for use in music instruments as stated earlier. Furthermore, because a metallic sound is not preferred, the materials had a relatively lower E/ρ than was expected.

4 Generation principle of sound and role of the mouthpiece

The principle of the occurrence of sound is schematically shown in Fig. 9 [9]. The air blown from the mouth makes a vibration wave at the mouthpiece.

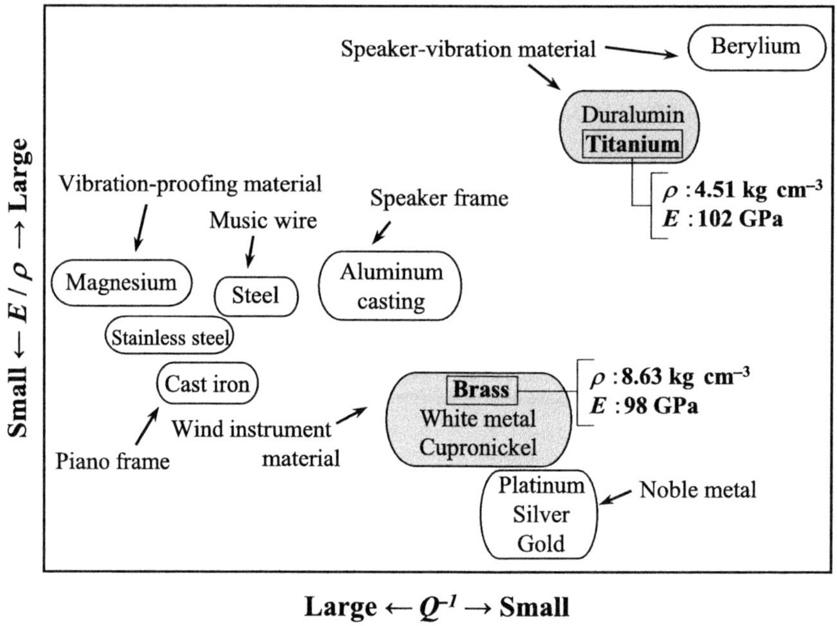

Fig. 8 Acoustic properties of various metallic materials. E, ρ, and Q^{-1} indicate elastic modulus, density, and damping of sound, respectively.

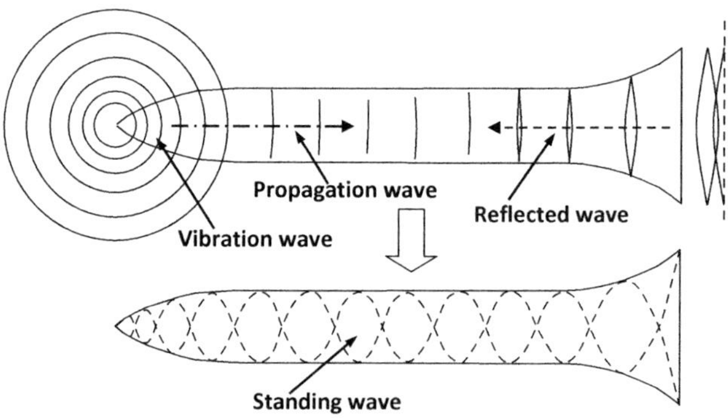

Fig. 9 Sound occurrence principle schematic drawing.

The vibration wave is reflected, which reflects inside the tube toward the bell. The reflected wave is resonant when traveling, and then a standing wave occurs. The standing wave is emitted outside the tube from the bell, which reaches the ears of the audience. Because the mouthpiece side of the brass instrument is closed, the sound pressure becomes maximum in gamut. Therefore, the mouthpiece is one of the important parts because it can change the sound from the music instrument in the most efficient manner due to changes in the weight according to the distribution of the thickness of the plate and physical properties according to the quality of the material and wrought process. The mouthpiece is consistent of the balance of the inside diameters of the rim, cup, and throat, and cooperation harmony of the weight, material quality, and other physical properties (see Fig. 10 [10]). It is reported that only 0.1 mm change in the inside diameter of the throat affects the feeling of the blowing; the inside diameter is, for example, 4 mm in the case of the Yamaha Co. standard model. In the case of Yamaha Co., the quality control of the inside diameter of the throat of the music instruments is kept to an error of ± 0.025 mm against the design criteria size.

5 Development of titanium alloys for trumpet mouthpieces

5.1 Young's modulus (E) and density (ρ)

Focusing on E/ρ as a factor, which affects the sound of musical instruments made of metallic materials, it is 4 for the brass and 22.6 for commercially pure Ti (CP Ti: JIS grade 2): $E = 102$ GPa and $\rho = 4.51$ g/cmm^3 as stated later. Therefore, the propagation rate of the sound is higher in CP Ti than in brass, and the metallic sound seems to occur when CP Ti is used for the brass instruments, although the metallic sound is favored depending on the music genre. It is necessary to reduce E of Ti and increase its density. Niobium (Nb) and tantalum (Ta) are judged to be nontoxic and allergy-free elements from Figs. 1–3. Their densities are greater than that of CP Ti; the densities of Nb and Ta are 8.56 and 16.65 g/cm^3, respectively.

Fig. 11 [11] shows the changes in elastic modulus (E) and density (ρ) of Ti-Nb and Ti-Ta system alloys as a function of Nb or Ta content, along with

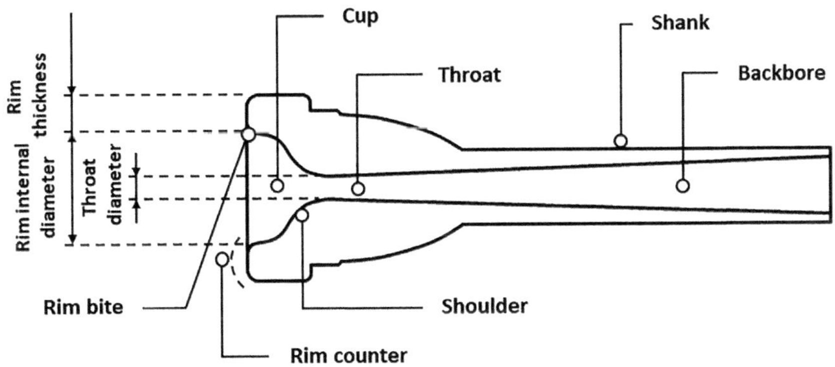

Fig. 10 Mouthpiece structure schematic drawing.

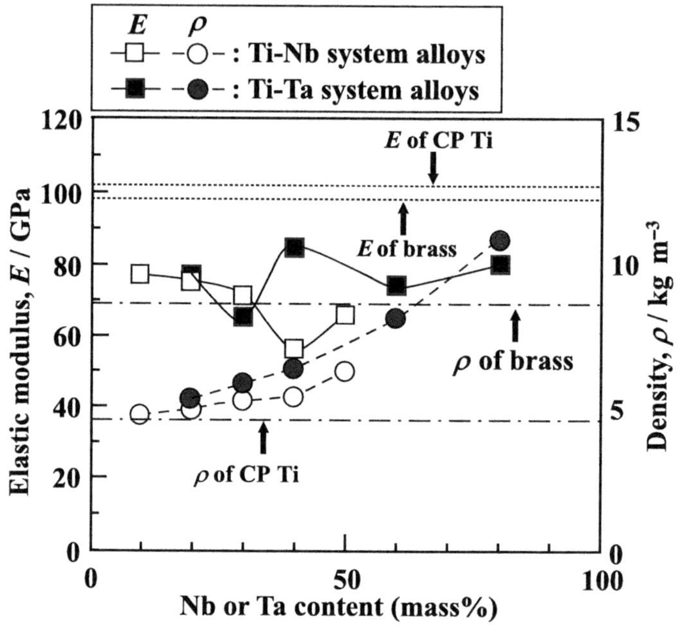

Fig. 11 Changes in elastic modulus (*E*) and density (*r*) of Ti-Nb and Ti-Ta alloys as a function of Nb or Ta content along with *E* and *r* of CP Ti and brass.

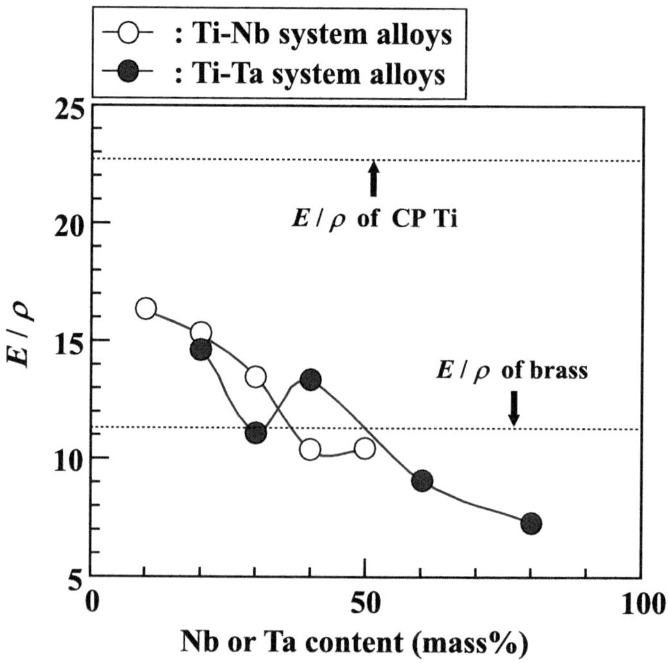

Fig. 12 Changes in *E*/*r* of Ti-Nb and Ti-Ta alloys as a function of Nb or Ta content along with *E*/*r* of CP Ti and brass.

E and ρ of CP Ti and brass. The E values of Ti-Nb and Ti-Ta system alloys exhibit the lowest values with a Nb content of 30 mass% and a Ta content of 30 mass%, respectively; approximately 56 and 65 GPa, respectively; and the E values of the brass and CP Ti are approximately 98 and 102 GPa, respectively. The ρ values of Ti-Nb and Ti-Ta system alloys increase linearly with increasing Nb and Ta contents, respectively.

Changes in the E/ρ values of Ti-Nb and Ti-Ta system alloys as a function of Nb or Ta content along with the E/ρ values of CP Ti and brass are shown in Fig. 12 [11]. The E/ρ values of Ti-Nb and Ti-Ta system alloys changes within 10.4–16.3 and 7.38–14.7, respectively, with increasing Nb and Ta contents, whereas those of the brass and CP Ti are 4 and 22.6, respectively. Therefore, the E/ρ value of Ti can be similar to or greater than that of the brass by controlling the content of Nb and Ta. Ti-60Ta and -80Ta are suitable for low tones whereas Ti-10Nb is suitable for high tones.

5.2 Resonant frequency and internal friction

The resonant frequency (f_R) of Ti-Nb and Ti-Ta system alloys, which are measured using a free resonance method shown in Fig. 13 [11], along with those of the brass and CP Ti, changes in the range of 3.01–3.62 kHz and 2.69–3.61 kHz with increasing Nb and Ta contents, respectively. The f_R values of the brass and CP Ti are approximately 3.01 and 4.58 kHz, respectively. Therefore, the f_R values of Ti-Nb and Ti-Ta system alloys can be equal to or greater than that of the brass by controlling Nb and Ta contents.

The changes in the internal friction ($Q^{-1}{}_R$) evaluated by free resonance method of Ti-Nb and Ti-Ta system alloys as a function of Nb or Ta content along with those of the brass and CP Ti are shown in Fig. 14 [11]. The $Q^{-1}{}_R$ value of Ti-Nb system alloys increases up to a Nb content of 20 mass% by increasing the Nb content, but decreases over a Nb content of 30 mass%. On the other hand, the $Q^{-1}{}_R$ value of Ti-Ta system alloy increases up to a Ta content of 40 mass% by increasing the Ta content, but decreases over a Ta content of 60 mass%. Fig. 15 [11] shows the X-ray diffraction profiles of Ti-Nb and Ti-Ta system alloys. The peaks of the α'' martensite and ω phases are observed in Ti-20Nb and Ti-30Nb. The peaks of the α'' martensite phase are observed in Ti-30 and Ti-40Ta. It has been reported that the $Q^{-1}{}_R$ values increase in Ti alloys, which comprise the α'' martensite or metastable β phases. It has been also reported that the $Q^{-1}{}_R$ value increases due to the twins and high-density dislocations in the α'' martensite phase. Therefore, Ti20Nb and Ti-40Ta exhibit high $Q^{-1}{}_R$ value due to the appearance of the α'' martensite phase in the β phase. The $Q^{-1}{}_R$ value of each alloy is located

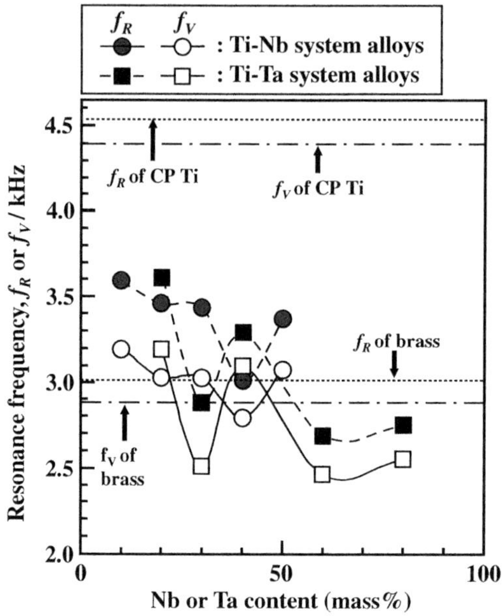

Fig. 13 Changes in resonance frequency (f_R) evaluated through the free resonance method of Ti-Nb and Ti-Ta alloys as a function of Nb or Ta content along with those of CP Ti and brass.

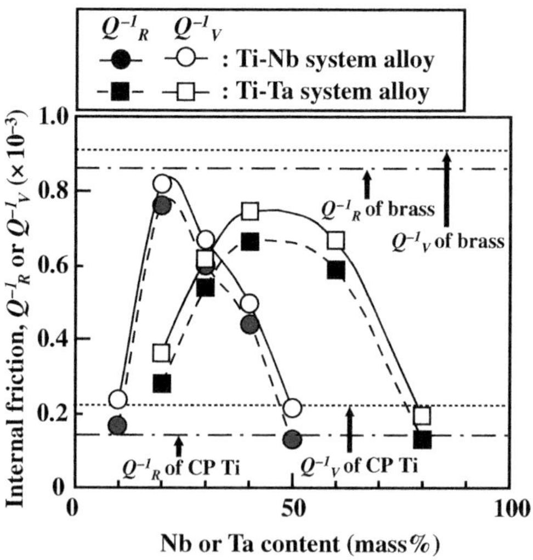

Fig. 14 Changes in internal friction (Q^{-1}_R) evaluated by resonance and vibration-damping methods of Ti-Nb and Ti-Ta system alloys as a function of Nb or Ta content along with Q^{-1}_R of CP Ti and brass.

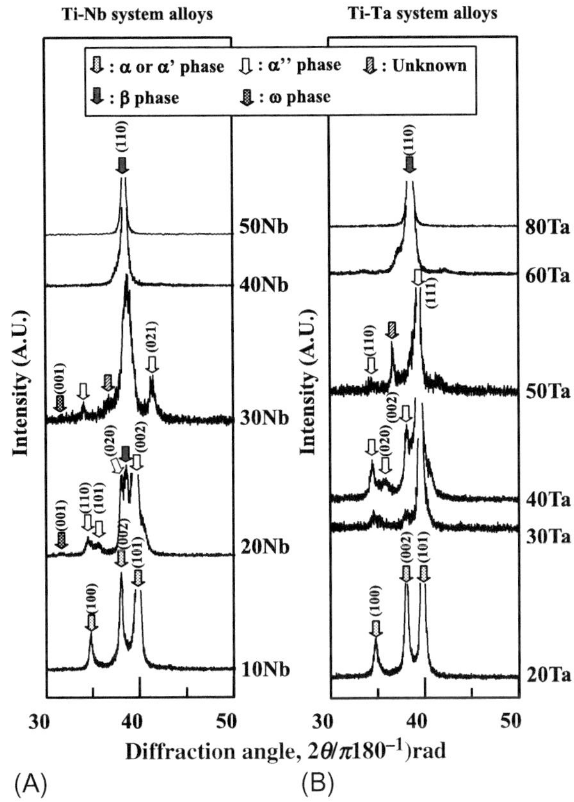

Fig. 15 X-ray diffraction profiles of (A) Ti-Nb and (B) Ti-Ta alloys.

in the range of those of the brass and CP Ti. The Q^{-1}_R value of Ti-20Nb is very close to that of the brass.

5.3 Relationship between E/ρ and f_R

The relationships between E/ρ and f_R of Ti-Nb and Ti-Ta system alloys are shown in Fig. 16 [11] along with E/ρ and f_R of the brass and CP Ti. The positive correlation can be recognized between E/ρ and f_R of each alloy. E/ρ and f_R of each alloy are located in the middle level or lower level of those of the brass and CP Ti. Among all the alloys, E/ρ and f_R of Ti-40Nb and Ti-30Ta are nearly equal to that of the brass. Therefore, E/ρ and f_R of Ti can be equal to that of the brass and controlled to be adapted to the low- or high-tone region by controlling the Nb and Ta content.

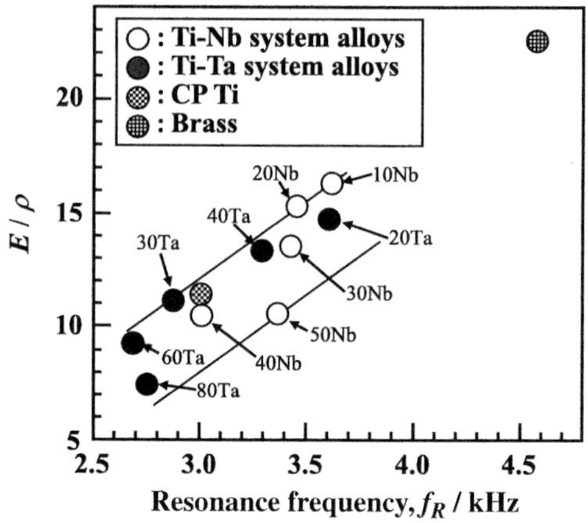

Fig. 16 Relationships between E/ρ and f_R of Ti-Nb and Ti-Ta alloys along with E/ρ and f_R of CP Ti and brass.

5.4 Corrosion resistance

Anode polarization curves of Ti-Nb and Ti-Ta system alloys are shown in Fig. 17 [11] along with those of CP Ti and brass in artificial saliva. The passivation current densities of Ti-Nb and Ti-Ta system alloys are approximately 40% lower than that of the brass, and equal to or a little bit lower than that of CP Ti. Therefore, the corrosion resistances of Ti-Nb and Ti-Ta system alloys are excellent. The dissolution of the constitutional elements of these alloys is inhibited; allergic problem is expected to be solved.

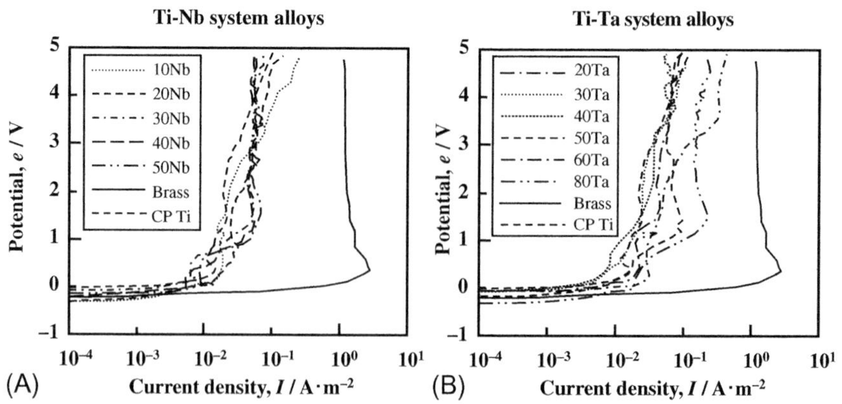

Fig. 17 Anode polarization curves of (A) Ti-Nb and (B) Ti-Ta alloys along with anode polarization curves of commercially pure Ti (CP Ti) and brass in artificial saliva.

6 Fabrication of mouthpieces of trumpets using the developed Ti alloys for brass instruments

Mouthpieces of brass instruments such as trumpets are generally made by machining. However, near net shape processing, such as casting, is suitable to reduce the cost of products and for Ti alloy product because the machinability of Ti and its alloys are poor. Among the casting processes, precision casting is suitable for fabricating trumpet mouthpieces, which have high size accuracy requirements.

The trumpet mouthpieces of YAMAHA Eric Miyashiro and YAMAHA Bobby Shew models for jazz were selected and fabricated by precision casting using Ti-10Nb because its E/ρ and f_R are in the middle level between those of the brass and CP Ti. Fig. 18 shows profiles of trumpets mouthpieces of YAMAHA Bobby Shew and Eric Miyashiro models made of Ti-10Nb by precision casting using zirconia mold material with sprue and without sprue. Fig. 19 shows profiles of trumpets mouthpieces of YAMAHA Bobby Shew and Eric Miyashiro models made of Ti-10Nb by precision

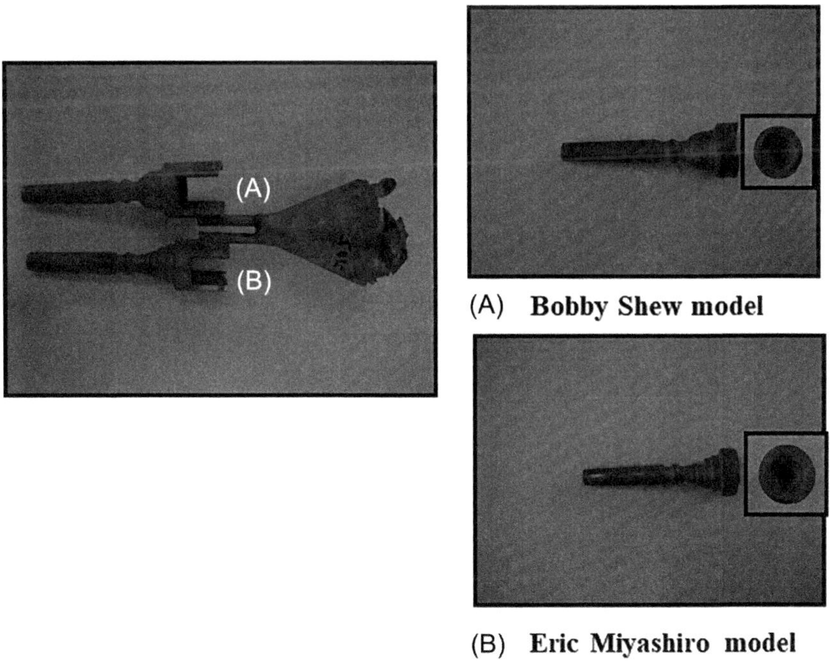

Fig. 18 Profiles of trumpets mouthpieces of YAMAHA (A) Bobby Shew and (B) Eric Miyashiro models made of Ti-10Nb by precision casting using zirconia mold material with sprue (left) and without sprue (right).

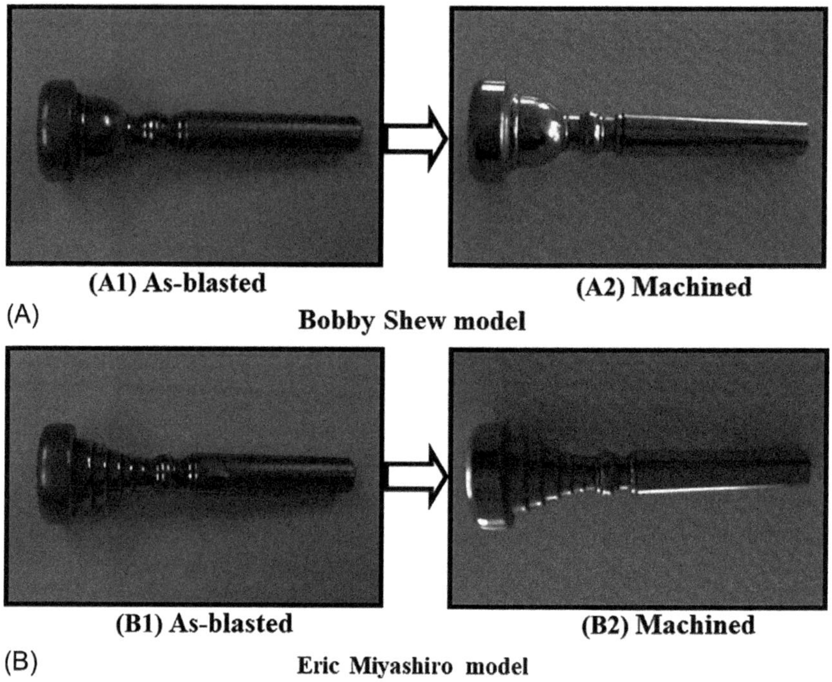

Fig. 19 Profiles of trumpets mouthpieces of YAMAHA (A) Bobby Shew and (B) Eric Miyashiro models made of Ti-10Nb by precision casting under (A1) and (B1) as-blasted conditions, and (B1) and (B2) under machined conditions.

casting under as-blasted and machined conditions. The trumpets mouthpieces of both models are well fabricated by the precision casting process.

7 Sound characteristics of a trumpet mouthpiece made of titanium alloy

Figs. 20 and 21 [11,12] show fast Fourier transforms (FFT) of sound-damping waves (left) and vibration-damping waves (right) of YAMAHA (a) Eric Miyashiro (Y_E) and (b) Bobby Shew (Y_B) models made of Ti-10Nb, CP Ti, and brass in low-pitch and high-sound regions. The sound-damping waveforms in low-pitch sound regions show no peculiarity in different materials (see Fig. 20, left). This trend is same as that of the sound-damping waves in high-pitch sound region of Y_E model, and in both low- and high-pitch sound regions of Y_B model (see Fig. 20A, right). However, for the vibration-damping wave in the low-pitch sound region (see Fig. 21A, right), the intensity peak of the waveform of each material becomes low, and the intensity peak of the waveform in the high-frequency region (near 7 kHz) of Y_E model made of Ti-10Nb is much closer to that made of the brass than that made of CP-Ti.

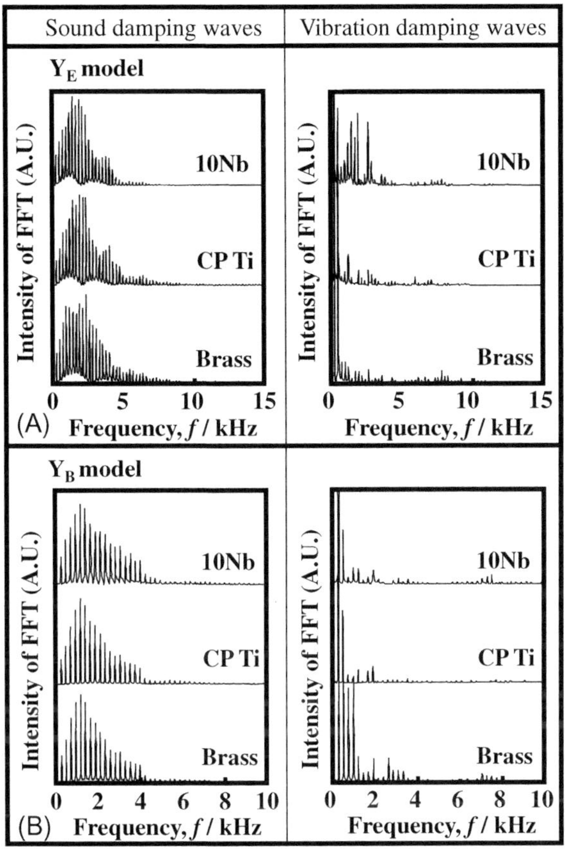

Fig. 20 Fast Fourier transforms (FFT) of sound-damping waves (left) and vibration-damping waves (right) of (A) Y_E and (B) Y_B models made of 10Nb, CP Ti, and brass in low-pitch sound region.

For the vibration-damping waveform of Y_E model made of each material in the high-pitch sound region, the same trend can be observed. Therefore, in the practical size trumpet mouthpiece, the vibration-damping waveform is also changed with the change in E/ρ, resonance frequency, and density. This trend is nearly the same as that of the vibration-damping waves in both low- and high-pitch sounds of Y_B model (see Fig. 20B, right and Fig. 21B, right). As stated earlier, the changes in the vibration-damping waveforms of Y_E and Y_B models are more obvious in the high-pitch sound region than in the low-pitch sound region because the vibration added to the mouthpiece is more obvious in the high-pitch sound region than in the low-pitch sound region. The other factor for this trend is also due to the Y_E and Y_B models, which are for jazz; it is easy to make the high-pitch sound.

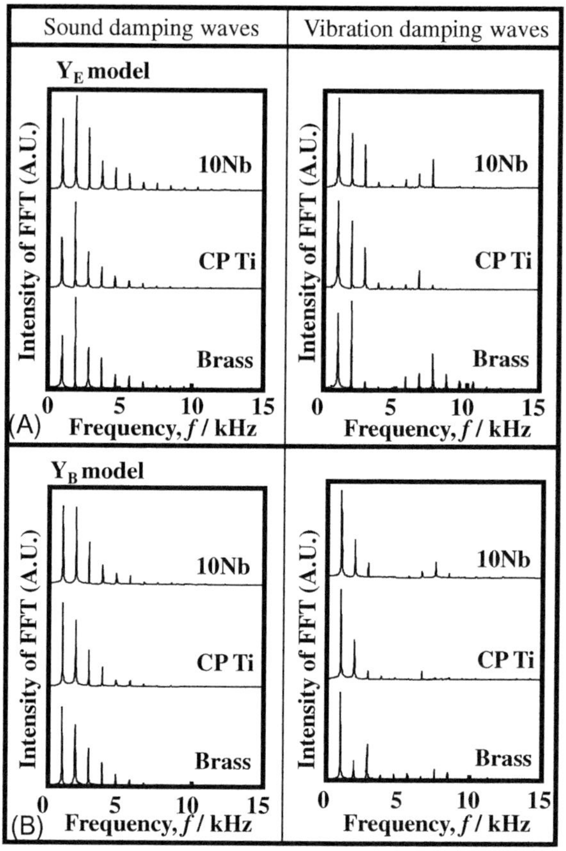

Fig. 21 FFT of sound-damping waves (left) and vibration-damping waves (right) of (A) Y_E and (B) Y_B models made of 10Nb, CP Ti, and brass in high-pitch sound region.

8 Summary

The sound characteristics of the practical size mouthpiece made of Ti can be controlled by changing the physical properties, such as E/ρ, by changing the content of the alloying elements, including nontoxic and allergy-free elements. The relationship between the vibration signal (vibration-damping wave) and the sound signal (sound-damping wave) is more obvious in the high-pitch sound region than in the low-pitch sound region. The mouthpieces of the brass instruments made of Ti alloys are expected to be more suitable in the high-pitch sound region than in the low-pitch sound region.

Acknowledgments

The author would like to express sincere thanks to Professors T. Akahori at Meijo University, Nagoya, Japan, and S. Horihata at Nihon University, Matsudo, Japan; Dr. J. Takeda at Ministry, Health, Labor, and Welfare, Japan, and Messrs. T. Suzuki, Suzuhiro Co., Ltd., and K. Sone, S. Niwata, and N. Fukuda, YAMAHA Co., Ltd., Hamamatsu, Japan, for their help with this article.

References

[1] H. Kawahara, Cytotoxicity of implantable metals and alloys, Bull. Jpn. Inst. Metals Mater. 31 (1992) 1033–1039.
[2] Y. Okazaki, Y. Ito, A. Ito, T. Tateishi, Effect of alloying elements on mechanical properties of titanium alloys for medical implants, J. Jpn. Inst. Metals Mater. 57 (1993) 332–337.
[3] S.G. Steinemann, in: G.D. Winter, J.L. Leray, K. de Goot (Eds.), Evaluation of Biomaterials, Wiley, New York, 1980, pp. 1–34.
[4] P.J. Uggowitzer, W.F. Bähre, M.O. Speidel, Metal Injection molding of nickel-free stainless steels, in: Proc. Adv. Powder Metall., Part Mater. 3, 1997. 18-113–18-21.
[5] http://www.kitanodrums.com/japanese.htm.
[6] https://www.musiciansfriend.com/brass-instruments/stomvi-5383-master-titanium-series-bb-trumpet.
[7] https://musicbliss.com.my/products/esp-f-50-titanium-malaysia-electric-guitar.
[8] https://ja.aliexpress.com/item/Titanium-Metal-Flute-Xiao-80cm-G-F-Key-Xiao-Flute-Transverse-Flute-not-Dizi-Professional-Metal/32410736183.html.
[9] T. Sata, Metallic materials for music instrument, Denkiseiko 64 (4) (1993) 269–275.
[10] M. Niinomi, J. Takeda, T. Akahori, H. Toda, T. Suzuki, K. Watanabe, M. Yamashita, K. Sone, Applications of titanium and its alloys for music instruments, J. Jpn. Light Metals 55 (12) (2005) 642–645.
[11] T. Akahori, M. Niinomi, M. Nakai, H. Tshutsumi, T. Nozaki, S. Hirohata, T. Suzuki, K. Sone, Development of allergy-free titanium alloys for brass instruments, J. Jpn. Light Matals 58 (11) (2008) 604–610.
[12] S. Horihata, M. Niinomi, T. Nozaki, T. Akahori, T. Suzuki, K. Sone, S. Niwata, N. Fukuda, Vibration and sound characteristics by shape in mouthpiece made of titanium alloy of trumpet, Trans. Jpn. Soc. Mech. Eng. C 79 (803) (2013) 2240–2249.

Index

Note: Page numbers followed by *f* indicate figures and *t* indicate tables.

C

D

M

N

O

9780128158203